T0317610

History of Anatomy

History of Anatomy

An International Perspective

R. Shane Tubbs
Seattle Science Foundation, Seattle, USA

Mohammadali M. Shoja
Tabriz University Neuroscience Institute, Iran

Marios Loukas
Department of Anatomical Sciences, St. George's University, Grenada

Paul Agutter
Theoretical Medicine and Biology Group, Glossop, UK

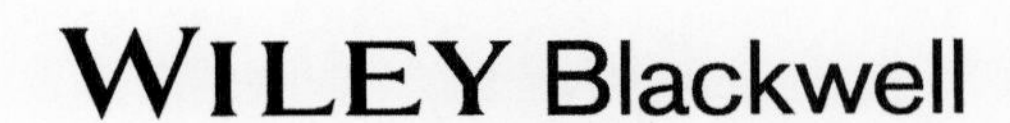

This edition first published 2019

Registered Office
John Wiley & Sons, Inc., 111 River Street, Hoboken, NJ 07030, USA

Editorial Office
The Atrium, Southern Gate, Chichester, West Sussex, PO19 8SQ, UK

For details of our global editorial offices, customer services, and more information about Wiley products visit us at www.wiley.com.

Wiley also publishes its books in a variety of electronic formats and by print-on-demand. Some content that appears in standard print versions of this book may not be available in other formats.

Library of Congress Cataloging-in-Publication Data

Names: Tubbs, R. Shane, author.
Title: History of anatomy : an international perspective / Dr. R. Shane Tubbs, Pediatric Neurosurgery, Children's of Alabama, Birmingham, AL, USA, Department of Anatomical Sciences, St. George's University, Grenada, Centre of Anatomy and Human Identification, University of Dundee, Scotland, UK, Dr. Mohammadali M. Shoja, Pediatric Neurosurgery, Children's of Alabama, Birmingham, AL, Dr. Paul Agutter, Director, Theoretical Medicine and Biology Group, Glossop, U.K., Dr. Marios Loukas, Department of Anatomical Sciences, St. George's University, Grenada.
Description: Hoboken, NJ : Wiley Blackwell, [2019] | Includes index. |
Identifiers: LCCN 2018041498 (print) | LCCN 2018042144 (ebook) | ISBN 9781118524343 (Adobe PDF) | ISBN 9781118524312 (ePub) | ISBN 9781118524251 (hardback)
Subjects: LCSH: Human anatomy–History. | BISAC: SCIENCE / Life Sciences / Human Anatomy & Physiology.
Classification: LCC QM11 (ebook) | LCC QM11 .T83 2019 (print) | DDC 612–dc23
LC record available at https://lccn.loc.gov/2018041498

Cover Design: Wiley
Cover Background: © ilbusca / Getty Images

Set in 10/12pt Warnock by SPi Global, Pondicherry, India

10 9 8 7 6 5 4 3 2 1

To my beautiful and loving wife Susan, who is always beside me – R. Shane Tubbs

To Susan Tubbs, for her care and moral support – Mohammadali M. Shoja

To the love of my life, my wife Joanna Loukas – Marios Loukas

I remain grateful to the staff of Edinburgh Medical School, who during the 1960s inspired me with respect for the traditions of medical practice and a lasting interest in the history of medical science – Paul Agutter

Contents

Preface

The aim of medical practice is to understand the causes of diseases and to use such understanding to cure those diseases or alleviate their effects. Diseases are regarded as perturbations of normal function in one or more parts of the body. Normal function and its disturbances cannot be understood unless the structures and interrelations of the parts of the body are known in detail. Therefore, knowledge of human anatomy is essential for medical practice and is considered a cornerstone of medical education; in particular, its importance as a foundation for surgery is self-evident.

Few people involved in the practice and teaching of medicine today would challenge these pronouncements, but although the belief they encapsulate appears universal, that belief is largely a product of post-Renaissance Europe. Not all cultures in human history have viewed diseases and their treatment in anything like the mechanistic cause–effect terms we take for granted today, and *a fortiori* knowledge of anatomy has not always been seen as fundamental to medicine. Indeed, anatomy received limited attention in most European medical schools in the centuries preceding the Renaissance.

Nevertheless, the internal as well as the external structures and organization of the body have excited interest and curiosity in virtually all times and places. Therefore, most societies throughout history have produced ideas about human anatomy, even if they are not necessarily related to medicine. Where writing developed, some of those ideas were recorded; anatomical descriptions were written by observers in ancient civilizations some four millennia before Europe began to emerge from the Dark Ages. It would thus be unforgivably ethnocentric to restrict a historical study of anatomy to developments in post-medieval Europe, important though these are. It would also blind us to a number of remarkable discoveries made in the remote past, including in China, India, and Egypt, some of which indirectly influenced the growth of knowledge in Europe. These considerations define the motivation for this book.

Our knowledge of human anatomy today, as encapsulated in Henry Gray's classic (first published in 1858), is regarded as objective and essentially complete; that is to say, it is value-neutral, transculturally valid, and fully detailed. Much of this knowledge was constructed in Europe between the sixteenth and twentieth centuries CE, but it can be traced to more ancient roots. Both the construction of present-day understanding and the tracing of these roots are fascinating studies for the historian, and we hope to convey that fascination in the following pages by presenting a wide variety of individuals from around the world who contributed to the field of anatomy.

The emergence of our modern understanding has been prolonged, slow, and often tortuous. There are three main reasons for this. First, the subject is inherently difficult; mammalian (including human) anatomy is very complicated. Most editions of *Gray's Anatomy*, for example, run to some 800 densely packed pages. Second, understanding of structure has tended to evolve hand in hand with understanding of function, and ideas about the functions of various internal organs have in past times been inspired by religious and other cultural beliefs and attitudes, rather than by the "scientific evidence" we now regard as the gold standard of credibility. Accounts of human anatomy, as well as "physiology," have therefore differed among cultures. Third, the only reliable source of anatomical knowledge – dissection – has been restricted or even forbidden in many societies. Galen, the great Greek physician of the second century CE, was famously obliged by Roman law to rely on dissections of pigs, dogs, and apes for his accounts of human anatomy, and the consequent errors in his writings were sustained for some 1500 years. Restrictions on dissection in European countries before, during, and after the Renaissance also limited the progress of anatomy, although they ultimately led to the emergence of "centers of excellence." During the early sixteenth century, for example, Vesalius traveled to many locations in Europe where he hoped to be free to dissect cadavers; only in Italy – specifically, Padua – was he able to dissect female corpses, a prerequisite for the compilation of his celebrated *Seven Books*. Subsequently, Padua became a center of anatomical learning, where the successors of Vesalius made a series of important advances in knowledge.

In this book, we present the gradual emergence of modern knowledge about human anatomy, largely through the biographies of selected contributors to the field, not all of them well known. These contributors made striking discoveries in a range of different cultural settings. By telling their stories, we hope to enrich the modern student's understanding of a fundamental aspect of medicine and surgery.

Lastly, earlier writings on the history of anatomy have reviewed this topic chronologically. Herein, we have chosen to celebrate regions of the world and their contributions to this discipline. Therefore, each geographic chapter reviews the lives of those in that part of the world who influenced and helped shape our current understanding of the human form. Instead of short glimpses, each anatomist is covered in some detail. No single text can include every contributor to this field. Therefore, we provide a smattering of many well-known – and some not-so-well-known – figures in the history of anatomy.

The authors

Acknowledgments

The authors would like to thank the following individuals who were instrumental in bringing this project together:

Hanna M, Alsaiegh NR, Louis RG Jr, Pinyard J, Vaid S, Curry B, Shea M, Shea C, Lutter-Hoppenheim M, Zand P, Cohen-Gadol AA, Bosmia AN, Patel TR, Watanabe K, Kato D, Ardalan MR, Linganna S, Chiba A, Riech S, Verma K, Chern J, Mortazavi M, Shokouhi G, Lanteri A, Ferrauiola J, Maharaja G, Yadav A, Rao VC., Salter EG, Oakes WJ, Carmichael SW, Malenfant J, Robitaille M, Schaefer J, Wellons JC, Housman B, Bellary S, Hansra S, Mian A, Bertino F, Shipley E, Veith P, Blaak C, Adeeb N, Deep A, Griessenauer CJ, Fukushima T, Steck DT, Malakpour M, Rozzelle CJ, Padmalayam D, Bellary SS, Walters A, Gielecki J, Song YB, Osiro S, Matusz P, Rompala O, Klaassen Z, Chen J, Dixit V, Bilinsky E, Bilinsky S, Abrahams P, Diamond M, Pennell C, Groat C, Vahedi P, Khalili M, Khodadoost K, Ghabili K, Akiyama M, Yalcin B, Saad Y, Hill M, Gribben WB, Tubbs KO, Lam R, Apaydin N, Eknoyan G, El-Sedfy A, Wartman C, Clarke P, Kapos T, Gupta AA, Spentzouris G, Kolbinger W, Chiba A, Trotz M, Jordan R, Downs E, Grater J, Gianaris N.

1

Africa

Few cultural transformations in human history have compared in magnitude with the emergence of the earliest civilizations such as those found in the cradle of humankind, Africa. During the Neolithic Period, productive agriculture in the flood plains of great rivers allowed local non-nomadic populations to expand, and this led to the building of the world's first cities. The consequences were wide-ranging and profound. They included organized warfare, which led to frequent serious injuries; the appearance of new diseases associated with high population densities; the elaboration of religion, including novel ideas about death and the afterlife; and the invention of writing, which was needed to record laws, food distribution patterns, and so on. War-related and other injuries promoted the practice of surgery, diseases led to the emergence of medicine, and these developments – together with certain burial practices – fostered knowledge of anatomy. Thanks to the invention of writing, some of this knowledge was recorded. Nowhere was this historical pattern more clearly exemplified than in ancient Egypt. The earliest surviving Egyptian writings pertinent to medicine and anatomy date back some 4000 years, but they seem to refer to texts around 1000 years older, which are no longer extant.

When the great Bronze Age empires declined and iron-based technology spread, long-distance trade increased, and ideas were carried along the trade routes. For instance, knowledge from ancient Egypt, as well as the former empires of the Middle East, was conveyed to the Greek city states during the first millennium BCE, contributing to the development of Hippocratic medicine. Two or three centuries later, this influence was reciprocated. The burgeoning of knowledge in Alexandria owed the most to Greek rationalism and naturalism, although the legacy of ancient Egypt was still apparent.

With the erosion of the Western Roman Empire and the concomitant decline of Alexandrian learning during the early centuries CE, Egypt ceased to make salient contributions to anatomical or other knowledge. Such learning as survived from the heyday of Alexandria was scattered far and wide – to be brought back together in the eighth and ninth centuries CE under the influence of the Caliphate of Baghdad (see Chapter 4). A few hundred years later, some of that corpus of knowledge was translated in Europe. The more ancient Egyptian knowledge, however, did not become known to Europe (or the rest of the post-Classical world) until the later nineteenth century, when it became possible to translate hieroglyphic writings into European languages. The quality and detail of that ancient knowledge surprised the translators, and still has the power to surprise the reader.

History of Anatomy: An International Perspective, First Edition. R. Shane Tubbs, Mohammadali M. Shoja, Marios Loukas and Paul Agutter.

1.1 Egypt

Historians and archeologists have spent much time researching and excavating ancient Egypt. Much of that work has focused on the visible aspects of ancient Egyptian society; however, research shows that little recognition has been given to Egypt's contribution to the anatomical sciences, even though the Egyptians were regarded as great physicians of their day. They were famously known for the process of embalming and mummification, which allowed them to have invaluable exposure to human anatomy. As devoted record-keepers, the ancient Egyptians chronicled a vast amount of their knowledge and experience on papyrus. Some of the papyri have shed light on what ancient Egypt knew about anatomy, including the Edwin Smith Papyrus (1700 BCE), the Ebers Papyrus (1500 BCE), and the Kahun Gynecological Papyrus (1825 BCE). These papyri contain a wealth of information on surgery, healing, skin diseases, stomach ailments, medicines, the head, dentistry, gynecology, and diseases of the extremities. Centuries later, during the Alexandrian Period, anatomy became a formally recognized discipline in the city of Alexandria, where it was led by two of the greatest anatomical scholars: Herophilus and Erasistratus. Although they did not have access to technologically advanced labs, their primitive studies and investigations produced the cornerstone on which the later study of anatomy was built. This section examines ancient Egypt's contribution to the anatomical sciences and how it influenced our current knowledge of human anatomy.

1.1.1 The Edwin Smith Papyrus

The Edwin Smith Papyrus is regarded as one of the most significant medical and anatomical references discovered in ancient Egypt. Dating back to 1700 BCE, it is one of the oldest known documents relating to anatomy and surgery. It was written by an anonymous author, who appears to have been attempting to salvage content from an older script dating back to 3000 BCE. An American living in Egypt, Edwin Smith, purchased the script in Luxor in 1862 under controversial circumstances. Some believe that he purchased it from Mustafa Agha, an Egyptian businessman, while others claim he obtained it from tomb raiders. The manuscript was donated to the New York Historical Society in 1906, and later translated by James Henry Breasted in 1930. It appears to have been intended as a surgical guidebook, starting with cases relating to the head and moving systemically down the body. The Edwin Smith Papyrus clearly illustrates the knowledge of the ancient Egyptians in anatomy, with 48 cases divided into *examination*, *diagnosis*, and *treatment* (Figure 1.1). The most helpful record and insight into the expertise of the Egyptians in anatomy is provided by the *examination* outline. The Egyptians' observational skills and systemic analysis were impressive, as they documented the anatomy associated with neurological deficits, musculoskeletal injuries, and the vascular and reproductive systems.

1.1.2 The Ebers Papyrus

Like the Edwin Smith Papyrus, the Ebers Papyrus was also purchased in Luxor by Edwin Smith in 1862. The Egyptologist George Ebers, for whom it is named, bought it from him in 1872. The Ebers Papyrus is by far the lengthiest of the medical papyri, and is dated by a passage to 1500 BCE, close to the date of the Edwin Smith Papyrus; however, other references in the text date the content of the script to 3000 BCE – again, like the Edwin Smith Papyrus. However, unlike the Edwin Smith Papyrus, the Ebers Papyrus is composed of medical guidance in reference to body systems. It incorporates references to many diseases and a comprehensible pharmacopeia. There is also a very illustrative and detailed picture of the cardiovascular system and its distribution, which will be discussed later.

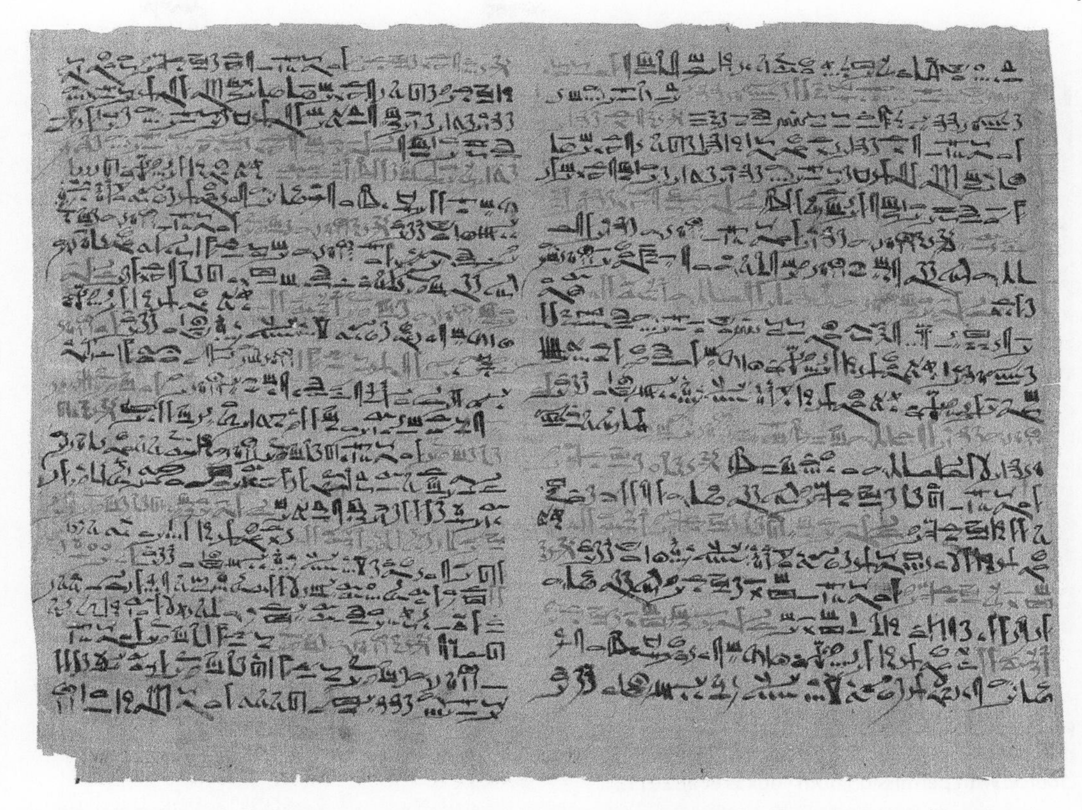

Figure 1.1 The Edwin Smith Papyrus, the world's oldest surviving surgical document. Written in hieratic script in ancient Egypt around 1600 BCE, the text describes anatomical observations and the examination, diagnosis, treatment, and prognosis of 48 types of medical problems in exquisite detail. Plates 6 and 7 of the papyrus, pictured here, discuss facial trauma. (*Source:* http://en.wikipedia.org/wiki/File:Edwin_Smith_Papyrus_v2.jpg.)

1.1.3 The Kahun Papyrus

The Kahun Papyrus is significant because it reveals the hypotheses and beliefs of the ancient Egyptians regarding gynecology and pelvic anatomy. It was discovered by Flinders Petrie in 1889 at the Fayum site of Lahun and dated back to the Middle Kingdom of 1825 BCE.

1.1.4 Egyptian Anatomy in General

In the foreword of the translation of the Edwin Smith Papyrus (page XVI), Breasted acknowledged the anatomical mastery of the ancient anatomist, as he wrote: "He knew of the cardiac system and was surprisingly near recognition of the circulation of the blood, for he was already aware the heart was the center and the pumping force of a system of distributing vessels. He was already conscious of the importance of the pulse and had probably already begun to count the pulse, a practice heretofore first found among the Greek physicians of the third century B.C. in Alexandria."

The ancient Egyptian anatomist was the embalmer. Embalmers held an honorable priestly position in Egyptian society, where their role was to prepare the dead for life beyond the grave, using extensive and elaborate techniques designed to preserve the human body. The embalmer's sole interest was to preserve the body ritually, but the processes used and

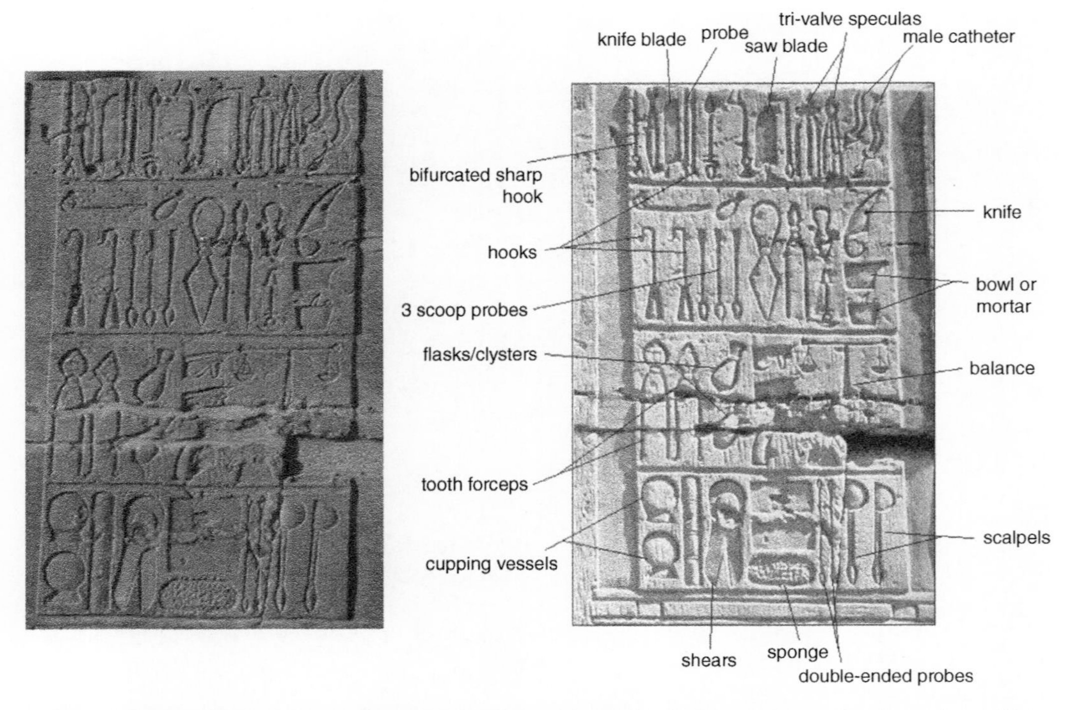

Figure 1.2 Relief in the outer wall of the Temple of Kom Ombo. The depressions represent the tools used by the ancient Egyptians. Some have still not been identified, and others have questionable functions. (*Source:* With permission from Loukas et al. 2011.)

developed for mummification led to discoveries that would later contribute significantly to the study of anatomy (Figure 1.2).

Much credit is also due to the ancient Egyptian physicians or surgeons, known as the *swnw*, from the hieroglyph for "arrow." Their use of limb prosthetics exemplifies their advanced knowledge for their time (Figure 1.3). This knowledge, transcribed into the Ebers and Edwin Smith papyri, was referred to by later physicians, not only for assistance with common ailments, but also to aid in diagnosing and treating a variety of conditions ranging from head injuries to trans-sphenoidal surgery. The detail seen in these papyri is evidence of the amount of knowledge the ancient anatomists were able to collect and record on the basis of their experiences.

1.1.5 Neurological Anatomy

The ancient Egyptian embalmers were experts at carefully removing and preserving organs from body cavities during their mummification ceremonies. Organs such as the brain were thought to have little overall significance for an individual in the realm of the afterlife. Surprisingly, ancient Egyptians were the first to apply an advanced technique of trans-sphenoidal access to the cranial vault and remove the brain without disfiguring the face. The use of this technique in ancient Egypt is evidenced by various studies on mummies and instruments found in archeological digs.

The Edwin Smith and Ebers papyri contain a plethora of cases relating to cranial and spinal injuries, as knowledge of how to treat these injuries was crucial during times of war.

Figure 1.3 Artificial big toe, found on the foot of an ancient Egyptian mummy. The toe has been dated to 1069–664 BCE and is on display at the Cairo Museum in Egypt. It is carved or made from wood and leather, and is attached to the foot by a sewn leather wrapping. (*Source:* Released from Copyright, http://www.experiment-resources.com/ancient-medicine.html, with permission from Loukas et al. 2011.)

The papyri address superficial lacerations, deep scalp wounds, skull fractures, and vertebral subluxation and dislocations. This in turn offers us a glimpse into the ancient Egyptians' repository of anatomical knowledge. In 1925, Breasted met Harvey Cushing (the father of American neurosurgery) and shared with him the first appearance in ancient medical literature of the word "brain," in the Edwin Smith Papyrus. Because of the emphasis on cranial and spinal injuries in these papyri, they are of great use to neurosurgeons today, as they provide a historical perspective.

Cases 4, 5, and 6 of the Edwin Smith Papyrus shed light on the anatomical knowledge the Egyptians possessed in treating severe head injuries accompanied by evidence of basilar skull fractures and meningismus. Cases 4 and 5 concern skull fractures and methods of diagnosis based on severity, while Case 6 introduces a skull fracture, dural laceration, and cerebrospinal fluid leakage. Further mention of the area of skull fracture indicates that the Egyptians had started to map the human skull. Case 7 makes mention of symptoms attributed to compound skull fractures penetrating the frontal air sinus, while Case 19 alludes to a zygomatic-temporal fracture or wound with subsequent deafness and aphasia associated with damage to Heschl's gyrus and to Broca's and Wernicke's areas. The detail and attention given to the treatment of these disorders allow us to appreciate the anatomical knowledge the Egyptians were able to assimilate.

The papyri also contain valuable information regarding vertebral and spinal cord injuries and subsequent symptoms, including paraplegia, persistent erections, seminal emissions, and urinary incontinence. The Egyptians were quick to make a connection between spinal cord injuries and paralysis or even death. Physicians were able to differentiate between different degrees of vertebral injury based on symptoms, as seen in Cases 31, 32, and 33 of the Edwin Smith Papyrus. Dislocation or a crushed vertebra produced the most severe symptoms, while a cervical vertebral sprain was diagnosed on the basis of stiff head rotation.

Later, during the Alexandrian Period, Herophilus (c. 335–280 BCE) is credited with the discovery and development of anatomical sciences in Egypt. He gave a description of the prolongation of the rhombencephalon, which he termed the "spinal cord" and believed was the source of the peripheral nerves. He also distinguished between motor and sensory nerves and proved the connection between loss of motor nerve function and paralysis. He described several cranial nerves, including the trigeminal, facial, auditory, and hypoglossal. He also identified the meninges and ventricles, and recognized the division between the cerebellum (*paraenkephalis*) and cerebrum (*enkephalos*). He was the first to assign to nerves the function of nerve impulses.

1.1.6 Cardiovascular Anatomy

"The beginning of the physician's secret: knowledge of the heart's movement and knowledge of the heart" is the opening statement on the anatomical-physiological section of the Ebers Papyrus. The Egyptians regarded the heart as the central organ in the human body; they believed that it was the motor organ and the seat of intelligence, emotions, and desire. During the mummification process, the heart was kept *in situ* and was meticulously preserved to ensure its continued well being, for its loss meant annihilation in the second life. The heart is depicted in hieroglyphics as a character with a black arrow, or sometimes a vessel with an inlet and outlet as seen in the Edwin Smith Papyrus. Also significant is the attempt to identify the base and apex of the heart and the bilateral symmetrical projections, namely the pulmonary artery and aorta, in ancient Egyptian drawings.

The Ebers Papyrus records even greater knowledge of internal medicine, including a treatise on the heart and the circulation. The first link made between the heart and the pulse is found in this papyrus: "From the heart arise the vessels which go to the whole body ... if the physician lay his finger on the head, on the neck, on the hand, on the epigastrium, on the arm or the leg, everywhere the motion of the heart touches him. Coursing through the vessels to all the members." It is also important to note that the Egyptians were the first to associate air with the blood and cardiovascular system.

The Ebers Papyrus goes into detail in mapping the circulation in the cardiovascular system, including the number of vessels reaching the nostrils, inferior temples, extremities, testicles, liver, lung, and bladder. However, its authors did not believe that these vessels were conduits for blood, but rather for the element of the organ they supplied. For example, "There are 2 vessels to his testicles; it is they which give semen" illustrates their belief that these vessels from the heart carried semen rather than blood.

During the Alexandrian Period, Herophilus' contribution to cardiovascular medicine was his most important work in terms of scientific achievement. He was the first to recognize the difference between arteries and nerves. Before Herophilus, nerves, tendons, arteries, and veins were all referred to as "cords," represented as an erect penis in hieroglyphics. Herophilus is also recognized for distinguishing between arteries and veins; he described the arteries as carrying blood from the heart and the veins as returning blood to the heart. He described the anatomy of the heart valves and, along with Erasistratus (c. 304–250 BCE), demonstrated how blood was prevented from flowing backwards under normal conditions. Erasistratus may be considered the founder of physiology. He named the trachea and described the heart valves and chordae tendinae.

1.1.7 Reproductive Anatomy

The Kahun Papyrus, a gynecological text, enables us to examine ancient Egyptians' concept of human reproduction. They had two theories in regard to conception. One view was that the man's semen originated from the spinal cord: this was the belief spread by Egyptian priests who worshiped bulls and perceived that the bull's phallus was an extension of the spine. The other hypothesis was that the heart was the source of the semen, as mentioned in the Ebers Papyrus in reference to the two vessels traveling from the heart to the testicles. The ancient Egyptians believed that menstruation ceased during pregnancy because the blood was diverted to sustain the embryo. An interesting note is that they did not understand the maternal role in reproduction, except in serving as an incubator for the fetus; however, they appreciated the vital role of the placenta as a source of nourishment.

Ancient Egypt did not utilize the last menstrual period date to confirm pregnancy. According to Hippocratic writings (460–377 BCE), they instructed the female to embed an onion bulb deep in her vagina overnight. If the onion's strong odor was detected on the patient's breath, it confirmed the diagnosis. It was later posited that the absorption of the onion's sulfuric compounds into the woman's blood via the submucosal blood vessels, which were engorged due to pregnancy, might have been responsible for this "onion breath." It is unlikely, however, that the ancient Egyptians understood the anatomical and physiological reasoning behind this method.

Other gynecological complications facilitated ancient Egypt's understanding of the female reproductive system. There are indications in the Ebers Papyrus of ancient Egyptians suturing postpartum perineal tears and lacerations. Records have also shown references to treatment of long-term complications of pregnancy such as a prolapsed uterus. More interestingly, mention is made of vesicovaginal fistulas.

There is no way of knowing how much of the surgical, medical, and anatomical knowledge of ancient Egypt is irretrievably lost. However, what survives of that knowledge, exemplified by the three papyri discussed here, is remarkable. Thanks mostly to their embalming practices and their pioneering work in surgery, the Egyptians of the second millennium BCE achieved a surprisingly accurate and detailed knowledge of human anatomy. However, although they identified and named a wide range of structures, their ideas about the functions associated with those structures (physiology) were very different from ours. During the "renaissance" of anatomical knowledge in Egypt, in the Alexandrian Period, Greek rationalism and naturalism led to more refined descriptions and to a grasp of physiology that we find easier to recognize. Indeed, our present-day physiology has descended, in part, from the surviving works of Herophilus and Erastristatus.

1.2 Algeria

1.2.1 Jean Baptiste Paulin Trolard

A hundred years after his death, Jean Baptiste Paulin Trolard's name endures in the medical literature primarily because of his work relating to the anastomotic veins of the cerebral circulation (Figure 1.4). Specifically, the great anastomotic vein, or the vein of Trolard, known to all neurosurgeons, records a portion of Trolard's contribution to neuroanatomy. Algeria has also remembered this influential colonist because of his life's work as a physician, professor, humanitarian, environmentalist, and French nationalist. Trolard fought deforestation, injustice, epidemics, and bureaucracy in northern Africa, and tragically died in the midst of these struggles.

1.2.1.1 Early Life

Trolard was born on November 27, 1842 in the town of Sedan (Ardennes), France. During this time, the French were involved in a long and violent conquest of Algeria, which began in 1830 and was completed in the early 1900s, at a cost of approximately one million Algerian lives. Tens of thousands of Europeans, including a large French contingent, settled and farmed much of Algeria, and populated the capital city of Algiers. This mass immigration also marked the arrival of Paulin Trolard.

Trolard studied medicine at the Algiers Preparatory College of Medicine and began his medical career as a municipal physician in Saint Eugene, a suburb of Algiers. In 1861, the College named him anatomy prosector, and in 1868 he defended his doctoral thesis in Paris, entitled "Research on the Anatomy of the Venous System of the Encephalon and Skull."

In 1869, Trolard became the Chair of Anatomy at the Algiers Preparatory College of Medicine, a post he held until 1910. As a professor, he had a very clear method of teaching that was free of unnecessary erudition, and he published several texts concerning the organization of medical studies.

Figure 1.4 Jean Baptiste Paulin Trolard (1842–1910). (*Source:* Courtesy of the Bibliothèque interuniversitaire de médecine et d'odontologie, Paris, France.)

1.2.1.2 Neuroanatomical Descriptions

In his doctorate thesis of 1868, and then in a subsequent publication in 1870, Trolard described one of the three major anastomotic veins of the cerebral circulation. Interestingly, in his early anatomical descriptions of his "great annectant" vein, which became known as the "great anastomotic vein" or the "vein of Trolard" (Figure 1.5), Trolard mentioned his contemporary Charles Labbé's (1851–1889) description of the vein that bears Labbé's name. This latter vessel forms the major communication between the superior and inferior cerebral veins, creating an anastomotic route between the superior sagittal sinus and the middle cerebral vein and, therefore, the cavernous sinus. It is normally present in the non-dominant hemisphere and found at the level of the postcentral sulcus. The vein of Trolard may be duplicated and is usually directed forward against the direction of flow of the superior sagittal sinus. Aran et al. indicated that injury to this large, superficial drainage vein, recognizable by the sixth or seventh month of fetal life, may result in venous hypertension, with manifestations that include seizure and focal neurological deficits. Labbé believed that adding the modifier "antérieure" to the vein of Trolard was appropriate:

> One can find, behind the vein described by Monsieur Trolard, another vein, almost equally important, like the former playing an anastomotic role between the sinuses and that has never been reported before. The last one should be called the greater anastomotic posterior cerebral vein.

Trolard also described a venous anastomosis at the base of the brain, which is the venous equivalent of the circle of Willis. This venous "circle" is composed of the anterior cerebral and communicating veins, the basal vein of Rosenthal, and the posterior communicating and lateral mesencephalic veins (Figure 1.6). While in Algiers, Trolard published various neuroanatomical papers on other topics, including olfaction, the middle meningeal veins, the joints of the spine,

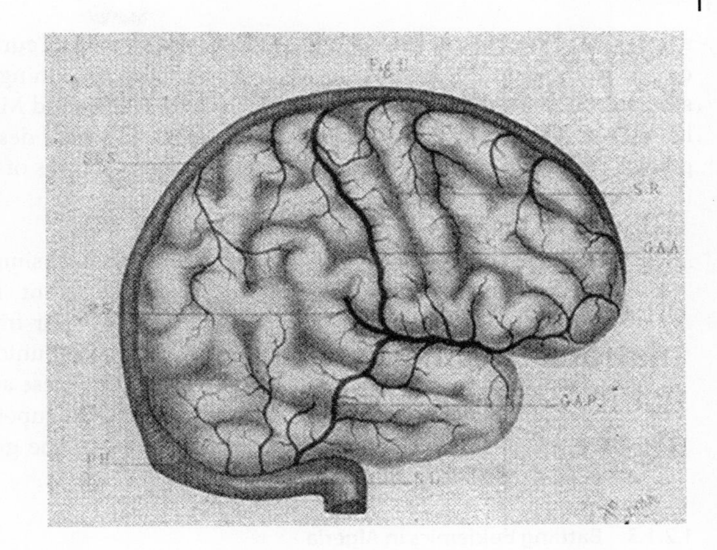

Figure 1.5 Drawing taken from Labbé's work on the superficial cerebral veins of the lateral brain, demonstrating the veins of Trolard (GAA) and Labbé (GAP). Interestingly, Trolard's original description of the vein that bears his name did not include figures.

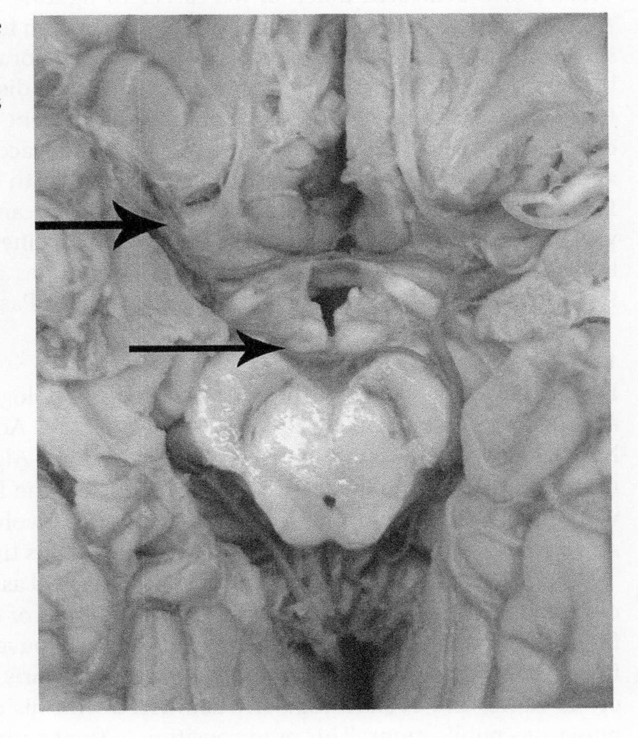

Figure 1.6 Cadaveric photo of the base of the brain. For reference, note the olfactory tracts and midbrain in the cross-section. Note the venous ring encircling the mammillary bodies and the floor of the third ventricle. The anterior cerebral veins are seen leaving the longitudinal fissure. Note the anterior communicating vein between the two anterior cerebral veins and the deep Sylvian vein (upper arrow). An anastomotic vein (lower arrow) is seen linking the basal vein of Rosenthal just posterior to the mammillary bodies. Note that the floor of the third ventricle has been perforated. (*Source:* With permission from Loukas et al. 2010.)

and the spinal cavity. He wrote that the vertebral vein acted like a dural venous sinus, as it surrounds the vertebral artery extracranially via multiple secondary channels linked by several septae and trabeculae. Trolard described a "net" or venous plexus, which now bears his name, and which surrounds the hypoglossal nerve while within the hypoglossal canal. He also correctly described the spinal dural cul-de-sac as terminating at the second sacral vertebrae, even in infants, which went against classical descriptions. Although Luschka studied the uncovertebral joints in 1858, Trolard is given credit for naming them in 1892. He described a ligament that connects the anterior spinal dura mater to the posterior longitudinal ligament and called it the "ligament a sacral anterius." The so-called "curtain sign" is due to the ligament of Trolard (sacrodural ligament of Trolard) and its partitioning of blood/tumor anterior to the spinal dural sac. Although Cruveilhier, Luschka, Henle, and Meyer had described the lacunae lateralis earlier, they carry the eponym "Trolard" and were described by him as the "lacs sanguins." Trolard is credited with first describing connections of regional veins to the arachnoid (Pacchionian) granulations and observing that

> The veins which open into the superior longitudinal sinus are described as opening into that canal in a direction contrary to the blood current. This is not quite true. The little veins, especially the anterior ones, open directly, or from before backwards, into the sinus. As to the two or three large veins, they communicate with the sinus from behind forward. But one must not overlook the fact that these anastomotic veins place in communication the superior longitudinal sinus with the superior petrosal and lateral sinuses. These anastomotic veins which are accessory to the great sinus must have the same direction that this last one had.

1.2.1.3 Battling Epidemics in Algeria

Trolard also dedicated much of his career to fighting contagious diseases and epidemics in Algeria. He was a major proponent of free vaccination for all indigent peoples and was responsible for procuring bacteriology and parasitology laboratories in Algiers. Even more remarkable was his work in combating rabies, a usually fatal disease at that time. Fascinated by Louis Pasteur's work on rabies and vaccination, Trolard went to Paris to visit him and to ask him to create an Institute in Algiers for the investigation of vaccination against rabies. Pasteur agreed, and in 1894, Professors Trolard and Henri Soulié, both of the Algiers School of Medicine and Pharmacy, founded the L'Institut Pasteur d'Algérie and established a rabies and smallpox vaccine department. Trolard went on to develop vaccines for anthrax and sheep pox. Although he lacked material resources to upgrade laboratories and develop research extensively, the Institute became the center of scientific research for Pasteur's methods in Algeria.

1.2.1.4 Conflict with France

For his efforts in studying the anatomy and physiology of the venous system, Trolard was awarded the Lallemand Prize (1800 francs) by the Academy of Sciences in Paris in 1893. Nevertheless, his commitment was to his work in Algeria. In addition to his professional responsibilities at the medical school, his duties at the Pasteur Institute, and his involvement with the League of Reforestation, Trolard was also involved in the local governance of Algiers, serving on the local governing council for much of his time in Algeria.

Despite his various positions, Trolard was described as "ennemi de la notoriété et dédaigneux des honneurs" (an enemy of notoriety and disdainful of honors). This led him into direct conflict with French authorities, particularly with the Gouvernement Général and the délégations financières with regards to his reforestation efforts. Paris disagreed with and was openly critical of his views, and Trolard regularly countered with his own scathing criticisms, often via his numerous publications. This predisposition to flout authority would eventually lead to the end of his career and, possibly, his death.

1.2.1.5 One Man's Legacy

Jean Baptiste Paulin Trolard, physician, anatomist, husband, and father, died in Saint-Eugène, Algeria on April 13, 1910, curiously soon after being "expelled" from Algeria. He had hoped to retire to his native France. Although he will forever be remembered by the anatomical structures bearing his name, specifically the greater anastomotic vein, his life meant much more to the countless people saved through his efforts, to the countless students he trained as physicians, and to the country he strove to make better. Algeria has remembered Trolard via a town, a church, and a street bearing his name.

Bibliography

Egypt

Breasted, J.H. (1992). The Edwin Smith Surgical Papyurs. In: *Neurosurgical Classics* (ed. R.H. Wilkins), 1–5. Rolling Meadows, IL: AANS.

Cappabianca, P. and de Divitiis, E. (2007). Back to the Egyptians: neurosurgery via the nose. *Neurosurg. Rev.* 30: 1–7.

Durant, W. (1939). *The Life of Greece, Vol 2. The Story of Civilization*. New York: Simon and Schuster.

Ebbel, B. (1937). *The Papyrus Ebers*. Copenhagen: Levinn and Munksgaard.

El Mahdy, C. (1989). *Mummies. Myth and Magic*. London: Thames and Hudson.

Fanous, A.A. and Couldwell, W.T. (2012). Transnasal excerebration surgery in ancient Egypt. *J. Neurosurg.* 116: 743–748.

Feldman, R.P. and Goodrich, J.T. (1999). The Edwin Smith surgical Papyrus. *Childs Nerv. Syst.* 15: 281–284.

Filler, A.G. (2007). A historical hypothesis of the first recorded neurosurgical operation: Isis, Osiris, Thoth and the origin of the djed cross. *Neurosurg. Focus.* 23: 1–6.

Haimov-Kochman, R., Sciaky-Tamir, Y., and Hurwitz, A. (2005). Reproduction concepts and practices in ancient Egypt mirrored by modern medicine. *J. Obstet. Gynecol. Reprod. Biol.* 123: 3–8.

Hamburger, W. (1939). The earliest known reference to the heart and circulation. *Am. Heart J.* 17: 259–274.

Kshettry, V.R., Mindea, S.S., and Batjer, H.H. (2007). The management of cranial injuries in antiquity and beyond. *Neurosurg. Focus.* 23: 1–8.

Lippi, D. (1990). An aneurysm in the Papyrus of Ebers. *Med. Secoli* 2: 1–4.

Longrigg, J. (1972). Herophilus. In: *Dictionary of Scientific Biography*, vol. 6 (ed. C. Gillespie), 316–319. New York: Charles Scribners Sons.

Loukas, M., Tubbs, R.S., Louis, R.G. Jr. et al. (2007). The cardiovascular system in the pre-Hippocratic era. *Int. J. Cardiol.* 120: 145–149.

Loukas, M., Hanna, M., Alsaiegh, N. et al. (2011). Clinical anatomy as practiced by ancient Egyptians. *Clin. Anat.* 24: 409–415.

Persaud, T. (1984). *Early History of Human Anatomy*, 44–49. Springfield, IL: Charles C. Thomas.

Sanchez, G.M. and Burridge, A.L. (2007). Decision making in head injury management in the Edwin Smith Papyrus. *Neurosurg. Focus* 23: 1–9.

Singer, C. (1957). *A Short History of Anatomy and Physiology from the Greeks to Harvey*, 2e. New York: Dover Publications.

Stiefel, M., Shaner, A., and Schaefer, S. (2006). The Edwin Smith Papyrus: the birth of analytical thinking in medicine and otolaryngology. *Laryngoscope* 116: 182–188.

Sullivan, R. (1997). Divine and rational: the reproductive health of women in ancient Egypt. *Obstet. Gynecol. Surv.* 52: 635–642.

Thurston, A.J. (2007). Pare and prosthetics: the early history of artificial limbs. *ANZ J. Surg.* 77: 1114–1119.

Von Staden, H. (1989). *Herophlius, the Art of Medicine in Early Alexandria*. Cambridge: Cambridge University Press.

Willerson, J.T. and Teaff, R. (1996). Egyptian contributions to cardiovascular medicine. *Tex. Heart Inst. J.* 23: 191–200.
Willius, F.A. and Dry, T.J. (1947). *A History of the Heart and Circulation*. Philadelphia: WB Saunders.
Wills, A. (1999). Herophilus, Erasistratus, and the birth of neuroscience. *Lancet* 354: 1719–1720.
Worth Estes, J. (1991). The medical skills of ancient Egypt. In: *Medicine, a Treasury of Art and Literature* (ed. A.G. Carmichael and R.M. Ratzans), 31–33. New York: Hugh Lauter Levin Associates Inc.

Algeria

Anonymous. Societies and academies. *Nature* 1893;49:216. http://books.google.com/books?id=BkoCBFtLx7AC&pg=PA216&lpg=PA216&dq=trolard+Lallemand+Prize++1893&source=bl&ots=vrj4NLqY61&sig=-jvT9kfa1qFGf0c5n43Q5Y0a-lk&hl=en&ei=KIURStuiCNfVlQf8rO2ICA&sa=X&oi=book_result&ct=result&resnum=1#PPA216,M1. Last accessed July 13, 2018.
Aran, E., Nogueira, N., Crespo, E. et al. (2004). Morphometric study with imaging techniques of Trolard's vein and its anastomosis to the superior longitudinal sinus. *Neurocirugia (Astur.)* 15: 372–376; disc. 376–377.
Bartels, R.H. and van Overbeeke, J.J. (1997). Charles Labbé. *J. Neurosurg.* 87: 477–480.
Davis, D. (2007). *Resurrecting the Granary of Rome: Environmental History and French Colonial Expansion in North Africa*. Cleveland: Ohio University Press.
Dobson, J. (1962). *Anatomical Eponyms*, 2e. Edinburgh: E&S Livingstone.
Ford, C. (2008). Reforestation, landscape conservation, and the anxieties of empire in French colonial Algeria. *Am. Hist. Rev.* 133: 341–362.
Franklin J. Il y a Cent Ans. http://nice.algerianiste.free.fr/pages/textes/trolard.htm. Last accessed July 13, 2018.
Henry, A.K. (1973). *Extensile Exposure*, 2e. Edinburgh: Churchill Livingstone.
Horne, A. (2006). *A Savage War of Peace: Algeria 1954–1962*. New York: Random House.
Le Minor, J. (2005). The anatomists in Algiers during the French colonial period (1830–1962). *Hist. Sci. Med.* 39: 385–396.
Loukas, M., Shea, M., Shea, C. et al. (2010). Jean Baptiste Paulin Trolard (1842–1910): his life and contributions to neuroanatomy. *J. Neurosurg.* 112: 1192–1196.
Morris, P. (2006). *Practical Neuroangiography*, 2e. Philadelphia: Lippincott Williams & Wilkins.
Petitjean, P., Jami, C., and Marie, A. (1992). *Science and Empires: Historical Studies about Scientific Development and European Expansion*. Boston: Kluwer Academic Publishers.
Rhoton, A.L. (2003). *The Cerebral Veins in Rhoton Cranial Anatomy and Surgical Approaches*. Schaumburg, IL: Congress of Neurological Surgeons.
Scapinelli, R. (1990). Anatomical and radiologic studies on the lumbosacral meningo-vertebral ligaments of humans. *J. Spinal Disord.* 3: 6–15.
Strachan, J. (2006). The pasteurization of Algeria? *Fr. Hist.* 20: 260–275.
Trolard, P. (1870). Recherches sur l'anatomie du système veineux du crane et de l'encéphale. *Arch Générales de Médecine* 15: 257–271.
Trolard, P. (1892). Les granulations de pacchioni les lacunes veineuses de la dure-mère. *J. de l'anat et de la physiol.* 28: 177–210.
Tubbs, R.S., Loukas, M., Shoja, M.M. et al. (2008a). The venous circle of Trolard. *Bratisl. Lek. Listy* 109: 180–181.
Tubbs, R.S., Loukas, M., Shoja, M.M. et al. (2008b). The lateral lakes of Trolard: anatomy, quantitation, and surgical landmarks. Laboratory investigation. *J. Neurosurg.* 108: 1005–1009.
Uhlenbrock, D. and Brechtelsbauer, D. (2004). Malformations of the spinal canal. In: *MR Imaging of the Spine and Spinal Cord* (ed. D. Uhlenbrock), 57–158. New York: Thieme.

2

Asia

Some of the earliest contributions to anatomy arose from a number of Asian civilizations, including China, Japan, and India. China's medical practices were centered around the concept of the two opposite forces of yin and yang, with various organs attributed to either one or the other. However, a lack of dissection – primarily due to superstition – led to many misconceptions about the body in this country. Much of Japan's knowledge of anatomy originated in concepts that arrived from China. However, the Japanese appear to have more readily embraced the exploration of the human form, with some influence from European traders and physicians. India's anatomic knowledge was influenced by both regional cultures and invading nations. It made significant and early advances in surgery, which necessitates some working knowledge of the human body.

2.1 China

Huang di was the third of the legendary "Five Emperors," known as the "Yellow Emperor"; he is said to have ruled from 2696 to 2598 BCE. He was the first sovereign of civilized China and is recognized as the common ancestor of the Han Chinese. He is also called "Xuanyuanshi," after Xuanyuan Hill, the place of his birth (in the northwest of present-day Xinzheng county, Henan province), and "Youxiongshi," after his capital (present-day Xinzheng county). Huang di is credited with civilizing the earth and inventing such things as the wheel, the compass, and coined money. Interestingly, he became a chief deity of Daoism during the Han Dynasty. His best known contribution to anatomy was the *Nei Jing* ("Internal Organs"), which forms the basis for traditional Chinese medicine and, as legend has it, was written as a narrative of questions between Huang di and his servant and chief physician, Qibo. Although traditionally entirely attributed to Huang di, several authors probably wrote the *Nei Jing* over a long period of time.

The *Nei Jing* divides the internal organs into five *zangs* (the yin, solid organs) and six *fus* (the yang, hollow organs). The *zangs* are the liver, heart, spleen, lungs, and kidneys; the *fus* are the gall bladder, stomach, large intestine, small intestine, bladder, and "burning spaces." There is controversy as to the function and location of the burning spaces. The *zangs* store but do not eliminate, while the *fus* eliminate but do not store. The *Nei Jing* says, "we read that the liver stores the blood, which contains the soul; that the heart stores the pulse, which contains the spirit; that the spleen stores nutrition, which contains the thoughts; that the lungs store the breath, which contains energy, and finally, that the kidneys store the germinating principle, which contains the will." By applying the five-element theory to the organs, Huang di attributed

History of Anatomy: An International Perspective, First Edition. R. Shane Tubbs, Mohammadali M. Shoja, Marios Loukas and Paul Agutter.

tastes, colors, and emotions to them. For example, the liver is sour, brown, and the center of anger; the heart is bitter, red, and the seat of happiness; the lungs are spicy, white, and the base of sorrow; and the kidneys are salty, black and the core of fright. These correlations play an important role in the diagnosis and treatment of patients in traditional Chinese medicine. If someone is deficient in kidney *qi* (life energy), it is recommended that they wear black and eat salty food.

Huang di also uses the analogy of the state to describe the functions of the organs and viscera:

> The heart is the king who directs the body, the lungs are the promulgators who carry out his orders. The liver is the general whose duty is to meditate carefully. The gall bladder is the central legal officer, who makes judgments. The pericardium is the minister who brings happiness. The spleen is the officer of granaries who creates the five tastes. The large intestine is the officer of communications. The small intestine is the receiving office in which digestion is performed. The kidney is the officer of vigor or strength who serves through his intellect. The three burning spaces constitute the sewage system from which all the canals drain into the bladder where the fluid is stored; after having been acted upon by the air, it is finally passed out.

Although the language and imagery used in the *Nei Jing* are lyrical, the concepts of the functions are surprisingly insightful. However, some descriptions are more concrete, such as that the "small intestine is attached to the spine posteriorly and the naval anteriorly." With the great stress on schematization (i.e., lack of detail), the picture of anatomy became very stylized.

A large emphasis is placed on the measurements of the organs and on their relationships to one another. It seems certain that the ancient Chinese anatomist had a strong curiosity in determining physical standards, as exemplified by the following distances found in the *Nei Jing* (translated into English units):

Width of the mouth: 2.5 inches
Distance from the teeth to the epiglottis: 3.5 inches
Weight of the tongue: 10 ounces
Width of the tongue: 2.5 inches
Length of the esophagus: 1.5 feet
Diameter of the rectum: 2.5 inches

Even if human dissection was not condoned at the time the *Nei Jing* was written, observations of the internal aspect of the human body were obviously being made – ostensibly, on the dead following battle or after trauma to the abdomen or chest.

2.1.1 Correcting the Errors in the Forest of Medicine

Wang Qingren (1768–1831) was a physician during the Qing Dynasty (1644–1912), when Chinese scholars were beginning to question traditional teaching and belief systems. Much of traditional anatomical teaching was still based on the *Nei Jing*. Division occurred between those who regarded the classic Chinese texts as infallible and indisputable and those who thought of them as simply documents that should be examined and discussed. This atmosphere no doubt had quite an influence on Wang. With his book, *Yi Lin Gai Cuo* ("Correcting the Errors in the Forest of Medicine"), he attempted, after 42 years of study, to depose the ancient teachings that had been passed down to him (Figures 2.1–2.4). *Yi Lin Gai Cuo* was first published in 1831, as a response to the *Nei Jing*. Wang touted direct observation

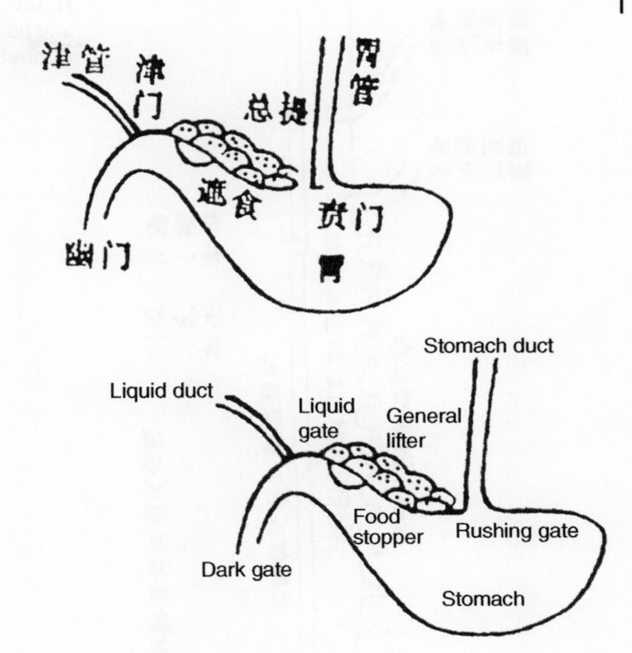

Figure 2.1 A depiction of the anatomy of the stomach from *Yi Lin Gai Cuo* ("Correcting the Errors in the Forest of Medicine").

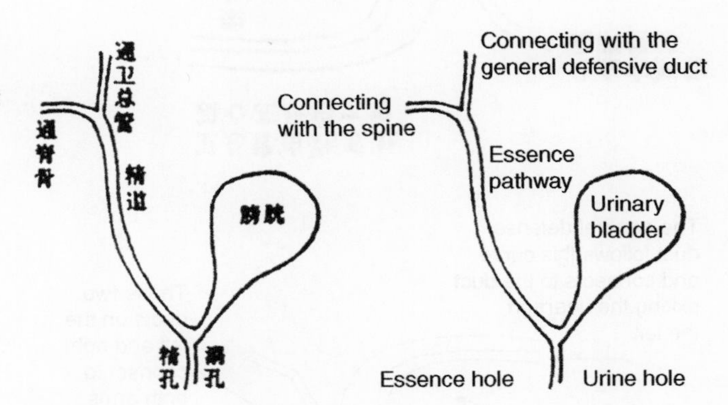

Figure 2.2 A depiction of the anatomy of the urinary system from *Yi Lin Gai Cuo* ("Correcting the Errors in the Forest of Medicine").

as the basis for his conclusions. He managed to circumvent the Confucian prohibition of body degradation in an ingenious yet unconventional way. He recounted:

> In 1797, when I was 29 years old … I was passing through the town of Daodi in Luan Prefecture [modern Hebei]. At the time an epidemic was killing eight or nine out of every ten of the children there. Poor families mostly used mats on which to bury the dead, the mats taking the place of coffins. The local custom was to bury the dead shallowly, so that dogs would eat them. The belief was that this would promote the live birth of the next child [of the same parents]. Because of this, every day in each charity graveyard there would be a hundred or so children with their abdomens torn, exposing their entrails. Every day I would urge on my horse when passing such places, at first always covering my nose. However, later I reflected that the reason the ancients had erroneously described the organs was because they had never seen them with their own eyes, so instead of avoiding the pollution, early every morning I would go to a charity graveyard and examine in detail the children with exposed organs.

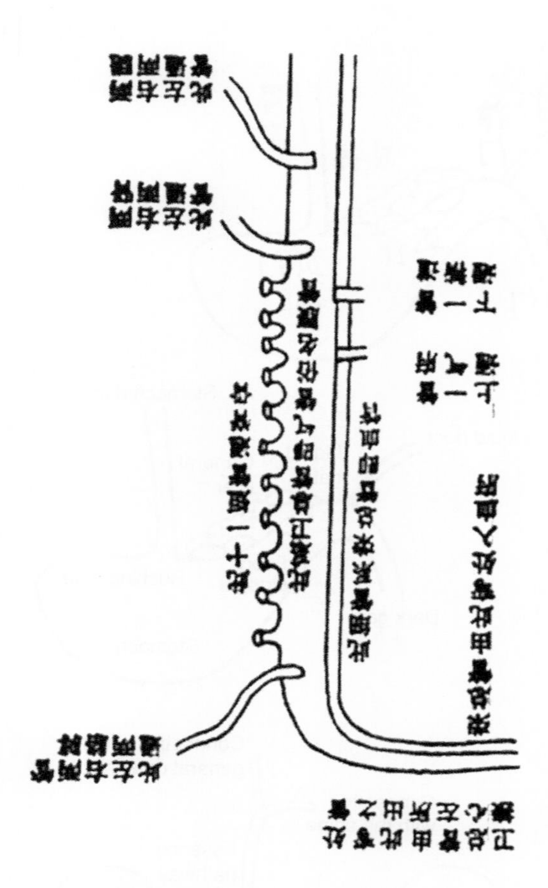

Figure 2.3 A depiction of the anatomy of the viscera and bowels (original Chinese) from *Yi Lin Gai Cuo* ("Correcting the Errors in the Forest of Medicine").

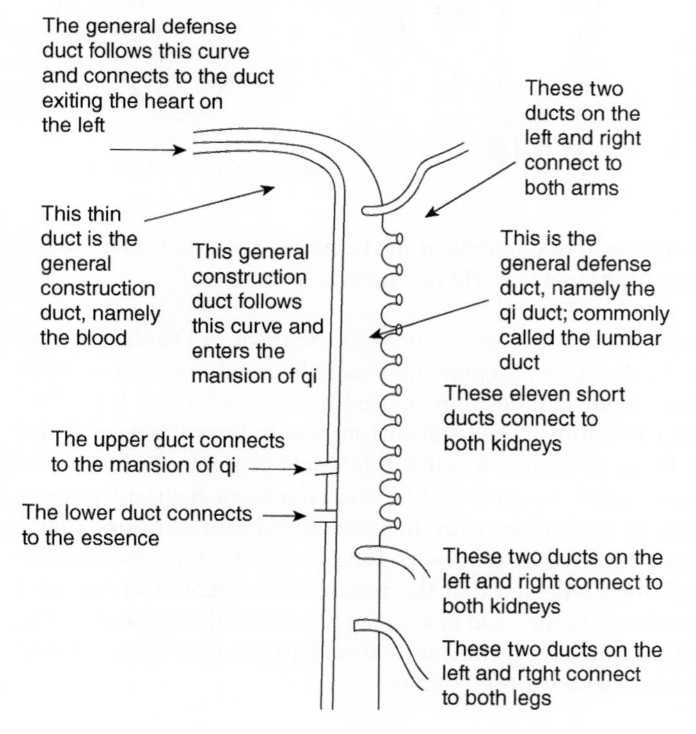
The general defense duct follows this curve and connects to the duct exiting the heart on the left
These two ducts on the left and right connect to both arms
This thin duct is the general construction duct, namely the blood
This general construction duct follows this curve and enters the mansion of qi
This is the general defense duct, namely the qi duct; commonly called the lumbar duct
These eleven short ducts connect to both kidneys
The upper duct connects to the mansion of qi
The lower duct connects to the essence
These two ducts on the left and right connect to both kidneys
These two ducts on the left and rtght connect to both legs

Figure 2.4 A depiction of the anatomy of the viscera and bowels (translated from Chinese) from *Yi Lin Gai Cuo* ("Correcting the Errors in the Forest of Medicine").

Although sporadic instances of dissection had occurred in Chinese history, the knowledge recorded was superficial and often incorrect. So intrigued was Wang with the study of anatomy, that he subsequently performed animal dissections on pigs and sheep. The human cadavers that he inspected contained no blood, which led him to erroneously believe that the thorax was a blood reservoir and that the heart did not transmit blood, but instead was a propeller of *qi*, and its vessels "*qi* pipes." However, his descriptions of the courses these vessels traveled were more or less correct. He described the carotid arteries as "'*qi* portals' that open into the throat ... *qi* pass into the heart, which transmits the *qi* through the 'main defensive *qi* pipe' [the aorta] down to the abdominal cavity where it is stored in the folds of the '*qi* reservoir' or mesentery." He also referred to the vena cavae as the "main constructive pipes." The concept of two forms of *qi* derived from digestion is part of the classical theory, in which there is pure, constructive *qi* and murky, defensive *qi*. Thus, Wang's differentiating between the aorta as the "main defensive *qi*" pipe and the vena cavae as the "main constructive *qi*" pipes demonstrates a primitive understanding of the arterial and venous systems. However, his interest in blood and its relationship to *qi* contributed to the understanding and treatment of a pathological condition referred to as "blood stasis" in traditional Chinese medicine. The bodies of the children that Wang examined all had large blood clots present in the thorax, and this led him to postulate that when the blood is stagnant to such an extent that the *qi* is unable to pass, a person will suffocate. Moreover, "if primordial *qi* is deficient, then it will be unable to reach the blood pipes and if the blood pipes have no *qi* in them, then blood is bound to stagnate and coagulate." With such concepts, Wang developed prescriptions that have become standard in modern traditional Chinese medicine practices, aimed at replenishing *qi* and invigorating the blood to prevent stasis and clotting.

Other anatomical findings recorded by Wang include his rebuke of the long-held belief that urine originates from excrement and that the lungs have 24 holes through them. Wang also, for the first time in Chinese writing, described the abdominal aorta, vena cavae, hepatic duct, pancreas, pyloric sphincter, and diaphragm. He averred that it was the brain and not the heart (as was previously thought) that was the seat of thought and memory; B.J. Andrews has stated, however, that this notion was already to be found in other medical works of the late Ming and early Qing Dynasties, apparently due to contact with Jesuit missionaries in the sixteenth century.

2.1.2 Hua Tuo

Regarded as China's first surgeon, Hua Tuo (also known as Yuan Hua of Hao County) (Figures 2.5 and 2.6) acquired a surgical acumen and a keen understanding of herbal medicine at an early age. Born around 108 CE in the Eastern Han Dynasty in present-day Haoxian, Anhui Province, Tuo's lofty reputation spread rapidly throughout the country, as the public praised his efforts to remedy even the most grueling conditions. As a child, he lived in poverty but explored various academic fields, including astronomy, geography, literature, history, and agriculture. He acquired his passion for medicine at the age of seven, upon witnessing the death of his father.

He is credited for spearheading the practice of laparotomy and organ transplant, using anesthetics, and he was the first Chinese surgeon to operate on the abdomen, including performing splenectomy and colostomy. Neurologically, Hua Tuo is said to have performed procedures to treat headache and paralysis, and to have suspected a brain tumor in one patient. Many physicians attribute to Tuo the ability to feel the pulse and evaluate the patient accordingly. Following a miscarriage, a woman approached Tuo, complaining of severe abdominal pain. Upon evaluating her pulse, he diagnosed her with a twin pregnancy, solidifying the presence of a dead fetus

Figure 2.5 Portrait of Hua Tuo.

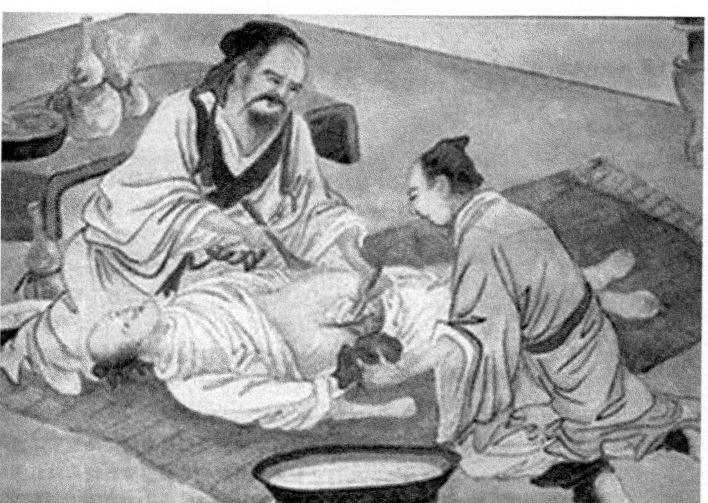

Figure 2.6 Woodblock by the Japanese artist Utagawa Kuniyoshi depicting Hua Tuo operating on General Guan Yu, as mentioned in the historical novel *Romance of the Three Kingdoms*.

inside her abdomen. With a mere session of acupuncture, he evoked the stillbirth. Furthermore, Tuo evaluated the position of the fetus to conjecture about the gender of the unborn child. His impact on medicine was so profound that the phrases "a second Hua Tuo" and "Hua Tuo reincarnated" were coined in his honor, in reference to physicians who demonstrate an equal caliber of surgical competence.

2.1.3 Conclusion

It is evident that some of the fundamental differences between Eastern and Western medicine can be traced to the relative importance placed on human dissection and the exploration and understanding of human anatomy. The Western tradition, starting with the ancient Greeks, put a great emphasis on dissection and the development of surgical and invasive medical techniques. The Chinese, bound by Confucian doctrine, developed methods of understanding and explaining the human body without the benefit of direct observation. This does not discount the contributions made to the study of anatomy through such works as the *Nei Jing* and *Yi Lin Gai Cuo*. Although highly conceptual and based on esoteric ideas of energy and harmony, the Chinese tradition's perspective on the interrelatedness and cooperation between all aspects of the human body teaches an important lesson that enables one to truly comprehend its functional anatomy. These writings made significant contributions to the understanding of human anatomy, both in early China and in nearby countries like Japan.

2.2 Japan

Early vague and mythical anatomical concepts derived from China prevailed for many centuries in Japan. Kajiwara wrote one of the earliest Japanese anatomical works in 1302. Anatomy was the first basic science to be established in the country, beginning simplistically during the 1600s and flourishing with the onset of the Meiji Restoration. As a result, Japan produced several of the most influential anatomists of the twentieth century.

The history of Japan is steeped in deep tradition. The country was historically closed to the West until the mid-1800s, but afterward began to accept and assimilate new ideas into its culture, including those surrounding the basic sciences. Durant describes the ethnic combinations that forged ancient Japan: the Ainu, who entered the country from the region of the Amur River (Heilong Jiang) in Neolithic times; the Mongols, who emigrated through Korea in the seventh century BCE; and the Malay and Indonesian cultures, which came to the islands from the south.

This section provides a discussion of the evolution of the study of anatomy in Japan, and how Japanese civilization has added to and promoted the anatomical sciences.

2.2.1 Feudal Japan (1186–1868)

In the feudal period, Japan was governed by powerful upper-class households (*daimyo*) and was militarily supported by the shogun, or supreme military leader; the emperor was simply a figurehead during this era. Important events of this period include the invasion of Japan by Kublai Khan in 1281 and the failed conquest of Korea by Hideyoshi in 1592. The onset of the Edo Period was marked by the beginning of the Tokugawa Shogunate in 1603, a regime that lasted until 1867. This feudal military dictatorship was established by Tokugawa Ieyasu (1542–1616) and was ruled by the shoguns of the Tokugawa family. This period is known as the

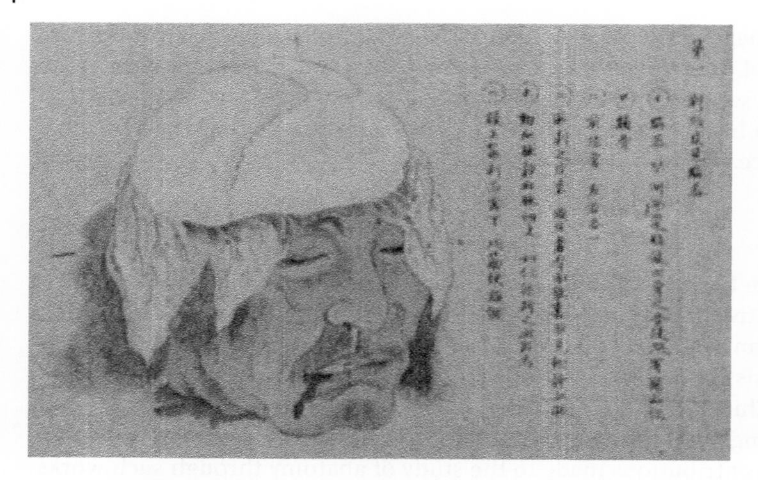

Figure 2.7 Painting of the exposure of the calvaria from the Late Edo Period.

Edo Period in reference to the capital city of Edo (now Tokyo). It was followed by the Meiji Restoration or Meiji Revolution (1868–1912), which signalled the end of Japan's feudalism.

In 1302, during the Kamakura Period, the priest-physician Shuzen Kajiwara (1266–1337) wrote the *Ton-i-sho,* one of the earliest anatomical works in Japan. The *Ton-i-sho* consisted of 50 volumes, one of which was devoted to human viscera (although it was not an accurate depiction).

During the Edo Period (1600–1867), medical and anatomical knowledge was heavily influenced by traditional Chinese medicine (Figures 2.7 and 2.8). In this era, the Tokugawa Shogunate insisted on isolating Japan from the rest of the world, allowing trade only with China and the Netherlands, through the port of Nagasaki. Some Portuguese and Spanish medical practitioners were also allowed entrance during this time. In fact, the Portuguese arrived in Tanegashima in southern Kyushu at a time that coincided with the publication of Vesalius' *Corporis Humani Fabrica* in 1543. The Portuguese arrival allowed an unprecedented direct contact between Europe and Japan. Five years later, Francis Xavier, a Jesuit missionary, arrived in the country. The introduction of Jesuit priests was important to the history of anatomy, because some of them practiced Western surgery, and thus studied the human body. Interestingly, Durant mentions that linguistic barriers beleaguered the missionary efforts of the Jesuits, and that the Jesuits considered the Japanese language to have been invented by the devil in order to prevent the preaching of the Gospel to the Japanese. This introduction of Christianity, however, was blocked in 1639 when the Tokugawa government implemented *sakoku* ("country in chains"), the closing of Japan to the rest of the world; the increasing influence of the Christian church had aroused the suspicion of many Japanese leaders.

Importantly, despite the induction of *sakoku,* an official decree of exchange with the Netherlands allowed Dutch physicians to bring in scientific literature, including anatomical texts. Interestingly, the Dutch study of surgery was referred to as *komo-ryu geka* (surgery of the red hair) or *oranda-geka* (Holland surgery). Bowers stresses that the "red" in the former term did not suggest hair color, but rather the diabolical nature of the group.

The Dutch influence spread when, in 1740, Shogun Yoshimune implemented a systematic study of Dutch, which gave rise to the Rangaku (scholars who studied Dutch), a group that included Genpaku Sugita (see later). The first anatomical text translated into Japanese was the *Pinax Microcosmographicus in Quo Certissimum Anatomae* by Johann Remmelin (Germany, 1667), translated by Ryoi Motoki (1628–1697) in approximately 1680. However, this Japanese

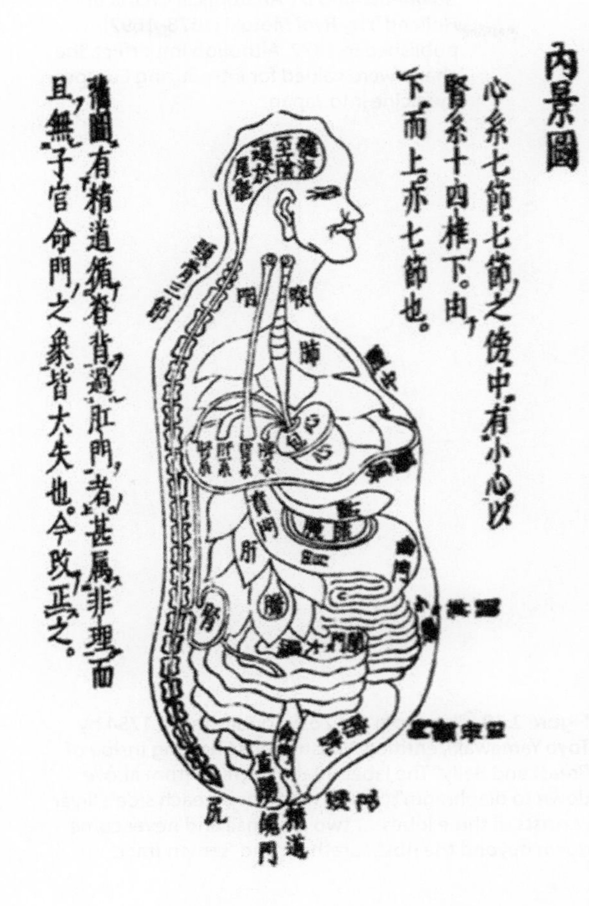

Figure 2.8 The "inner landscape" of the human body, based on historic Chinese concepts. (*Source:* Courtesy of the Kyushu University Library).

translation, entitled *Oranda Zenku-Naigai-Bungo-Zu* ("Anatomical Charts of Holland"), was not published until 1772, 75 years after Motoki's death (Figure 2.9). Ogawa verified that Motoki performed the translation almost one century prior to the publication of the *Kaitai-Shinsho* (see later).

In 1713, Terashima Ryoan compiled the *Wakan Sansai Zue* ("Illustrated Sino-Japanese Encyclopedia"). This publication was influenced by traditional Chinese medicine, but disseminated the simplistic anatomical drawings known as *Gozo roppu zu* ("drawing of the five viscera and six entrails"). The *gozo,* or five viscera, were the liver, heart, spleen, lung, and kidney (Figure 2.10). The *roppu,* or six entrails, were the gall bladder, small intestine, stomach, large intestine, urinary bladder, and *sansho* (for which there is no corresponding term in Occidental languages). The function of these internal organs was steeped in Confucian teachings, however. For example, the testes were thought to produce bone marrow, milk was formed from female blood in the spleen and stomach, and the reproductive organs ventilated through the ears by means of channels that passed from the spleen along the right side of the body.

Kataoka et al. mention that in 1741, Toshuku Negoro drew human skeletons based on his observations of the corpses of two criminals who had been burned at the stake. Negoro, an ophthalmologist and artist from Kyoto, may have been a student of the *Ko-i-ho* school. His drawings were incorporated into an anatomical book (*Zobutsu Yotan,* published in 1781) written by the noted Japanese physician and philosopher Baien Miura (1723–1789), who had an interest in anatomy.

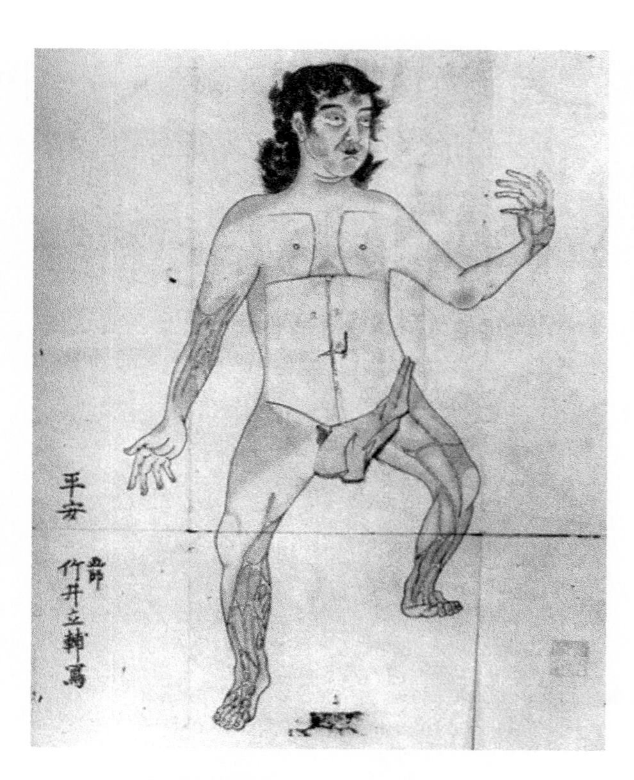

Figure 2.9 Drawing from *Oranda Zenku-Naigai-Bungo-Zu* ("Anatomical Charts of Holland") by Ryoi Motoki (1628–1697), published in 1772. Although imperfect, the charts were valued for introducing European medicine into Japan.

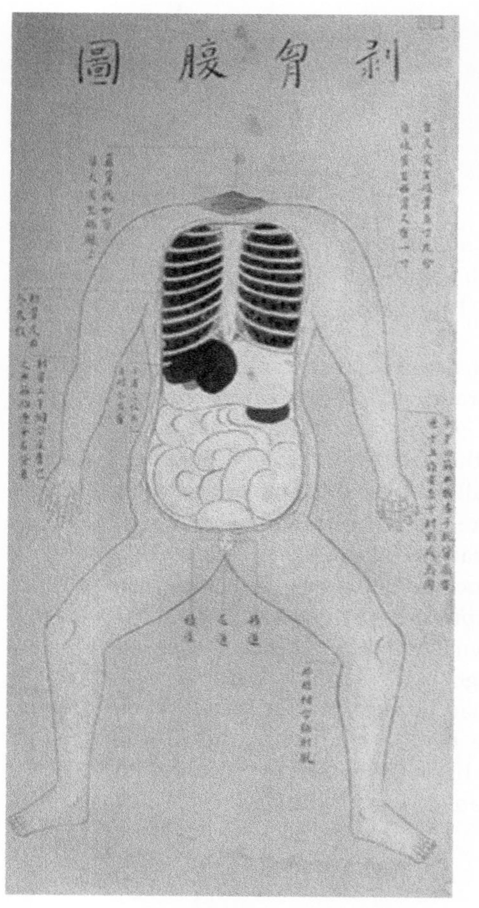

Figure 2.10 Plate from the *Zoshi*, published in 1754 by Toyo Yamawaki, entitled, "Illustrations Showing Inside of Breast and Belly." The labels read: "sternum from above down to diaphragm"; "lungs, nine ribs on each side"; "liver consists of three lobes … two are small and never come down beyond the ribs"; "urethra"; and "semen tract."

Kataoka et al. also recall that a primitive dissection was performed for the first time in Kyoto (and Japan, for that matter) under the supervision of Toyo Yamawaki (1705–1762), a pupil of Konzan Goto (1659–1733), who went on to write *Zoshi* ("Record of the Viscera") in 1759. Toyo was a member of the Kohoha (Ancient Practice) School, and in his publication he refuted the traditional teachings of the viscera, as taught by the Goseiha School. Bowers notes that Yamawaki used Johan Vesling's (1598–1649) text *Syntagma Anatomicum* as a reference guide. Low observes that Yamawaki participated in dissections as early as 1754, but that he was unable to persuade other Japanese physicians to free themselves from many erroneous Chinese perceptions of the human body. However, Ogawa stated that some individuals defended Yamawaki's observations; in fact, a commemorative stone marks the location of Yamawaki's dissection in Kyoto. In 1773, Riken Nakai (1732–1817), friend and sympathizer to Confucianists in Osaka and to the famous philosopher Goryu Asada (1734–1789), wrote *Esso-rohitsu* to describe Asada's views on the anatomy of the human body. This text was published one year prior to the *Kaitai-Shinsho* (see later).

2.2.2 Shinnin Kawaguchi

In 1770, Shinnin (Nobuto) Kawaguchi and Gengai Ogino published *Kaishihen* ("Notes on Autopsy"), an original description of the process of opening a cadaver (Figure 2.11). Some have considered this work to be as important to Japanese anatomy as Toyo Yamawaki's *Zoshi*. Kawaguchi studied with Doi Kurisaki, a proponent of European surgery.

In 1770, following an execution, Kawaguchi was given two cadavers and a head to dissect on-site, and these became the fifth and sixth human dissections recorded in Japan. In addition, Kawaguchi became the first person in Japan to dissect the brain. Unlike Yamawaki, Kawaguchi performed dissections himself. Kawaguchi's teacher was opposed to the publication of his notes on the dissections, as he did not want to contradict Yamawaki's findings. Regardless, Kawaguchi commissioned figures from Shunmei Aoki and published his text.

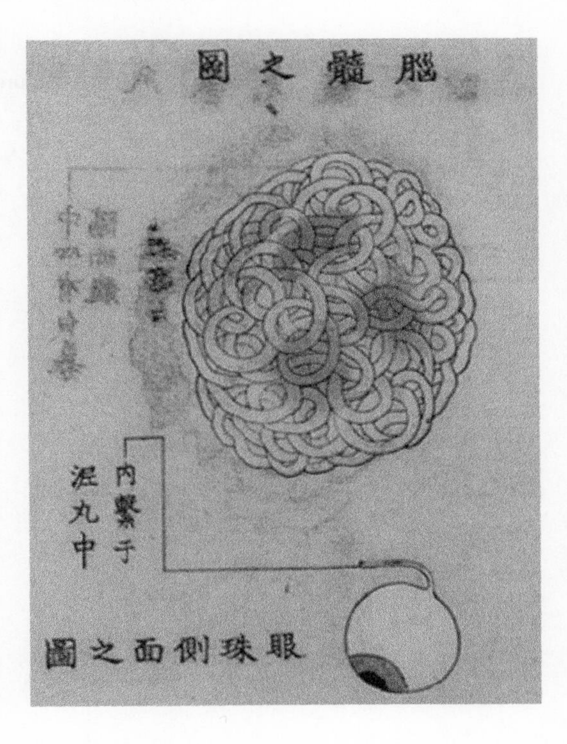

Figure 2.11 Brain and eye in Kawaguchi's *Kaishihen*.

2.2.3 Genpaku Sugita

Genpaku Sugita was born in 1733 (d. 1817) (Figure 2.12). He was the son of the official doctor to Lord Wakasa, and he studied medicine under the surgeon Nishi Gentetsu. After establishing himself as a medical practitioner, he began studying Dutch. In 1774, he and his colleagues, including Ryotaku Maeno (1723–1803), Hoshu Katsuragawa (1750–1808), Genjo Ishikawa (1744–1816), and Junnan Nakagawa (1739–1786), translated the 1734 Dutch anatomical text *Ontleekundige Tafelen* by Johann Adam Kulmus (1689–1745) (originally published in 1722 as *Anatomische Tabellen* and dedicated to the Dutch anatomist Frederik Ruysch, 1638–1731). After 11 draft translations (and approval by shogunate officials and court nobles, to ensure that there would be no significant political repercussions), they published the translation as the *Kaitai-Shinsho* ("A New Book of Anatomy") (Figures 2.13–2.15).

The impetus behind this translation was the 1771 dissection of a female body following an execution at Kotsugahara in Edo. Sugita recalled his thoughts after witnessing the dissection:

> On our way home, Maeno, Nakagawa and myself talked about the strong impression this made on us. We were ashamed of having lived so long in such a complete ignorance and served our lords day after day as physicians without the slightest idea of the true configuration of the body, whereas this should have been considered the foundation of our art. (Horiuchi, 2003)

The art within the *Kaitai-Shinsho* was unprecedented in Japanese anatomical works, and it originated from a variety of sources. The artist Odano Naotake (1749–1780) (Figure 2.16), an Akita samurai and remote relative of the Aoyagi clan, rendered copies of the artwork used in Kulmus' text, although the last four plates were derived from *Ontleding des Menschelycken Lichaams* (1690), written by the Dutch anatomist Govert Bidloo (1649–1713). After learning Western art techniques, Naotake went on to establish the Akita Ranga (Dutch drawing method). The Japanese artist Eisho Yoshio also created special engravings for the *Kaitai-Shinsho*.

Figure 2.12 Painting of Genpaku Sugita (1733–1817).

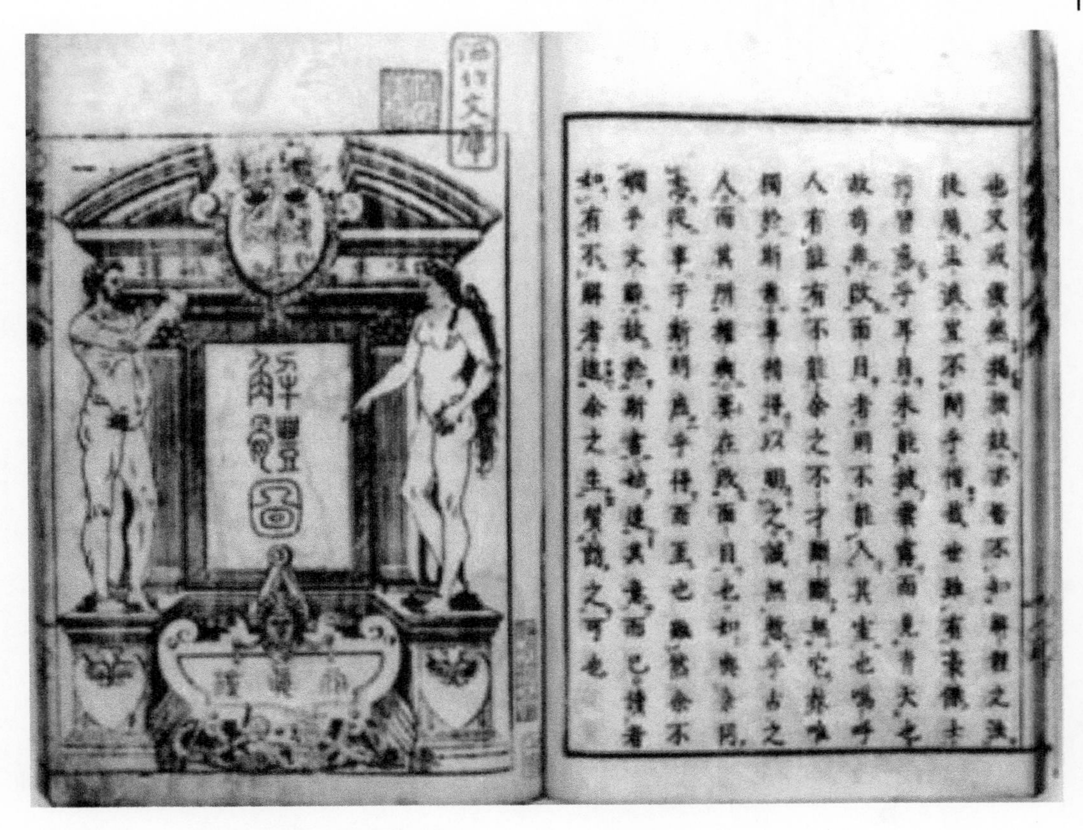

Figure 2.13 Cover page of the *Kaitai-Shinsho*.

The *Kaitai-Shinsho* was the first European work to be printed and published in Japan. This book became a nidus for increased Japanese interest in Western medicine and science. Eventually, it was also printed in Chinese, so that its influence could spread. Notably, many new words for anatomical structures were created in Japanese at this time; there were no equivalent Chinese or Japanese words for many of the Dutch anatomical terms. In fact, Sugita is said to have stated, "If the reader [of the *Kaitai-Shinsho*] cannot understand some parts, please come to me to make it sure while I am alive." The *Kaitai-Shinsho* has been referred to as the "Japanese Vesalius," and a copy was presented to the Shogun.

Soon after (c. 1800), in Osaka, Kagami Bunken (1765–1829) produced *Fujin Naikei No Ryaku Zu* ("Illustrations Showing an Internal View of the Female Body"). Shortly after that, a student of Sugita, Genshin Utagawa (1769–1834), published his three-volume *Ihan Teiko* in 1805, which described the essence of the Western view of anatomy. Ogawa posits that this latter work was the most popular among the translated medical texts until the end of the Tokugawa Period.

2.2.4 Ryotaku Maeno

Ryotaku Maeno (1723–1803), listed earlier as a co-contributor to the *Kaitai-Shinsho*, was educated by his physician uncle Zentaku Miyata. At the age of 47, Maeno became the pupil of Konyou Aoki (1698–1769), who was the Shogunate's book-magistrate. He became fluent in Dutch and played the greatest role in the translation of the *Kaitai-Shinsho*. Interestingly, however, Maeno's name was not included on the cover page of the book. One account states that

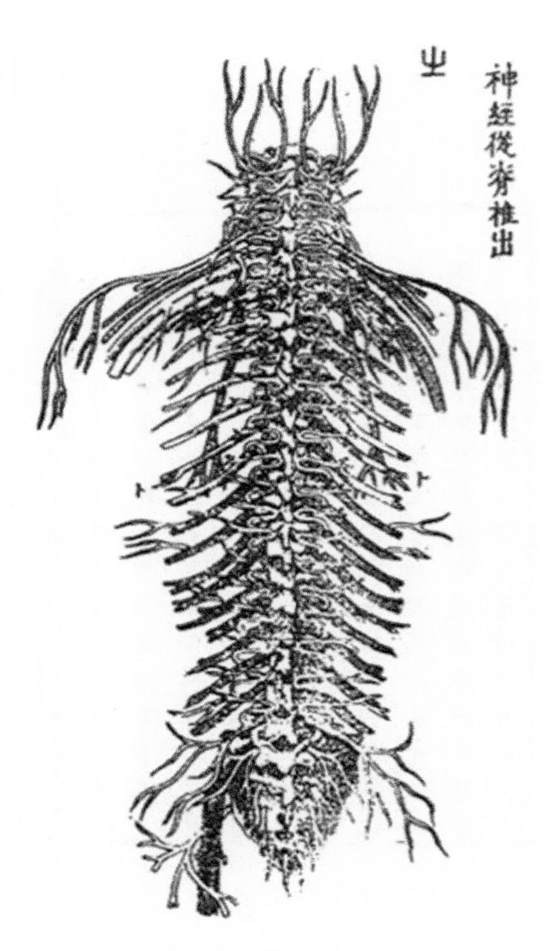

Figure 2.14 Illustrations from the *Kaitai-Shinsho*.

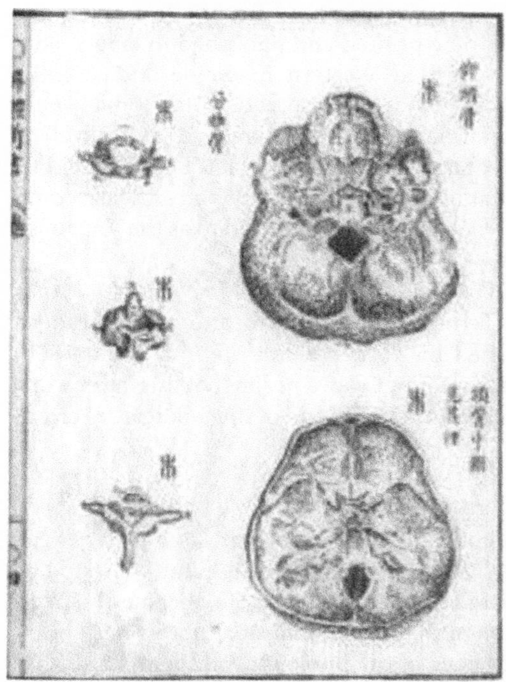

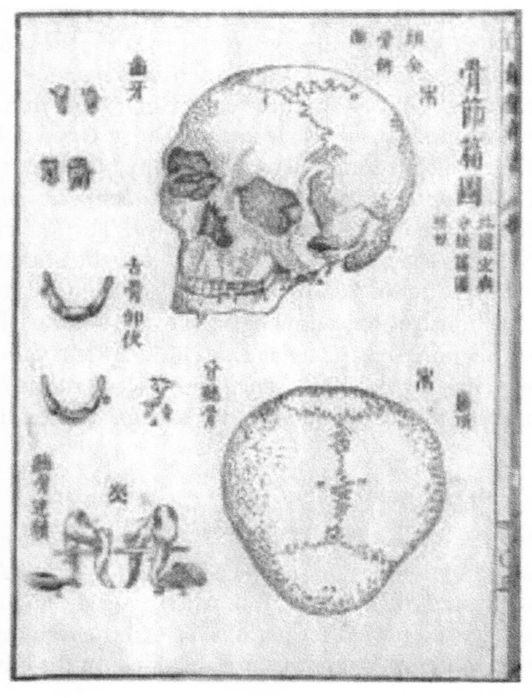

Figure 2.15 Illustrations from the *Kaitai-Shinsho*.

Figure 2.16 Bust of Odano Naotake (1749–1780).

while on his way to study in Nagasaki, Maeno prayed for fulfillment of his studies and vowed not to study to raise his own name; hence, he chose not to be included as an author of the *Kaitai-Shinsho*.

2.2.5 Shoteki Fuseya

Shoteki Fuseya (1745–1811) was a practitioner of Chinese medicine who converted to studying texts based on Western anatomy, such as the *Kaitai-Shinsho*. Although he performed many physiological experiments on animals, he also made detailed observations on female anatomy. Through observation and experimentation, he found that the kidney filtered the blood to produce urine, and that this urine was transferred to the urinary bladder via the ureters. Fuseya also studied the automaticity of the frog heart and the neuromuscular junction. Such observations were published in his *Oranda-Iwa* ("Dutch Medical Study") in 1805. Interestingly, it was almost two decades after Fuseya's experiments that William Bowman proposed his secretory theory for urine formation.

2.2.6 Otsuki Gentaku

An important anatomical text was published in 1826 by Otsuki Gentaku (1757–1827). This book, the *Chotei-Kaitai-Shinsho* (*chotei* meaning "revised"), corrected many of the errors in translation within the *Kaitai-Shinsho* and was the most complete Japanese anatomical book during the Tokugawa Period. As a child, Otsuki studied the Dutch language under Maeno and Sugita; he surpassed his teachers to become the greatest authority in the Rangaku. In fact, Sugita asked him to revise the *Kaitai-Shinsho*. Otsuki went on to establish a private school in Edo called the Shirando.

2.2.7 Koji Harada and Ryoetsu Hoshino

In the late Edo Period, approximately 10 wooden skeletons (mokkotsu) were constructed for the study of medicine. These skeletons, made between 1792 and 1820, were important because during this period of Japanese history, cadavers and real skeletons were prohibited for use in medical education. During this same period, practitioners of Eastern medicine denounced the credibility and usefulness of cadavers and skeletons. They argued that one would not acquire further knowledge of bodies by studying dead organs, since the interpretation of such complicated matter is beyond human ability.

In 1792, Koji Harada crafted the first wooden skeleton at the request of the physician Ryoetsu Hoshino. Hoshino, who lived in Hiroshima, had studied the temporomandibular joint of a skull that he had found, and had identified a method for relocating this joint when dislocated. He successfully used this technique on a patient, and became convinced that a model of the human skeleton could be of great use to physicians. After repeated requests to the feudal government of Hiroshima, he was finally given two corpses, which he and other physicians dissected. They collected the bones and constructed a wooden model based on them. The Hoshino skeleton narrowly escaped destruction by the atomic bomb in 1945; fortunately, it had been moved outside of the city for care.

2.2.8 Human Dissection

As already mentioned, although human dissection was occasionally allowed in Japan (for the purpose of confirming the anatomy as described by Chinese medicine), it was generally considered contrary to Confucian teachings, because it damaged the body given by one's parents. Shintoism also associated those handling the dead with impurity. This mindset was contrary to that of European perspectives during the same period; European anatomists believed that the mysteries of the universe could only be answered by studying the minimized version of the universe, the human body. They fervently insisted that studying cadavers was essential under Christian doctrine, since it allowed one to understand the nature of God, as it is stated in the Bible: "God made man in his own image."

Lower than slaves in Japan were the *eta* (now known as the *burakumin*), literally meaning "full of dirt." This caste of pariahs was considered despicable and unclean by Buddhist Japan, and acted as butchers, tanners, scavengers, and "dissectors." Low elaborates that the dissections mediated by the *eta* were carried out primarily on beheaded criminals, were hardly experimental in nature, and thus did not advance science. Instead, they were often carried out for monetary gain, in order to collect human bile for medicinal purposes. Physicians observed and organized such procedures, but they were not directly involved in the acts (Figure 2.17). Low notes that one Japanese physician mentioned that he had seen various internal body structures and did not know what they were, but that they had appeared in all of the bodies that he had seen dissected.

In 1759, Kuriyama Koan, a student of Yamawaki, dissected the corpse of a 17-year-old girl who had been accused of adultery and faced crucifixion. Prior to her death, the girl agreed to have her body dissected, and thus her sentence was "reduced" to a beheading. Koan made accurate descriptions of the female reproductive system based on this dissection.

Another important contributor to the Japanese introduction of modern Western medicine, including anatomy, was the German physician and scientist Philipp Franz Balthasar von Siebold (1796–1866), who was sent to Dejima in 1823. Siebold commented that it would be impossible for Japanese physicians to understand Western medicine as long as human dissection was banned. He was eventually expelled from Japan in 1830 after being accused of treachery. Recently, however, Japan has honored Siebold's influence by issuing a commemorative stamp in

Figure 2.17 Scroll dating to the Late Edo Period depicting a dissection by an *eta*, under the observation of a physician.

Figure 2.18 Bust of Philipp Franz Balthasar von Siebold (1796–1866), from his homeplace in Nagasaki.

1996, and by placing a bust at the site of his historic residence in Narutaki, Nagasaki (Figure 2.18). Another notable European contributor to Japanese medical knowledge during this time was J.L.C. Pompe van Meerdervoort (1829–1908), a Dutch physician who taught in Nagasaki and performed dissections for educational purposes, probably for the first time in Japan.

2.2.9 Seishu Hanaoka

Seishu Hanaoka (1760–1835) was an important figure in the advancement of surgery in Japan. He was a practitioner of both traditional Japanese medicine and Dutch-imported Western surgery. During Hanaoka's time, Japan was under *sakoku.* This meant that his achievement of performing surgery with narcotherapy could not be shared with the rest of the world until after it was lifted in 1854. Hanaoka was successful in the development of a surgical anesthetic, tsusensan or mafutsusan, which was composed of aconite and Korean morning glory (Mandarage and Torikabuto roots). Both Hanaoka's wife and his mother allowed him to experiment on them with the anesthetic during its different stages of development. Hanaoka can be credited with performing the world's first mastectomy with the use of an anesthetic in 1805, more than four decades before the West was to begin experimenting in the science of anesthesia (Figures 2.19 and 2.20). The *sakoku* policy played a large role in this 40-year delay. It has been documented that throughout his career, Hanaoka performed somewhere in the realm of 150 mastectomies for the treatment of breast cancer.

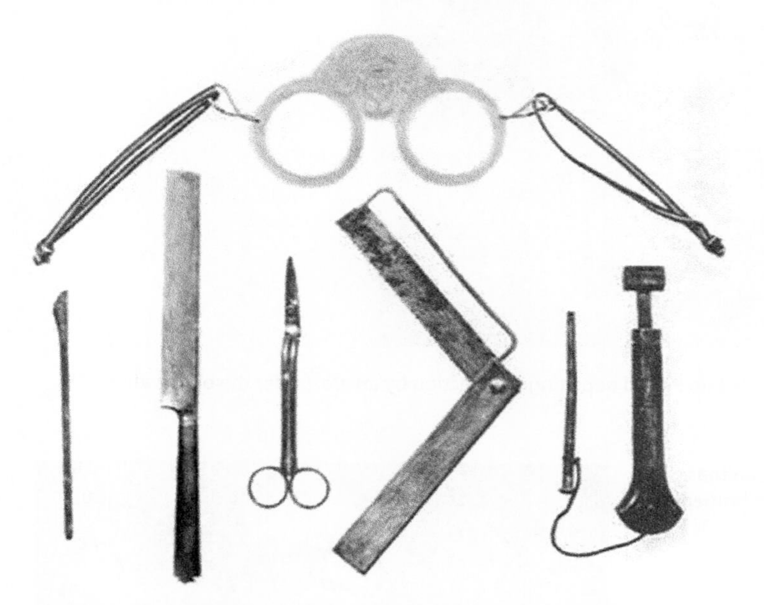

Figure 2.19 Surgical instruments used by Seishu Hanaoka (1760–1835).

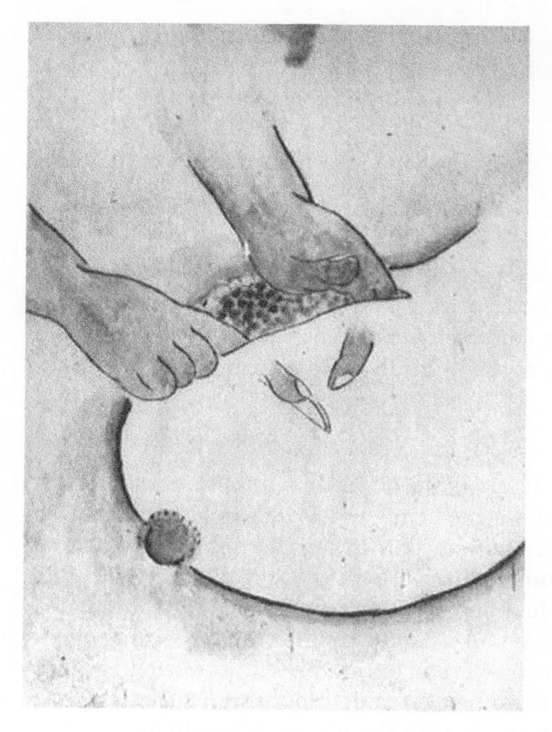

Figure 2.20 Illustration of the method of mastectomy drawn by Akaishi (1794–1857), who was a pupil of Hanaoka.

At the same time as Hanaoka was making his important advancements, works by a French surgeon, Jean Louis Petit, were published, which tied together different proposed aspects of the mastectomy into a more united model. Petit's idea stressed the removal of axillary lymph nodes, but also included removal of pectoralis fascia, adipose tissue, and part of the pectoralis muscle if it was involved, while preserving as much skin as possible. Dr. William Stewart Halsted later made this method of mastectomy (now known as the radical mastectomy) famous.

2.2.10 Buntaro Adachi

Buntaro Adachi (1865–1945) (Figure 2.21), a Japanese physician, anatomist, and anthropologist, is best known for his detailed descriptions and classifications of vascular anatomic variations in humans. His most famous studies are *Das Arteriensystem der Japaner* ("The Arterial System of the Japanese") (Figure 2.22), published in 1928, and *Das Venensystem der Japaner* ("The Venous System of the Japanese") (Figures 2.23 and 2.24), published in 1933 and 1940. These books, written in German, the academic language of early twentieth-century Japan, are still regarded as masterpieces of vascular anatomy.

Adachi was born on June 15, 1865 in what is now Izu City, Shizuoka Prefecture, Honshu. He was the eldest son born to his father Chozo and his mother Suga. In his youth, Adachi's uncle, Kiyoshi Inoue (his mother's brother), who was a medical doctor, supported his academic interests. This mentorship resulted in Adachi attending and graduating from Tokyo Imperial University (currently, Tokyo University) in1894, when he was 29 years old. After his graduation, he studied under Yoshikiyo Koganei (1859–1944), an anatomist and anthropologist, and the first Japanese to give a formal lecture on anatomy in Japan. Koganei studied anatomy with Wilhelm Waldeyer in Strasbourg and Berlin from 1880 to 1885. In order to uncover the origins

Figure 2.21 Photograph of Buntaro Adachi (1865–1945).

Anatomie der Japaner

von

PROF. DR. BUNTARO ADACHI

I

Das

Arteriensystem

der

Japaner

von

DR. BUNTARO ADACHI

Professor der Anatomie an der Kaiserlichen Universität zu Kyoto

unter Mitwirkung von

DR. KOTONDO HASEBE

früherem Extraordinarius für Anatomie an der Kaiserlichen Universität zu Kyoto
zurzeit Professor der Anatomie an der Kaiserlichen Universität zu Sendai

In zwei Bänden

Mit 539 Abbildungen im Text und auf vier farbigen Tafeln sowie mit etwa 700 Tabellen

Band I: A. pulmonalis, Aorta—Arcus volaris profundus
Band II: Aorta thoracalis—Arcus plantaris profundus

Kyoto 1928

Verlag der Kaiserlich-Japanischen Universität zu Kyoto
In Kommission bei „Maruzen Co.", Kyoto und Tokyo
Gedruckt von „Kenkyusha" in Tokyo

Figure 2.22 Cover of Adachi's *Das Arteriensystem der Japaner.*

of the Japanese race, he returned to Japan and became Professor of Anatomy at Tokyo Imperial University. He studied the ancient and modern skeletons of the Japanese minority ethnic group the Ainu, and suggested that this group originated from early humans (i.e., Jomon people). Adachi himself would publish on similar topics, such as his *Nihon sekki-jidai zugai* ("Crania from the Stone Age in Japan"). In 1895, Adachi began teaching at the University of Okayama Medical School. During this time, he studied variations between Japanese skulls and skulls from neighboring countries, as well as the variations of the bones and muscles of the Japanese, and was prolific in publishing both articles and anatomical texts, such as his *Nihonjin domyaku keito* ("Studies on the Anatomy of the Japanese").

In 1899, the Japanese government sent Adachi to study anatomy in Strasbourg, Germany. There, he studied under Gustav Schwalbe (1844–1916), who was Professor of Anatomy at the University of Strasbourg. Schwalbe was a famous anatomist who had discovered the medial vestibular nucleus, the nerve of the femoral artery, and the iridocorneal ring and intervaginal space of the optic nerve. Schwalbe was also known as an expert anthropologist, owing to his research on Neanderthal skulls. For five years, Adachi immersed himself in the study of anatomy in Germany. The Japanese government paid for his first three years of study abroad and he covered his own expenses for the additional two. During this time, Adachi visited

Figure 2.23 Cover of Adachi's *Das Venensystem der Japaner.*

Anatomie der Japaner

II

Das

Venensystem

der

Japaner

Von

DR. BUNTARO ADACHI

Professor (emer.) der Anatomie an der Kaiserlichen Universität Kyoto
Mitglied der Kaiserlichen Akademie

Zweite Lieferung

Mit 27 Abbildungen im Text und
74 Abbildungen auf 23 teilweise farbigen Tafeln
sowie mit etwa 120 Tabellen

Vv. thoracicae longitudinales
und Vv. intercostales

V. cava caudalis

Kyoto 1940
Druckanstalt KENKYUSHA Tokyo

many European museums and investigated their osteologic collections. He also studied the Mongolian spot (birthmark over the lumbosacral junction), which was thought by many Japanese to be the result of coitus during pregnancy. He found that the pigment in this spot was located in the epidermis and corium and occurred in variable amounts in different races. He studied the body odor of both Europeans and the Japanese, finding the smell of Germans to be unpleasant.

It is assumed that the anthropologic studies of Koganei and Schwalbe affected Adachi's research. Anthropology in those days focused primarily on the shape of the skeleton. However, Adachi felt it was also necessary to study the differences in soft tissues between races. Later in life, he summarized his thinking as follows:

> Anthropology was hitherto the field of study that treats only differences of external appearance such as color of the skin, shape of the face, character of the hair, or skeleton. However, if there are variations in the external appearance and skeleton among races, there also should be variations in all the soft tissues such as the muscles that cover the bones, vessels and nerves supplying these muscles, and the viscera as well. Hence, I want to establish a new research field called "anthropology of the soft tissues." Thinking on

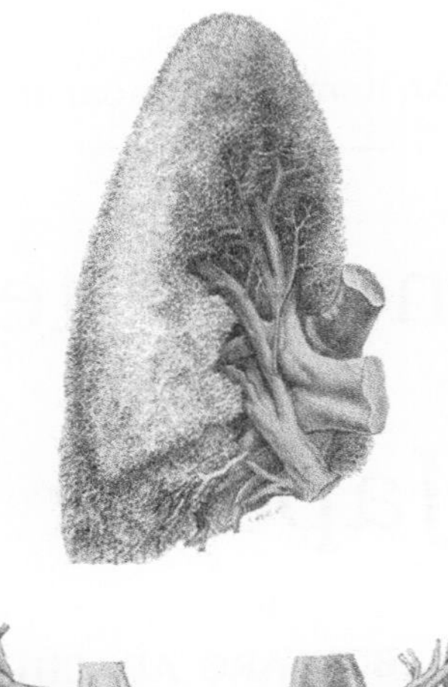

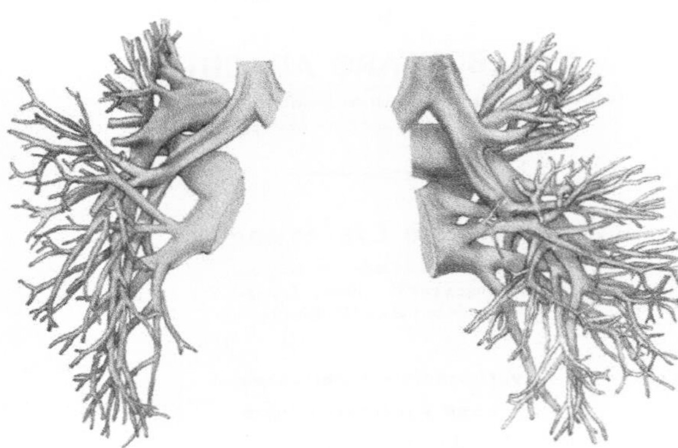

Figure 2.24 Color drawing from *Das Venensystem der Japaner*, noting variations in the arrangement of the pulmonary vessels and bronchi of the lungs.

these matters, present human anatomy is not human race anatomy, but is European-specific anatomy derived from research on the European race. Knowledge of specific anatomy is needed for each race. Thus, Japanese anatomists first need to establish Japanese-specific anatomy. These are the reasons I majored in anatomy and it is the starting point of my work.

Adachi became Associate Professor of Anatomy at Kyoto Imperial University in 1900, and then Professor of Anatomy in 1904, just after returning from abroad. Upon his return to Japan, he worked to establish Japanese-specific anatomy. He dissected large numbers of Japanese cadavers and examined soft-tissue variations. He studied variations of arteries and veins in the whole body statistically. He published few articles in the 24 years after his return to Japan, and it is assumed that he spent most of his time collecting and analyzing massive amounts of data on vascular variations. He also studied the lymphatic system (*Anatomie der Japaner: Das Lymphgefasssystem der Japaner* – "The Lymphatic System of the Japanese"), muscle variations (*Nihonjin kinhakaku no tokei* – "Statistics on the Muscles of the Japanese"), and the anthropology of cerumen.

In 1925, Adachi retired from Kyoto Imperial University, as was mandatory. Two years later, in 1927, he became a principal of what is now Osaka Medical College. Adatchi was the first president of this college, and there is today a statue of him near the main gate of the academic building. During his anatomy lectures at Osaka Medical College, Adachi always advised his students that they should never give up until they achieved their dream. During World War I, he went to great efforts to protect his research data from destruction – specifically, data regarding thoracic duct variations. After the war, he published the compilations of his long-term studies.

Adachi wished to publish more of his studies; however, with unstable world conditions such as the aftermath of World War I (1914–1918), the significant economic depression of the 1920s, and the Great Kanto Earthquake that struck Tokyo in 1923, this plan was never realized. Nevertheless, based on the dissections of hundreds of bodies, he published his *Anatomie der Japaner: Das Arteriensystem der Japaner* in 1928. This was followed by the two parts of his *Anatomie der Japaner: Das Venensystem der Japaner* in 1933 and 1940 (Figure 2.25). These three books were sent to medical schools all over the world and received much attention among anatomists and anthropologists. He was elected an honorary member of the Japan Academy

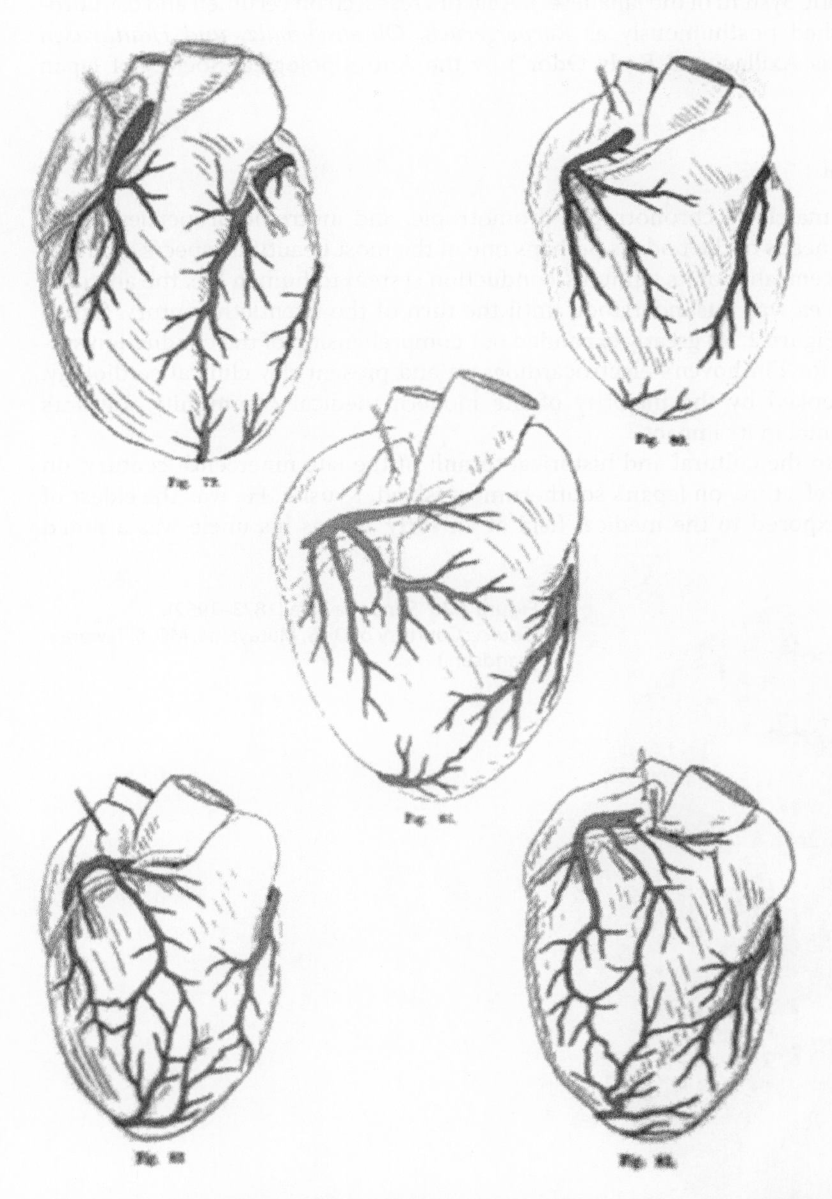

Figure 2.25 Drawing from Adachi's *Anatomie der Japaner: Das Venensystem der Japaner* demonstrating the variations in the cardiac veins from an anterior view.

and similar academic societies in Germany, Austria, Russia, Poland, and Argentina. He was given the Imperial Prize for his study on arterial variations by the Japan Academy (Nippon Gakushi-in) in 1930, and was elected as a member of the Leopoldina Nationale Akademie der Wissenschaften in 1941. He continued studying after retirement, and compiled his research on the variations of the thoracic duct. In 1945, Adachi died from a cerebral hemorrhage at his home in Kyoto at the age of 80 years. It is said that he was studying at his desk when he died. After his death, his successor, Takusaburo Kihara, completed his study on the variations of the lymphatic system, and in 1953 he published *Anatomie der Japaner: Das Lymphgefasssystem der Japaner* ("The Lymphatic System of the Japanese"). Adachi's research on cerumen and osmidrosis axillae was published posthumously as *Korpergeruch, Ohrenschmalz, und Hautdrusen* ("Cerumen, Osmidrosis Axillae, and Body Odor") by the Anthropological Society of Japan in 1981.

2.2.11 Sunao Tawara

The heart's ability to match its chronotropic, dromotropic, and inotropic properties to the dynamically changing needs of the body is perhaps one of the most beautiful aspects of physiology, but despite the centrality of its stimulus-conduction system to human life, the anatomy and function of this area was misunderstood until the turn of the twentieth century. Sunao Tawara (1873–1952) (Figure 2.26) greatly expanded our comprehension of the conduction system, setting the stage for Einthoven's electrocardiogram and present-day clinical cardiology. While arguably overlooked by the majority of the modern medical community, Tawara's research was monumental in its impact.

Tawara was born into the cultural and historical tumult of the late nineteenth century, on July 5, 1873, in Oita Prefecture, on Japan's southernmost island, Kyushu. He was the eldest of 10 children. He was exposed to the medical field at an early age, as his uncle was a noted

Figure 2.26 Sunao Tawara (1873–1952). (*Source:* Courtesy of Dr. S. Murayama, MD, S. Tawara's grandson.)

physician in the Kyushu city of Nakatsu. His uncle adopted him when he was 15, providing him with additional opportunities to pursue medicine. The younger Tawara's academic success was well noted. At the age of 16, he moved to Tokyo to study English and German at the Dai Ichi High School, and he went on to become the top student at the Imperial University of Tokyo, receiving his undergraduate degree from the renowned institution in 1901. Tawara then began his medical education at the Imperial University, studying both internal medicine and dermatology, before deciding in 1903 to continue his studies in Marburg, Germany with the noted pathologist Dr. Karl Albert Ludwig Aschoff (1866–1942) at the Pathological Institute of Philipps–Universität Marburg. Tawara spent the next three years working with the famed German (himself remembered today via the Aschoff bodies that are pathognomonic of rheumatic fever), before returning to the Imperial University in 1906. He was well received by academia, and became the Vice-Chair Professor of Pathology at Kyushu University after earning his medical degree from the Imperial University in 1908. Tawara went on to become dean of the Pathology Department at Kyushu and the president of Japan's Pathological Association, before he retired in 1933. He succumbed to dementia in 1952 at the age of 78.

Tawara's great discoveries in cardiac anatomy occurred during his years of collaboration with Aschoff. From 1903 to 1906, Tawara published two works in Germany, the latter being his landmark treatise on the heart's conduction system, *Das Reizleitungssystem de Saugetierherzens. Eine anatomich-histologische Studie uber das Atrioventricularbundel und die Purkinjeschen Faden* ("The Conduction System of the Mammalian Heart: An Anatomico-histological Study of Atrioventricular Bundle and the Purkinje Fibers") (Figure 2.27).

As is the case in much of scientific discovery, Tawara hit upon the conducting system in a somewhat unanticipated manner. Aschoff had asked him to examine the bundle of His with the hope of gaining insight into the histopathology of valvular and myocardial cardiac failure, but Tawara was unable to discover any notable morphological changes characteristic of this cardiomyopathy. However, the systematic dissections and observations of the bundle of His required by the cardiomyopathy study enabled him to gain a deeper understanding of the heart's conducting system. Several months before his departure from Marburg, Tawara wrote a letter to Aschoff asking for his help in completing the cardiomyopathy manuscript. Tawara additionally stated his desire to continue studying the conduction tissue of the heart, with the intention of penning a monograph on the subject to be dedicated to his father, to

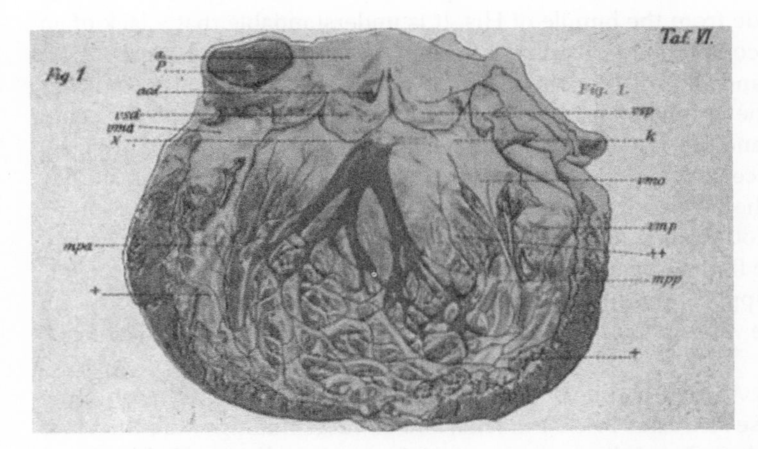

Figure 2.27 Picture from Tawara's monograph *Das Reizleitungssystem de Saugetierherzens. Eine anatomich-histologische Studie uber das Atrioventricularbundel und die Purkinjeschen Faden*, showing the tree-fascicular arrangement of the left bundle branch in humans. (*Source:* Tawara 1906.)

whom he felt deeply indebted. Tawara's letter emotionally touched Aschoff, and the student's wishes were granted. Continuing onwards with his conduction research, Tawara was able to correctly observe the heart's conducting system under the dissecting microscope. He completely dissected the "AV connecting system" by exploring both proximal and distal to the bundle of His in humans, cattle, rats, rabbits, guinea pigs, dogs, and cats. In so doing, he dissected out the Pukinje fibers, the bundle of His, the bundle branches, and the atrioventricular node, discovering not only the atrioventricular node ("knotten"), but also the connections between these structures.

Tawara's work, by providing a complete visualization of the ventricular conducting system, allowed scientists to view the heart's wiring in a proper fashion. Each of the system's individual components (with the notable exception of the AV node) had been described previously. In fact, they were discovered in a nearly retrograde manner, with Jan Evangelista Purkinje first observing his terminal fibers in 1839, then Wilhelm His Jr. discovering the His bundle in 1893, Retzer and Braeunig supplying independent commentaries on the bundle branches in 1904, and Tawara reporting his discovery of the AV node in 1906. However, the bundle of His and Purkinje fibers had only been viewed in isolation, and thus their form and function were mistaken. The role of Purkinje fibers, for example, was not understood prior to Tawara's dissections demonstrating them to be part of the "AV connecting system." Similarly, observations of the delay between atrial and ventricular contractions and the work of Wilhelm His Jr. prompted physiologists to believe that the bundle of His was a slow conduction system that directly excited myocytes near the base of the heart, leading to the erroneous conclusion that the base of the heart began contracting before the apex. However, Tawara recognized that the bundle of His was a fast conduction system connected to a previously undescribed nodal structure, presumably by employing what would eventually be called the "Criteria of Aschoff and Monckeberg," which stated that cardiac conduction tissue must be: (i) histologically distinct from surrounding tissue; (ii) always traceable from one serial section to another; and (iii) insulated by fibrous tissue. As the bundle of His and the bundle branches satisfy all three criteria, Tawara appreciated these connected structures as fast conduction systems, as they in fact are. Tawara's description of the bundle of His's insulation made it clear that this structure could not depolarize the heart's base (as was previously thought). The AV node, by contrast, fulfills only two of the Aschoff–Monkenberg criteria, since insulating fibrous connective tissue does not surround the node. In fact, Tawara used the lack or presence of insulating connective tissue to demarcate the AV node from the bundle of His. It is understandable that a lack of an insulating layer would slow conduction, providing an ostensible electrophysiological reason for the delay between atrial and ventricular contraction. Likewise, Tawara was able to recognize the specialized role of the terminal Purkinje fibers in cardiac contraction, since he could visualize the system *en bloc*, and this visualization (which he equated to the structure of a tree) allowed him to properly discern the timing and locations of myocyte contractions as they relate to the contraction of the cardiac apex prior to the base. This was how Sunao Tawara, a pathologist by training – not to mention a student – was able to solve a centuries-old physiological mystery. These findings were not just interesting in the pure academic realms, as they had definite clinical applications in the development and interpretation of electrocardiograms. Consequently, the AV node was – for a time – referred to as the "AV Node of Tawara" in his honor.

Unfortunately, Tawara and Aschoff chose to issue their findings in a dense German tome published in 1906 by the house of Gustav Fischer in Jena, and Tawara's findings were therefore scantily circulated within the broader medical community. His noted modesty probably played a role in his historical overshadowing as well. Indeed, the eponymous use of Tawara's name has virtually disappeared within medical parlance, and broader interest in his works might have been extinguished altogether were it not for the single-handed effort of Koso Suma. Nonetheless, Tawara's work was well appreciated in some academic communities (such as among cardiovascular physiologists) and in certain parts of Japan: a monument commemorating the scientist stands in Fukuoka City, and one of the streets in the Kyushu University Medical Center is named after him. With the 100th anniversary of his publication recently passing, curiosity over his historic contributions has risen across Japan. In spite of his humility, perhaps Tawara's works will gain greater renown among modern disciples; perhaps they will even influence modern students as they once influenced Sir Arthur Keith, who stated, "With the discovery of the conducting system of Tawara, heart research entered a new epoch."

2.3 India

Healing traditions and medical practices are inextricably tied to human history, with the oldest known civilizations – Egypt, China, Mesopotamia, and others – having healing traditions that increased human knowledge concerning medical sciences, particularly anatomy. As one of these ancient civilizations, India has a rich history and tradition of medicine. The foundation for modern Indian Ayurvedic medicine can be identified in ancient texts, some of which are said to predate the Christian era by 4000 years. The developmental history of ancient India can be divided into three periods: the Vedic (c. 1500–500 BCE), Brahmanic (600 BCE–1000 CE), and Mughal (1000 CE until the eighteenth century). The four Vedas, considered the oldest Sanskrit literature and the basis for Hindu religion, were compiled during the Vedic Period. They contain rituals, hymns, and incantations, and the Atharva Veda scripture focuses on health and medicinal practices.

The Vedic philosophies form the basis of the Ayurvedic tradition, considered to be one of the oldest known systems of medicine. They are founded on two sets of texts, the *Sushruta Samhita* and the *Charaka Samhita*. The *Sushruta Samhita* was written by the famous physician and surgeon Sushruta in the sixth century BCE. Sushruta taught at the University of Benares (alternatively, Kasi or Varanasi), situated on the Ganges River, and is best known for his tome of surgical wisdom, practices, and tools. In Sushruta's work, considerable thought is given to anatomical structure and function, as he was a proponent of human dissection; his texts include a systematic method for the dissection of a human cadaver. Charaka lived in the middle of the second century BCE, and was associated with the northwestern part of India and the ancient university of Taksasila. His six-volume book, the *Charaka Samhita*, focuses on the maintenance of health. The *Charaka Samhita* is philosophical and ethical in its considerations, and includes an Oath of Initiation that is similar to the Hippocratic Oath. However, it does not deal much with anatomy, so will not be examined further in this section.

2.3.1 Sushruta

Sushruta (Figure 2.28) is believed to have been born in the eastern part of India, near Bihar. Known as the Father of Indian Surgery, he was the first man to practice rhinoplasty in India. His dates of birth and death are a controversial subject among medical historians, as his famous work, the *Sushruta Samhita*, has not survived, and exists only in the form of distant revisions and copies.

Sushruta's *Samhita* emphasized surgical matters, including the use of specific instruments and types of operations, and his work concerns significant anatomical considerations of the ancient Hindu. Furthermore, there is compelling evidence that knowledge of human anatomy was

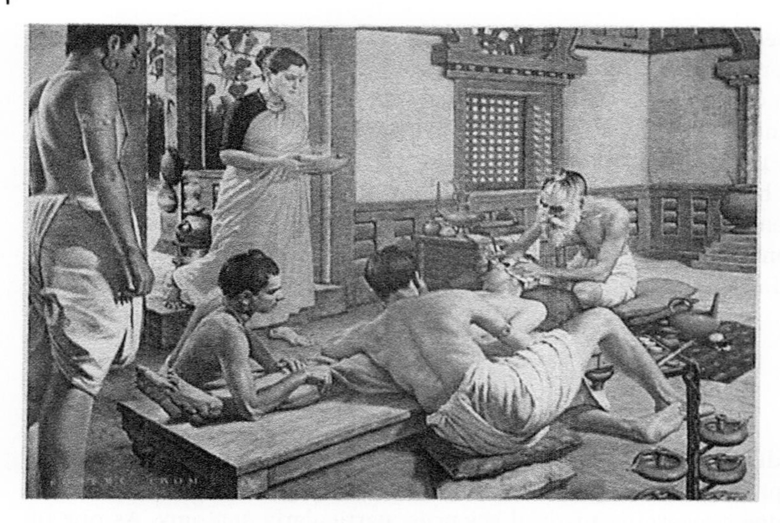

Figure 2.28 Sushruta depicted performing an otoplastic operation. Friends and relatives steady the patient, drugged with wine, as the great surgeon sets about fashioning an artificial earlobe. He will use a section of flesh cut from the patient's cheek; it will be attached to the stump of the mutilated organ, treated with homeostatic powders, and bandaged.

revealed by inspecting the surface of the human body and through human dissection, as Sushruta believed that students aspiring to be surgeons should acquire good knowledge concerning the structure of the body.

The advances in surgery made by the ancient Indians are significant considering the obstacles that impaired the study of anatomy. According to Hindu tenets, the human body is sacred in death. Hindu law (Shastras) states that no body may be violated with a knife and that deceased persons older than two years must be cremated in their original condition. Sushruta was able to bypass this decree and achieve remarkable knowledge concerning human anatomy by using a brush-type broom to scrape the skin and flesh without actually touching the body.

Sushruta's description of anatomical specimens included over 300 bones, types of joints, ligaments, and muscles in various parts of the body. Critics suggest that his overestimation of this number may be due to the child cadavers he studied; it is possible that he counted individual parts of bones that had not yet fused. Despite his erroneous accounts of the skeleton, he offered an in-depth understanding of bones, muscles, joint, and vessels, far exceeding what was previously known.

2.3.1.1 The *Sushruta Samhita*

Arguably the oldest known surgical textbook is the *Sushruta Samhita*. The first translation from Sanskrit was made into Arabic in the late eighth century; it was later translated into Latin, German, and English. The most recent English translation is by Kaviraj Bhishagharatan, first published in 1910, with a later edition in 1963. The *Sushruta Samhita* is divided into two parts: the Purva-tantra and the Uttara-tantra. The Purva-tantra is subdivided into five books, the Sutrasthana, Nidana, Sarirasthana, Chikitasathanam, and the Kalpastham, together totaling 120 chapters. The Sutrasthana concerns basic medical science and pharmacology; the Nidana addresses disease processes; the Chikitsasthanam, forming the bulk of the text, devotes 34 chapters to surgical procedures and postoperative management; and the Kalpasthanam offers eight chapters on toxicology. It was Sushruta's belief that to be a skillful and erudite surgeon, one must first be an anatomist: the Sarirasthana is thus composed of 10 chapters concerning human anatomy.

The following sections offer translations from the *Sushruta Samhita* describing various procedures as they relate to anatomical regions of the human body and the preparation of a human cadaver for dissection.

2.3.1.2 Dissection Preparation

As already discussed, the use of humans for dissection was in opposition to the religious law of the time. However, dissection is an essential tool for true understanding of human anatomy. The following concerns the method that Sushruta developed to enable him to work within the confines of this law:

> Therefore for dissecting purposes, a cadaver should be selected which has all parts of the body present, of a person who had not died due to poisoning, but not suffered from a chronic disease (before death), had not attained a 100 years of age and from which the fecal contents of the intestines have been removed. Such a cadaver, all whose parts are wrapped by any one of "munja" (bush or grass), bark, "kusa" and flax, etc. and kept inside a cage, should be put in a slowly flowing river and allowed to decompose in an unlighted area. After proper decomposition for seven nights, the cadaver should be removed (from the cage) and then dissected slowly by rubbing it with the brushes made out of any of usira (fragrant roots of plant), hair, bamboo or "balvaja" (coarse grass). In this way, as previously described, skin, etc. and all the internal and external parts with their subdivisions should be visually examined.

2.3.1.3 Head and Neck

The Hindu Laws of Manu that governed life in ancient India dictated that the punitive measure for the crime of adultery was for the offender to have his nose cut from his face. The *Sushruta Samhita* offers a procedure for repairing such damage, equivalent to a modern-day "free flap" used during reconstructive surgery (Figures 2.29 and 2.30).

> First the leaf of a creeper, long and broad enough to fully cover the whole of the severed or clipped part, should be gathered; and a patch of living flesh, equal in dimension to the receding leaf, should be sliced off [from down upward] from the region of the cheek and, after scarifying it with a knife, swiftly adhered to the severed nose. Then the cool-headed physician should steadily tie it up with a bandage decent to look at and perfectly suited to the end for which it has been employed. The physician should make sure that the adhesion of the severed parts has been fully affected and then insert two small pipes into the nostrils to facilitate respiration, and to prevent the adhered flesh from hanging down. After that, the adhered part should be dusted with [hemostatic] powders; and the nose should be enveloped in Karpasa cotton and several times sprinkled over with the refined oil of pure seasmum.

The piercing of earlobes and subsequent enlarging of the hollow was a widely practiced talisman for warding off evil. However, this often resulted in tearing of the lobes. Sushruta instructed that a pedicle flap reconstructive procedure be carried out, in which a graft of skin was taken from an adjacent area (carefully leaving its vascular supply intact), rotated to the defective area, and reattached.

> A surgeon well versed in the knowledge of surgery should slice off a patch of living flesh from the cheek of a person so as to have on its ends attached to its former seat (cheek). Then the part, where the artificial ear lobe is to be made, should be slightly scarified

Figure 2.29 The famous Indian rhinoplasty. "Note the scar on the forehead." (*Source:* Reproduced in the October 1794 issue of the *Gentleman's Magazine of London*, Urban BL, 64:8912.)

> (with a knife) and the living flesh, full of blood and sliced off as previously directed, should be adhered to it (so as to resemble a natural ear lobe in shape). The flap should then be covered with honey and butter and bandaged with cotton and linen and dusted with the powder of baking clay.

Sushruta gave considerable time to ophthalmic study, as conditions such as cataracts were common in this region. His description of the eye is couched in terms of five basic elements: earth (Bhu), fire/heat (Agni), air (Vayu), fluid (Jala), and void (Akasa). The extraocular muscles are the solid earth; the heat/fire is the blood flowing through the vessels; air forms the iris and pupil; fittingly, the vitreous region is attributed to the fluid element; and the lacrimal ducts are derived from the void. Sushruta delineated five anatomical divisions (Madalas) of the eye: eyelashes, eyelid, sclera, choroid, and pupil. The following is the procedure for removing cataracts:

> This procedure is auspiciously performed primarily in the warm season ... [Preoperatively] the skin is rubbed with a pledget of cotton saturated with an oily medicine followed by a heated bath. The patient is given a light refreshment. The sick room is fumigated with vapors of white mustard, bdellium, nimva leaves, and the resinous gums of shala trees (in order to rid the area of insects and the diseases they harbor) ... Incense of cannabis is used in addition to wine for sedation ... The patient sits on a high stool with the surgeon facing him. The hands are secured with proper fastenings. The patient is asked to look at his own nose while the surgeon rests his little finger on the [bony margin of the outer angle of the orbit], holding a Yava Vaktra Salaka between his thumb, index, and middle finger. The left eye should be pierced with the right hand, and vice versa. The eye is entered at the junction of the medial and lateral two-thirds of the outer portion of the sclera. If a sound is produced following the gushing of a watery fluid, the needle is in the correct place, but if bleeding follows the puncture, it means that it is misplaced. The eye is then sprinkled with breast milk. Care is taken to avoid blood vessels in the region.

Figure 2.30 Picture of Sushruta examining a patient. Note the palpation of the radial pulse.

The tip of the needle is then used to incise the anterior capsule of the lens. With the needle in this position, the patient is asked to blow down the nostril, while closing the opposite nare. After this, lens material (Kapha) is seen coming alongside the needle. When the patient is able to perceive objects, the needle is removed ... [Postoperatively,] indigenous roots, leaves, and ghee are applied with a lined bandage. The patient then lies flat and is asked not to sneeze, cough, or move. The eye is examined every fourth day for ten days. If the whitish material recurs, the same procedure is repeated.

2.3.1.4 Pelvis and Perineum

Hindu religion places much emphasis on reproduction and sexual energy. Male urogenital issues received substantial attention in the *Sushruta Samhita,* and the health of women was also addressed. Sushruta advocated the use of dilators, irrigating syringes, and catheters. The following is a method of managing a urethral stricture, dilation, and urethroplasty:

> In the case of Niruddhaprakasha (stricture of the urethra), a tube open at both ends made of iron, wood, or shellac should be lubricated with clarified butter and gently introduced into the urethra. Thicker and thicker tubes should be made to dilate in this manner and emollient food should be given to the patient. As an alternative, an incision should be made into the lower part of the penis avoiding the sevani (raphe) and it should be treated as an incidental ulcer.

2.3.1.5 Conclusion

The studies of Sushruta had a notable impact on the fields of anatomy, medicine, and surgery, particularly plastic surgery. Sushruta's seminal work, the *Sushruta Samhita,* forms the basis of the Ayurvedic tradition, which is still widely practiced today. The contributions of ancient civilizations to modern understanding are well appreciated, and ancient India is no exception. Abundant knowledge can be gleaned from the texts of this time.

Bibliography

China

Andrews, B.J. (1991). Wang Qingren and the history of Chinese anatomy. *J. Chin. Med* 35: 30–36.

Bynum, W.F. and Porter, R. (1993). *Companion Encyclopedia of the History of Medicine Vol 1.* London: Routledge.

Calkins, C.M., Franciosis, J.P., and Kolesari, G.L. (1999). Human anatomical science and illustration: the origin of two inseparable disciplines. *Clin. Anat.* 12: 120–129.

Clark, E. (1913). Anatomy in the Far East. *Anat. Rec.* 7: 237–245.

Dikötter, F. (1992). *The Discourse of Race in Modern China.* Stanford, CT: Stanford University Press.

Ho, P.Y. and Lisowski, F.P. (1997). *A Brief History of Chinese Medicine and its Influence.* Singapore: World Scientific.

Hoizey, D. and Hoizey, M.J. (1993). *A History of Chinese Medicine.* Vancouver: UBC Press.

Hsieh, E.T. (1921). A review of ancient Chinese anatomy. *Anat. Rec.* 20: 97–127.

Hsu, H.Y. and Preacher, W.G. (1977). *Chen's History of Chinese Medical Science.* Long Beach, CA: Oriental Healing Arts Institute.

Hume, E.H. (1924). The contributions of China to the science and art of medicine. *Science* 59: 345–350.

Kaptchuk, T.J. (2000). *The Web That Has No Weaver: Understanding Chinese Medicine.* New York: McGraw-Hill Professional.

Lassek, A.M. (1958). *Human Dissection. Its Drama and Struggle.* Springfield, IL: Charles C. Thomas.

Loukas, M., Tubbs, R.S., Louis, R.G. Jr. et al. (2007). The cardiovascular system in the pre-Hippocratic era. *Int. J. Cardiol.* 120: 145–149.

Maoxing, N. (1995). *The Yellow Emperor's Classic of Medicine: A New Translation of the Neijing Suwen with Commentary.* New York: Shambhala Publishers.

Morse, W.R. (1934). *Chinese Medicine.* New York: P.B. Hoeber.

Persaud, T.V.N. (1984). *Early History of Human Anatomy. From Antiquity to the Beginning of the Modern Era*. Springfield IL: Charles C. Thomas.
Shoja, M.M., Tubbs, R.S., Shokouhi, G., and Loukas, M. (2010). Wang Qingren and the 19th century Chinese doctrine of the bloodless heart. *Int. J. Cardiol.* 145: 305–306.
Tubbs, R.S., Loukas, M., Kato, D. et al. (2009). The evolution of the study of anatomy in Japan. *Clin. Anat.* 22: 425–435.
Tubbs, R.S., Riech, S., Verma, K. et al. (2011). China's first surgeon: Hua Tuo (c. 108–208 AD). *Childs Nerv. Syst.* 27: 1357–1360.
Wang, Q.R. (2007). *Yi Lin Gai Cuo. Correcting the Errors in the Forest of Medicine*, 1. Translated by Chung Y, Oving H, Becker Se. Fletcher, NC: Blue Poppy Press.
Waye, J.D. (1973). A short account of Chinese medicine. In: *Theories and Philosophies of Medicine*, 2e. New Delhi: Institute of History of Medicine and Medical Research.

Japan

Adachi, B. (1896). Ueber die anthropologische Angiologie der Japaner; Einleitung und Arterien der oberen Extremitäten. *Ztschr. d. med. Gesellsch.* 10: 1231–1257.
Adachi, B. (1897). Die Blutgefässe der Japaner. *Ztschr. d. med. Gesellsch.* 11: 1039–1054.
Adachi, B. (1900). Muskelvarietäten. *Ztschr f Morphol u Anthropol.* 2: 221.
Adachi, B. (1902a). Hautpigment beim Menschen und bei den Affen. *Ztschr f Morphol u Anthrop* VI: 1–132.
Adachi, B. (1902b). Ueber den Penis der Japaner. *Ztschr f Morphol u Anthropol.* 5: 350–356.
Adachi, B. (1903). Hautpigment beim Menschen und bei den Affen. *Zeitschr f Morphol Anthropol.* 6: 1–131.
Adachi, B. (1904). Häufigeres Vorkommen des Musculus sternalis bei Japaner. *Ztschr f Morphol u Anthrop* 7: 133–141.
Adachi, B. (1928). *Anatomie der Japaner 1: Das Arteriensystem der Japaner*. Kyoto: Kaiserlich-Japanischen Universitat zu Kyoto.
Adachi B. Anatomie der Japaner 2: Das Venensystem der Japaner; 1. Einleitung, Material, Methode und allgemeine Vorbemerkungen, Vv. pulmonales, Vv. cordis (S. Mochizuki), V. cava superior, Vv. anomymae. Kyoto: Kaiserlish-Japanischen Universitat Kyoto. 1933.
Adachi B. Anatomie der Japaner 2: Das Venensystem der Japaner; 2. Vv. thoracicae longitudinales und Vv. intercostales. Kyoto: Kaiserlich-Japanischen Unicersitat Kyoto. 1940.
Adachi, B. (1985). Ueber die Spina trochlearis. *Ztschr d med Gesellsch.* 9: 177–179.
Adachi, B. and Akaza, S. (1900). Ueber die Lage des Augapfels bei den Japaner. *Mitt d med Gesellsch zu Tokyo.* 14: 94–99.
Adachi, B. and Fiyisawa, U. (1899). Untersuchungen über die knöcherne Orbita der Japaner. *Mitt d med Gesellsch.* 13: 890–919.
Adachi, B., Ogawa, T., and Terada, K. (1981). *Korpergeruch, Ohrenschmalz und Hautdrusen: eine anthropologische Studie.* Tokyo: Anthropological Society of Nippon.
Anderson, R.H., Ho, S.Y., and Becker, A.E. (2000). Anatomy of the human atrioventricular junctions revisited. *Anat. Rec.* 260: 81–91.
Arita, M. (2002). History of cardiology in the last 100 years: Japanese contribution to the studies on atrioventricular node (Tawara). *Nippon Naika Gakkai Zasshi.* 91: 841–845.
Baba, H. (2007). Estimating how and why Dr. Okuda made a complete wooden human skeleton in the Edo era, Japan. *Anat. Sci. Int.* 82: 31–37.
Bowers, J.Z. (1965). *Medical Education in Japan.* New York: Harper & Row.
Bowers, J.Z. (1970). *Western Medical Pioneers in Feudal Japan.* Baltimore: Johns Hopkins Press.

Bowers, J.Z. (1973). The development of the basic sciences in Japan. *Clin. Med.* 8: 229–237.
Cowdry, E.V. (1920). Anatomy in Japan. *Anat. Rec.* 12: 67–95.
Durant, W. (1954). *Our Oriental Heritage. Part 1*. New York: Simon and Schuster.
Evans, B. (1958). *The Natural History of Nonsense*. New York: Vintage Books.
Fodstad, H., Hariz, M.I., Hirabayashi, H., and Ohye, C. (2002). Barbarian medicine in feudal Japan. *Neurosurgery* 51: 1015–1025.
Freédéric, L. (2005). *Japan Encyclopedia*, trans. Roth K. Boston: Harvard University Press.
Hiki, S. (2001). Surgeons who contributed to the enlightenment of Japanese medicine. *World J. Surg.* 25: 1383–1387.
Horiuchi, A. (2003). When science develops outside state patronage. Dutch studies in Japan at the turn of the nineteenth century. *Early Sci. Med.* 8: 148–171.
Huard, P. (1958). How Japan adopted western anatomy (French). *Concours Med.* 80: 4861–4864.
Inoue, Y. (1981). Kouhi zuisou (Japanese). *Nihon Iji Shinpou* 2974: 61–64.
Inoue, Y. and Inoue, F. (1985). Nichidoku igakukai no kyosei Adachi Buntaro (Japanese). *Chishiki* 40: 296–307.
Izumi, Y. and Isozumi, K. (2001). Modern Japanese medical history and the European influence. *Keio J. Med.* 50: 91–99.
Japanese Society of Medical History (1959). *Catalogue of the Historical Writings and Materials in the Early Stage of the Development of Modern Medicine in Japan*. Kyoto: Benrido Co.
Kataoka, K., Suzaki, E., and Ajima, N. (2007). The Hoshino wooden skeleton, the first wooden model of a human skeleton, made during the Edo era in Japan. *Anat. Sci. Int.* 82: 38–45.
Keith, A. (1950). *An Autobiography*. London: Watts & Co.
Kozai, T. (2007). History of collecting cadavers in Japan (Japanese). *Kaibogaku Zasshi.* 82: 33–36.
Le Minor, J.M. and Sick, H. (2002). The chair of anatomy in the Faculty of Medicine at Strasbourg: 350th anniversary of its foundation (1652–2002). *Surg. Radiol. Anat.* 24: 1–5.
Loukas, M., Linganna, S., Chiba, A., and Tubbs, R.S. (2008). Sunao Tawara, a cardiac pathophysiologist. *Clin. Anat.* 21: 2–4.
Loukas, M., Tubbs, R.S., Mirzayan, N. et al. (2011). The history of mastectomy. *Am. Surg.* 77: 566–571.
Low, M.F. (1996). Medical representations of the body in Japan: gender, class, and discourse in the eighteenth century. *Ann. Sci.* 53: 345–359.
Magari, S. (1989). The history of basic lymphology in Japan. *Jpn. J. Lymphol.* 12: 1–12.
Mestler, G.E. (1954). A galaxy of old Japanese medical books with miscellaneous notes on early medicine in Japan. Part I. Medical history and biography, general works, anatomy, physiology and pharmacology. *Bull. Med. Libr. Assoc.* 42: 287–327.
Michel, W. (2000). *"Inner Landscapes" – Japan's Reception of Western Conceptions of the Body. Medicine in Japan and Germany*. Bonn: Japan Society for the Promotion of Science/Deutsche Gesellschaft der JSPS-Stipendiaten.
Mochizuki, S. (1944). Adachi Buntaro hakushi Goethe shou wo uku (Japanese). *Kaibogaku Zasshi..* 1989 22: 197–200.
Motomiya, K. and Orley, R. (1999). Adachi Buntaro no hitoto gyousekinitsuite (Japanese). *Nihon Igakushi Zasshi* 45: 270–271.
Ogawa, T. (1975). The beginnings of anatomy in Japan. *Okajimas Folia Anat. Jpn.* 52: 59–72.
Olry, R. and Lellouch, A. (2003). The arterial system of the Japanese anatomist Buntaro Adachi. *Hist. Sci. Med.* 37: 89–94.
Olry R, Motomiya K. Dutch influence on Japanese neuroanatomy in the eighteenth century: From Johann Kulm's Ontleedkundige Tafelen to Gempaku Sugita's Kitai Shinsho. Presented at the International Society for the History of the Neurosciences and the European Club on the History of Neurology, Leiden, The Netherlands, June 1997.

Osborn, H.F. (1916). Gustav Schwalbe. *Science* 44: 97.
Sakai, S. (1987). Adachi Buntaro hakushi to chosaku (Japanese). *Kagaku-Igakushiryo Kenkyu* 63: 1–4.
Sakai, T. (2007). Anatomical education in the late Meiji era – Lu Xun, Doctor Fujino and their contemporaries (Japanese). *Kaibogaku Zasshi* 82: 21–31.
Sakula, A. (1985). Kaitai Shinsho: the historic Japanese translation of a Dutch anatomical text. *J. R. Soc. Med.* 78: 582–587.
Screech, T. (1997a). *The Lens within the Heart: The Western Scientific Gaze and Popular Imagery in Later Edo Japan*. Honalulu: University of Hawaii Press.
Screech, T. (1997b). *Edo no Jintai O Hiraku (Opening the Edo Body)*. Tokyo: Sakuhinsha.
Shimada, K. (1996). Scarce anatomical textbook by professor Buntaro Adachi (Japanese). *Kaibogaku Zasshi* 71: 133–134.
Shimada, K. (2003). Meijiki no kaibougakusho (Japanese). *Keitaikagaku* 7: 1–6.
Shimada, K. (2007). Beginning of modern anatomy in Japan from the perspective of anatomical bibliographies of the Meiji era (Japanese). *Kaibogaku Zasshi* 82: 9–20.
Shimizu, S. (1989). Topographical anatomy of the atrioventricular node: findings by macro-microscopic dissection under dissecting microscope. *Nippon Kyobu Geka Gakkai Zasshi.* 37: 227–233.
Shirasugi, E. (2007). Envisioning the inner body during the Edo period in Japan: Inshoku yojo kagami (rules of dietary life) and boji yojo kagami (rules of sexual life). *Anat. Sci. Int.* 82: 46–52.
Silverman, M.E., Grove, D., and Upshaw, C.B. Jr. (2006). Why does the heart beat? The discovery of the electrical system of the heart. *Circulation* 113: 2775–2781.
Suma, K. (2001). Sunao Tawara: a father of modern cardiology. *Pacing Clin. Electrophysiol.* 24: 88–96.
Tawara, S. (1906). *Das Reizleitungssystem de Saugetierherzens. Eine anatomich-histologische Studie uber das Atrioventricularbundel und die Purkinjeschen Faden*. Jena: Verlag von Gustav Fischer.
Tawara, S. (2000). *The Conduction System of the Mammalian Heart: An Anatomico-Histological Study of the Atrioventricular Bundle and the Purkinje Fibers*. London: Imperial College Press.
Terada, K. (1981). Adachi Buntaro no jinruigaku (Japanese). *Kagaku* 51: 464–466.
Tomita, T. (1999). Der Entdecker des Atrioventrikularbundles Dr. Sunao Tawara. *Nachbemerkungen Nippon Ishigaku Zasshi.* 45: 503–514.
Tubbs, R.S., Loukas, M., Kato, D. et al. (2009). The evolution of the study of anatomy in Japan. *Clin. Anat.* 22: 425–435.
van Gulik, T.M. and Nimura, Y. (2005). Dutch surgery in Japan. *World J. Surg.* 29: 10–17.
Yasuda, K. (2007). Seiyou igaku no denrai to doitsu igaku no sentaku (Japanese). *Keio Igaku* 84: 69–84.
Yeo, I.S. and Hwang, S.I. (1994). A historical study on the introduction and development of anatomy in Japan (Korean). *Uisahak.* 3: 208–219.

India

BBC. 2006. The religious capital of Hinduism. http://news.bbc.co.uk/2/hi/south_asia/4784056.stm. Last accessed July 13, 2018.
Bhishagratna, K.L. (1963). *An English Translation of the Susruta Samhita*. Varanasi: Chowkhamba Sanskrit Series Office.
Das, S. (1983). Susruta of India, the pioneer in the treatment of urethral structure. *Surg Gynecol Obstet* 157: 581–582.

eVaranasiTourism.com. History of India. http://www.evaranasitourism.com/history-of-varanasi-india/index.html. Last accessed July 13, 2018.

Hoernle, A.F. (1907). *Studies in the Medicine of Ancient India. Part I. Osteology or the Bones of the Human Body*. Oxford: Claredon Press.

Keswani, N.H. (1970). Medical education in India since ancient times. In: *The History of Medical Education* (ed. C.D. O'Malley), 329–366. Berkley, CA: University of California Press.

Lannoy, R. (1999). *Benares Seen from Within*. Washington, DC: University of Washington Press.

Loukas, M., Lanteri, A., Ferrauiola, J. et al. (2010). Anatomy in ancient India: a focus on the Susruta Samhita. *J. Anat.* 217: 646–650.

Mahabir, R.C. (2001). Ancient Indian civilization: ahead by a nose. *J. Investig. Surg.* 14: 3–5.

Micozzi, M.S. (2006). *Fundamentals of Complementary and Integrative Medicine*, 3e. Missouri: Saunders Elsevier.

Mukhopadhaya, G. (1929). *History of Indian Medicine*. Calcutta: Calcutta University Press.

Persaud, T.V.N. (1984). *Early History of Human Anatomy*. Springfield, IL: Charles C Thomas.

Persaud, T.V.N. (1997). *A History of Anatomy. The Post-Vesalian Era*, 1e. Springfield, IL: Charles C Thomas.

Porter, R. (1998). *The Greatest Benefit to Mankind. A Medical History of Humanity*, 1e. New York: W.W. Norton & Co.

Raju, V.K. (2003). Susruta of ancient India. *Indian J. Ophthalmol.* 51: 119–122.

Ruthkow, I.M. (1961). *Great Ideas in the History of Surgery*. Baltimore: William & Wilkins Co.

Singhal, G.D. and Guru, L.V. (1973). *Anatomical & Obstetrical Considerations in Ancient Indian Surgery*. Banaras: Banaras Hindu University Press.

3

Europe

The countries of Western Europe began to attain stability after c. 1000 CE. In the succeeding centuries, universities were established in many major cities. Teaching in these universities relied on ancient texts translated into Latin, mostly from Arabic. The cultural forces responsible for these developments, and for the dramatic intellectual and technological innovations associated with the European Renaissance of the fifteenth century and its aftermath, have been discussed in a number of publications. Inter alia, they led to advances in medicine and associated studies, particularly anatomy.

The systematic teaching of medicine in Europe was pioneered in the eleventh century in the medical school of Salerno, which was run by the Church. The Salerno teachers utilized texts from the Byzantine Empire as well as the Islamic world, which were assembled as the *ars medicinae*. During the twelfth century, many ancient Greek and Islamic medical texts were translated from Arabic into Latin at other centers in southern Europe. These included the Hippocratic Corpus, the works of Akindus, Haly Abbas, Abulcasis, and many other Islamic scholars, and most especially the writings of Galen and Avicenna's *Al-Qanun fi al-Tibb* ("The Canon of Medicine"). These texts, particularly Avicenna's *Qanun*, were to dominate medical teaching in Europe's new universities until, and indeed after, the Renaissance.

The universities inculcated a *tradition* of learning: each new generation of students benefited from the work of those who had gone before, so every student was taught on the basis of an accumulated and ever-growing body of information. There were no precedents for this in human history, as far as is known, and it was central to the increasing dominance of Western European thought. As the following sections will show, every European country in which new contributions to anatomical knowledge were made had its own tradition, extending up to the present day, although there was extensive cross-fertilization among them.

Medical teaching spread from Salerno to the new universities as they were founded, so the traditions of medical education were established first in Italy and France, then in Holland and England, and afterwards in other parts of Europe. For many years, this teaching was imitative rather than innovative, particularly in regard to anatomy. Surgery, the aspect of medical practice most dependent on detailed anatomical knowledge, was largely disregarded in the universities and was left mainly in the hands of uneducated practitioners, so for several generations the study of anatomy occupied a lowly position in the medical curriculum.

The anatomist Mondino de Liuzzi (c. 1270–1326 CE) did much to change this. Mondino personally performed systematic human dissections, rather than employing a demonstrator. His dissections were permitted by the Vatican; the cadavers he used were probably from executed criminals. Mondino was born into a family of apothecaries in Bologna, and he was the first to incorporate the formal study of anatomy into medical education in Europe.

History of Anatomy: An International Perspective, First Edition. R. Shane Tubbs, Mohammadali M. Shoja, Marios Loukas and Paul Agutter.

His handbook of dissection, *Anathomia corporis humani*, the first of its kind, was written in 1316. Its importance in medieval and Renaissance teaching is indicated by the fact that it was printed in Padua more than a century and a half later; by then, Padua had become a key center for anatomical study. Although Mondino described his personal experiences of dissection, he ensured that they were consistent with the anatomical teachings of Avicenna and other Islamic scholars, based on their development of the work of Hippocrates, Aristotle, and (especially) Galen. He was not an innovator, but – albeit inadvertently – he opened the door to innovation.

The Renaissance was characterized by an increasing willingness to re-evaluate and even to challenge the teachings of the ancients, including Galen, though many generations were to pass before the works of Galen and Avicenna ceased to be major influences on European thought. Anatomists of the Renaissance, including Leonardo da Vinci (1452–1518) and Alessandro Achillini of Bologna (1463–1512), acknowledged the influence of Mondino, an unquestioning Galenist. Nevertheless, these anatomists gradually corrected the errors in the Galenist tradition, paving the way for modern medical teaching. The publication in 1543 of the *Seven Books on the Fabric of the Human Body* by Andreas Vesalius (1514–1564) was pivotal in this process, but as much as a century later the eminent Parisian doctor of medicine Jean Riolan (1577–1657) was still defending Galenist teaching against William Harvey (1578–1657). Interestingly, Renaissance innovations in human anatomy followed more or less the same pathway as the establishment of university medical schools three centuries earlier; they began in Italy and France and spread to England and Holland, and then to the rest of Europe.

European contributions to knowledge, including the study of anatomy, were distinctive in many ways. We have already emphasized the development of *traditions* of learning in the universities. Another salient European contribution was the establishment of overseas empires in the centuries following the Renaissance. Collectively, these empires spread over most of the world. As a result, the bodies of knowledge developed in European countries – as well as their beliefs, values, and technologies – were introduced to peoples with entirely different histories and cultures. In particular, the rest of the world acquired a tradition of medical knowledge and practice – and anatomical study – from European imperialism. We have already seen examples of the consequences in the case of Japan.

3.1 Austria

3.1.1 Heinrich Obersteiner

Heinrich Obersteiner (1847–1922) (Figure 3.1) was an influential neuroscientist in the nineteenth century. He was the founder of Vienna's Neurological Institute. Born November 13, 1847, Obersteiner was the son and grandson of Viennese physicians. He gained early exposure to medicine, becoming fascinated with the central nervous system through experience at his father's private psychiatric clinic. It was therefore no surprise when he decided early in his life to pursue medical studies, enrolling at the University of Vienna. During this time (1865–1870), the medical school was home to several famous members of the "Second Vienna School of Medicine," namely Rokitansky, Hyrtl, Skoda, and Oppolzer, who served as part of the teaching faculty. As a medical student, Obersteiner developed a keen interest in research, much of which was channeled toward the nervous system. This particular curiosity was encouraged by his mentor, the physiologist Ernst Wilhelm von Brucke (1819–1892), who accepted Obersteiner as a coworker in the neurohistology laboratory at Vienna University. Notably, Obersteiner's first paper on the development of tendons was published only two years after his admittance to medical school.

Figure 3.1 Portrait of scientist and physician, Heinrich Obersteiner (1847–1922). (*Source:* With permission from Marburg 1923.)

3.1.1.1 Academic Career and Research

Obersteiner's career in academia began only a few years after he finished medical school. He joined the faculty of the University of Vienna in 1873, where he was initially made dozent (lecturer) on the anatomy and pathology of the nervous system. This marked the inception of a rigorous and highly successful teaching career. Young graduates and budding neurologists from all over the world flocked to his lectures and sought out his tutelage and expertise. Much of the attention he attracted was thought to be due to his personal charm, outstanding abilities as a teacher, and excellent command of English. In 1880, he was promoted to extraordinarius (associate professor), in 1898 to ordinarius (professor), and in 1906 he had the title hofrat (privy councilor) bestowed upon him. Over these three decades, Obersteiner carried out intensive scientific research, which contributed many original ideas that became a part of his well-known textbook and of the field of neurology. He covered many aspects of neurology and psychiatry, and was one of the first scientists to promote the idea of practical neuropsychiatry, noting that organic diseases often display a psychogenic component. Following this idea, he diversified his research pursuits, focusing not only on morphological and pathological studies of the nervous system, but also on areas such as the comparative psychology of the various senses and hypnosis. Interest in the latter was facilitated by his inheritance (from his parents) of a private sanatorium for mental illness, where he was able to conduct most of his psychological experiments. He was among the first to use hypnotism in psychiatry, and he eventually introduced the concept to the public in 1885 with an address to Vienna's "Scientific Club." After limited success with this practice, he proposed that only one-third of any given population can obtain benefits from hypnosis, and that researchers seeking explanations for hypnotic phenomena should use only healthy people, since the clinically insane were usually resistant. Later, in 1891, he incorporated the idea of psychotherapy as a means of "calming and diverting" patients, thus allowing for some hypnotism and suggestion on nervous patients. On the basis of his previous experience, he acknowledged that this new practice was still inaccessible for psychotic patients.

Obersteiner's contributions to neuropathology were wide-ranging, including comparative studies of the cerebellum, spinal cord anatomy, progressive paresis, epilepsy, and intoxication psychoses. He devoted much of his time to examination of the normal histology, anatomy, and pathology of the cerebellum (Figures 3.2–3.4). He was the first to identify lipochrome, a pigment in ganglion cells that is a product of catabolism. He also found this pigment in glial cells, and hypothesized that the corpora amylacea were probably derived from them. He was one of the first to investigate the problem of general paralysis, correlating it with the presence of syphilis. In collaboration with his scholar Otto Redlich, he was able to demonstrate that tabes dorsalis involved pathology of the posterior spinal roots. This observation resulted in a publication about the pathogenesis of this disease. He was the first German-speaking neurologist to describe status epilepticus, and he was able to reproduce epilepsy experimentally. He also proved that the condition had a genetic component and that it might in some cases be related to trauma. He introduced the term "allochiria" to denote a "confusion of sides" or mislocation of sensory stimuli to the opposite side of the body.

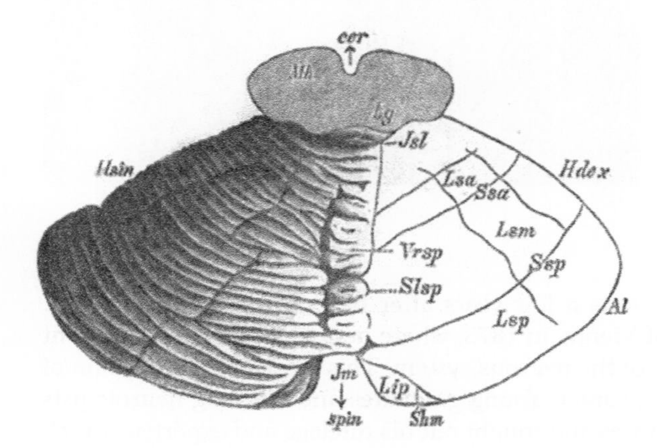

Figure 3.2 Cerebellum, dorsal view. Original illustration in the first edition of Obersteiner's textbook. (*Source:* From Obersteiner 1890.)

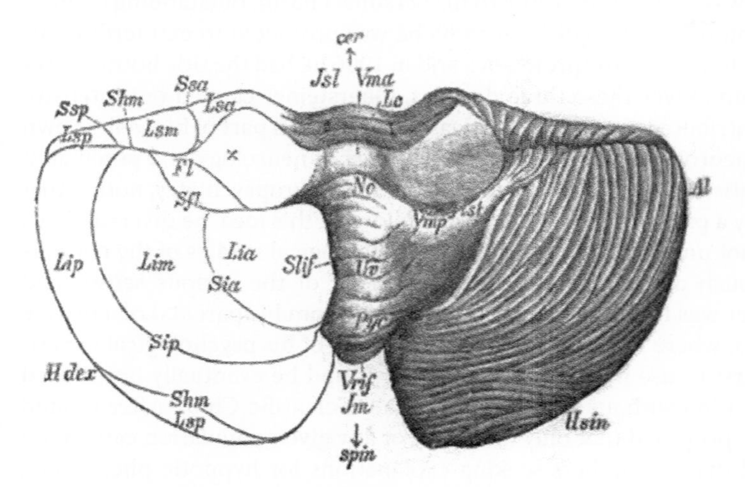

Figure 3.3 Cerebellum, ventral view. Original illustration in the first edition of Obersteiner's textbook. (*Source:* From Obersteiner 1890.)

3.1.1.2 His Book

From his days as a medical student, Obersteiner displayed a peculiar interest in the morphology of the nervous system. One of the greatest manifestations of this curiosity was his textbook, *Introduction to the Study of the Structure of the Central Nervous Organs in Health and Disease* (Figure 3.5), first published in German in 1888. Almost instantaneously, it became a standard reference for neurologists in training, providing a systematic syllabus of all the organic knowledge of neurology at that time. As a result of its popularity, it was translated twice into English, twice into Russian, and once each into French and Italian. Over the next 25 years, the German edition was revised four times.

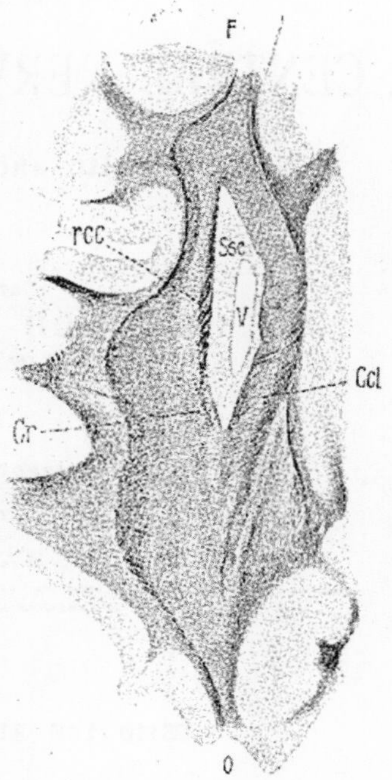

The textbook provided the most up-to-date and in-depth knowledge of neurology at the time, allowing scholars to study the brain and spinal cord in a structured and logical manner through macroscopic and microscopic aspects, as well as to review neuronal pathways of the pyramidal tracts (Figure 3.6). He was able to provide this information with great clarity and simplicity, tackling material that had previously been considered conceptually difficult. An exemplary instance may be noted with the calling to attention of Obersteiner's representation (Figure 3.7) of the protoplasmic processes of a Purkinje's cell. Much of the content was based largely on Obersteiner's own research findings, and he integrated many original observations into the subject matter with remarkable modesty and a characteristic tendency toward "objectification" and anonymity.

Figure 3.4 Horizontal section of the brain of a five-month-old infant. This illustration was meant as a means of locating and describing the subcallosum in humans. (*Source:* From Obersteiner & Redlich 1902.).

In 1887, after Obersteiner's book was written but before it was published, Sigmund Freud wrote a review of it for the *Wiener Medizinische Wochenschrift*, a notable medical journal in Vienna. Freud, who was in his early career as a neuroanatomist, compared Obersteiner's compilation to previous texts of a similar nature, noting that previous authors, in contrast to Obersteiner, had not addressed the finer structure of the brain and the course of its fibers. He stated that Obersteiner's book was "a reliable guide for anyone seeking to acquire a good anatomical understanding of the nervous system."

The book received many other acclaimed reviews, and is sometimes referred to as Obersteiner's greatest contribution to the neurological world. It encompassed a comprehensive review of the normal and pathological anatomy of the nervous system. This established a new field of basic and clinical neuromorphology and helped set the foundation for the eventual development of neurology as a medical discipline. Its tremendous educational and publishing success led to its becoming, according to Marburg, "the textbook of the entire neurological world."

3.1.1.3 The Neurological Institute

Obersteiner was among the first to conceive the idea of neurological sciences as a specialized branch of medicine. After he was designated extraordinarius at the University of Vienna in 1880, he pursued a long-desired goal of developing his own Institute for the specialization of his research in neuroanatomy. At the time, research of this sort was only performed at the neurohistological laboratory in a nearby pharmacological institute.

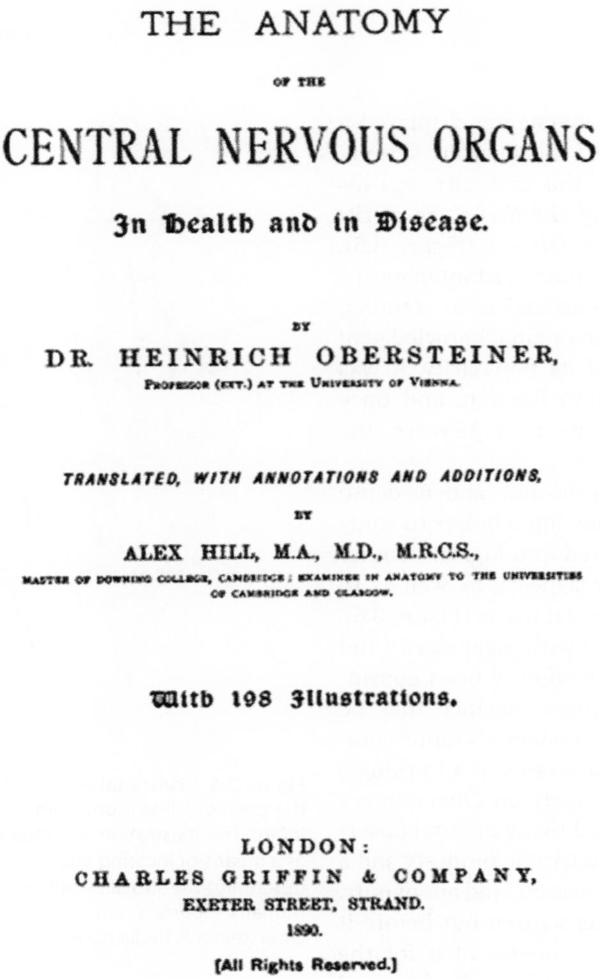

THE ANATOMY

OF THE

CENTRAL NERVOUS ORGANS

In Health and in Disease.

BY

DR. HEINRICH OBERSTEINER,

PROFESSOR (EXT.) AT THE UNIVERSITY OF VIENNA.

TRANSLATED, WITH ANNOTATIONS AND ADDITIONS,

BY

ALEX HILL, M.A., M.D., M.R.C.S.,

MASTER OF DOWNING COLLEGE, CAMBRIDGE; EXAMINER IN ANATOMY TO THE UNIVERSITIES OF CAMBRIDGE AND GLASGOW.

With 198 Illustrations.

LONDON:
CHARLES GRIFFIN & COMPANY,
EXETER STREET, STRAND.
1890.

[All Rights Reserved.]

Figure 3.5 Title page of the 1890 English version of Obersteiner's textbook. (*Source:* From Obersteiner 1890.)

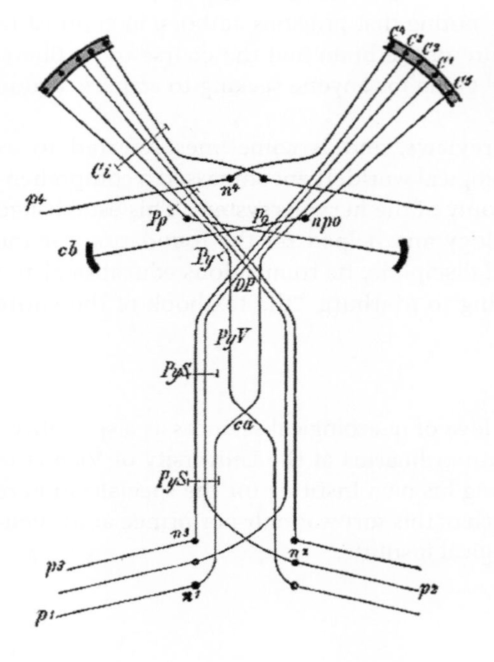

Figure 3.6 Schematic of the pyramidal tracts, as per Obersteiner. PyS, lateral pyramidal tracts; PyV, anterior pyramidal tracts. (*Source:* From: Obersteiner 1890.)

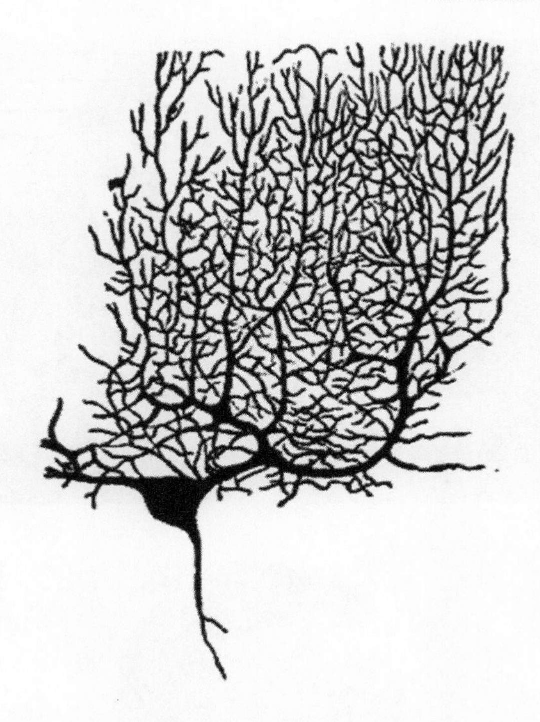

Figure 3.7 Illustration of a Purkinje's cell exposed by a vertical section and stained in Golgi's method. (*Source:* From: Obersteiner 1890.)

In 1882, after personally donating half a million crowns to the Austrian Ministry of Culture, Obersteiner founded the University Institute for Anatomy and Physiology of the CNS. In 1900, it was officially renamed the Neurological Institute. Initially, he was granted only a large cabinet in the corridor of Rokitansky's Institute, and he donated many of his own private books to incoming scholars for reference. His generosity continued as the Institute grew, resulting in a unique library of about 40 000 volumes. He also funded the costs of laboratory supplies and other essentials up until his retirement as director in 1919.

The Institute was considered to be the world's first interdisciplinary brain research center, and for many years was the leader in neuroscience research and teaching. It served as a venue for international meetings and as a model for brain research institutes that were founded subsequently. Graduates from all over the world went to study there, including highly noted Viennese neurologists: von Economo, von Frankl-Hochwart, Karplus, Redlich, Schlesinger, and Spitzer. Otto Marburg was Obersteiner's most devoted and outstanding student, and succeeded him as Director of the Institute (Figure 3.8). In the time that Obersteiner served as director, 22 volumes of work were produced and over 500 scientific papers and monograms were published. The Institute became one of the few of its kind to be accepted by the International Association of Academies of Science as a "Specialized Institute for Brain Research." It stands as a lasting monument to Obersteiner's tremendous and exceptionally important contributions to the development of the fields of neurology and neuroanatomy.

3.1.2 Emil Zuckerkandl

The Austrian anatomist and pathologist Emil Zuckerkandl (1849–1910) (Figure 3.9) is remembered by several eponyms: the organ of Zuckerkandl (para-aortic chromaffin body), Zuckerkandl's fascia (posterior layer of the renal fascia), Zuckerkandl's gyrus (subcallosal area), Zuckerkandl's tuberculum (lateral projection of the thyroid gland), Zuckerkandl's operation

Figure 3.8 Heinrich Obersteiner and Otto Marburg in front of the Neurological Institute in 1912. (*Source:* With permission from Seitelberger 1992.)

Figure 3.9 Photograph of Professor Emil Zuckerkandl (1849–1910).

(perineal prostatectomy), Zuckerkandl's dehiscence (fissures in the ethmoid bone), and concha of Zuckerkandl (a rare nasal concha). He was a favorite pupil of both Josef Hyrtl (1810–1894) and Carl von Langer (1819–1887) at the Vienna School of Anatomy, and succeeded to the Chair of Anatomy there. Zuckerkandl was noted for his sharp observational powers and his critical mind, and he made significant contributions to normal and pathological anatomy. He strove to make anatomy "subservient" to the patient, and was thus an early pioneer of clinical anatomy; he is credited with the comment, "anatomy is the war map for the operations of the physician." This concept gave rise to many surgical subspecialities, and his anatomical descriptions serve as a basis for our current morphological understanding. He had written over 164 publications by the time of his death, leaving a tremendous legacy with regard to anatomical knowledge.

Zuckerkandl was born on September 1, 1849 in Raab, today Györ, Hungary, to a Jewish family. He was the older brother of Otto Zuckerkandl (1861–1921), a urologist and university professor in Vienna. He married Berta Szeps (later Berta Zuckerkandl-Szeps), a famous Austrian journalist, in 1889. Berta was the daughter of Moritz Szeps, the editor-in-chief of *Morgenpost* in Vienna, and sister-in-law of the French Prime Minister Georges Clemenceau (1841–1929), himself a physician who graduated in the United States. She was an influential personality in Jewish society in Vienna from the last decades of the Austro-Hungarian Empire until World War II. Their house was a *salon* for such famous figures in the arts and sciences as the sculptor Auguste Rodin (1840–1917), the painter Gustav Klimt (1862–1918), the architect Otto Wagner (1841–1918), the writer Hermann Bahr (1863–1934), the playwright Arthur Schnitzler (1862–1931), and the composer Gustav Mahler (1860–1911).

Zuckerkandl began to study medicine at the University of Vienna in 1867, and obtained his medical degree in 1874. His discourses were so influential that he soon became one of the favorite pupils of the anatomist Joseph Hyrtl at the Vienna School of Anatomy, of whom he said, "he spoke like Cicero and wrote like Heine." Hyrtl made him a demonstrator, and later an assistant. On Hyrtl's recommendation, Zuckerkandl became a prosector in Amsterdam (University of Utrecht) in 1870. In 1873, he returned to Vienna to become assistant to the Pathological-Anatomical Chair under Carl Freiherr von Rokitansky (1804–1878). He also worked with Carl von Langer, Hyrtl's successor at the Vienna School of Anatomy, which was the world's anatomy center for more than half a century. He became an assistant professor in 1879, and in 1882 was called to Graz to follow Julius Planers (1827–1881) as Professor of Anatomy there. In 1888, following Langer's death, Zuckerkandl succeeded to the Chair of Anatomy and continued his career as Professor of Descriptive and Topographical Anatomy in Vienna. He was a member of the Akademie der Wissenschaften from 1898. Between 1902 and 1905, he revised and edited Carl Heitzmann's *Atlas der descriptiven Anatomie des Menschen* ("Atlas of Descriptive Anatomy of Man"), a two-volume work that included 600 illustrations. His notable assistant and colleagues were Heinrich Albrecht (Austrian physician, 1866–1922), William Stewart Halsted (American surgeon, 1852–1922), Georg Lotheissen (Austrian surgeon, 1868–1941), and Bernhard Aschner (Austrian gynecologist, 1883–1960). He died in Vienna on May 28, 1910.

Zuckerkandl worked in almost all fields of morphology, making advances in topographical and comparative anatomy. He particularly contributed to the normal and pathological anatomy of the nasal cavity and to the anatomy of the facial skeleton, hearing organs, teeth, blood vessels, brain, and chromaffin system. His first detailed descriptions and illustrations of the paranasal sinuses, in *Normale und pathologische Anatomie der Nasenhöhle und ihrer pneumatischen Anhänge* (1882), set the standard for generations of anatomists and surgeons, and provided the basis for such nasal operations as endoscopic sinus surgery. Zuckerkandl used the term "hiatus semilunaris" to signify the two-dimensional cleft between the posterior margin of the uncinate process and the anterior face of the ethmoid bulla ("bulla" is also a term first used by Zuckerkandl), and "ethmoid infundibulum" for the depression extending forward from the hiatus semilunaris both inferiorly and superiorly into the lateral nasal wall. He coined the phrase "physiological septal deviation," noting that the nasal septum is often not straight between the lateral nasal walls. The anterolateral process of the nasal septum is known as the ventral process of Zuckerkandl. The first report of a sphenochoanal polyp and nasolabial cyst is attributed to Zuckerkandl.

In 1883, in a report titled "Uber den fixationsapaparat der nieren" ("Supporting structures of the kidney"), Zuckerkandl stated that "if one divides the anterior fascia of the quadratus lumborum, then one sees another membrane, which is the posterior wall of the renal capsule [the fascia retrorenalis], then the adipose tissue, then the kidney itself." For such an original description, the eponym "Zuckerkandl's fascia" is often used to indicate the posterior layer of the renal fascia. However, he was unable to recognize the anterior renal fascia individually, but saw it as a continuation of the parietal peritoneum. Dimitrie Gerota (1867–1939) cited Zuckerkandl in a paper 15 years later, and clarified the presence of an anterior renal fascia. The cardiac vein of Zuckerkandl is the anterior cardiac vein that begins between the ascending portion of the aortic arch and the pulmonary artery. The epitrapezium bone of the wrist is known as Zuckerkandl's bone. In 1891, Zuckerkandl demonstrated that the periodontal membrane is supplied with nerves and blood vessels from the interdental and interalveolar arteries and drains into the interdental veins and lymph vessels, which form a bundle surrounded by loose connective tissue in a relatively wide interdental canal, which opens at the alveolar crest. He demonstrated several anatomical variations, including the aberrant origins of the middle meningeal artery from the ophthalmic and internal carotid arteries. This latter variant was also reported by Hyrtl.

Zuckerkandl is probably best remembered for his descriptions of the chromaffin tissue found in a para-aortic location: Zuckerkandl's bodies, glands, or paraganglia, near the inferior mesenteric artery (Figure 3.10). These enclaves are the largest concentrations of extraadrenal paraganglia of sympathetic origin, and account for up to 5% of all neuroblastomas. Zuckerkandl termed them "Nebenorgane."

Other structures that bear his name include Zuckerkandl's fascia (the posterior layer of the renal fascia), Zuckerkandl's gyrus (gyrus paraterminalis) or convolution (subcallosal area, the ventral continuation of the transparent septum pellucidum anterior to the lamina terminalis and anterior commissure that continues anteriorly to the cingulate), Zuckerkandl's tubercle (the posterolateral projection of the thyroid gland and the remnant of the ultimobranchial body, which is an anatomical landmark used for localizing the recurrent laryngeal nerve during thyroidectomy), Zuckerkandl's operation (perineal prostatectomy), Zuckerkandl's dehiscence (small fissures occasionally seen in the lamina papyracea of the ethmoid bone, which may allow

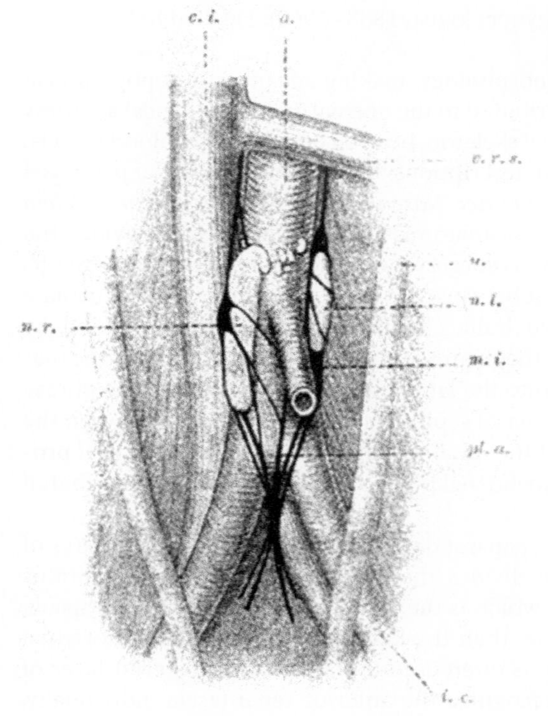

Figure 3.10 Drawing from Zuckerkandl's paper (1901) depicting the "organs of Zuckerkandl" (n.l.). For reference, note the inferior mesenteric artery (m.i.).

infections to spread into the orbit and then intracranially), and the concha of Zuckerkandl (a rare, small nasal concha situated above the supreme nasal concha). In 1876, Zuckerkandl discovered the vestibular aqueduct. The diagonal band of Broca is also known as the olfactory radiation of Zuckerkandl. Zuckerkandl described an olfactory fasciculus that becomes detached from the main portion of the column of the fornix and passes down in front of the anterior commissure to the base of the brain, where it divides into two bundles, one joining the medial stria of the olfactory tract, the other the subcallosal gyrus, and through it the hippocampal gyrus. He is also known for identifying the fasciculus cortico-dentatus, which is a diffuse system of unmyelinated fibers that arise with the alvear paleo-ammonic fibers, originating from the pyriform cortex. These fibers pass directly into the dentate gyrus. The emissary vein, which may connect the anterior part of the superior sagittal sinus to veins of the nasal cavity via the foramen cecum, is referred to as the vein of Zuckerkandl.

It is important to note that the anatomist Emil Zuckerkandl mentioned herein is different from the biologist Emile Zuckerkandl, who was born in Vienna in 1922 and founded, together with his colleague Linus Pauling, the theory of the molecular evolutionary clock.

3.2 Czechoslovakia

3.2.1 Vincent Alexander Bochdalek

Vincent Alexander Bochdalek (1801–1883) (Figure 3.11) was a distinguished physician associated with the first descriptions of congenital diaphragmatic hernias. Often mistaken for a German because of the lack of a Czech accent on the letter "a" of his last name, Bochdalek was born on February 11, 1801, in Litoměřice, a town at the junction of the Rivers Labe and Ohře in the northern part of what is now the Czech Republic (approximately 70 km northwest of Prague). His father was a Czech gamekeeper and his mother was a Slovak of Polish descent from the village of Skripov in northern Moravia. From 1816 to 1820, Bochdalek attended the Gymnasium in Opava. He then studied in Vienna, but later returned to the Carolinum in

Figure 3.11 Photograph of Vincent Alexander Bochdalek (1801–1883).

Prague, where he studied medicine with Rokitansky, Skoda, Krombholz, and Ilga. As a medical student, his talent for dissection enabled him to work in the laboratory, and he graduated with a degree in medicine in 1833. Bochdalek's interest in the medical dilemmas of his time was made evident in his graduate degree dissertation, "Instruction to Practical Autopsy of the Human Brain with Special Attention to the Cerebellum," which was published as part of a monograph containing 19 defended theses.

In Prague, Bochdalek spent most of his career in the dissecting room, and competed for a position in 1836 to succeed Professor Ilgovi. However, he was unsuccessful in becoming Professor of Pathological Anatomy, and the position was awarded to Josef Hyrtl (1810–1894). Bochdalek originally had doubts about the position because of his family responsibilities, so he applied for the pathological prosector position in Prague General Hospital. He was reluctant to pursue a career in anatomy because of the societal stigma associated with dissection and working with cadavers that existed in Prague at the time. In October 1837, Professor Nadherny noticed Bochdalek's potential and appointed him as a prosector, which he accepted without a salary. In 1838, Bochdalek received 1000 zlaty for his work in the dissecting room, which at the time was considered a high wage.

Bochdalek was also appointed to the faculty of Carolinum, where he became a highly respected professor. He was obsessed with his work, and with dissection. When human cadavers were lacking, he would dissect animal ones. In 1829, along with Professor Krombholz, he was the first Czech to dissect a dog with rabies. In addition, he dissected livestock that had died from the plague.

Owing to the societal views in Prague at the time regarding cadaveric studies, Bochdalek often had difficulty in finding a location for his dissection laboratory. He initially worked out of the Carolinum, but was soon evicted. He then obtained a room for his dissection in the General Hospital of Prague, but encountered more difficulty in maintaining his laboratory. He was always ostracized by hospital clinicians and, despite the heat, was forced to close the windows of his laboratory during the summer because of the unbearable odor. In addition, hospital employees stole and consumed the alcohol and other materials he used to preserve the bodies, diluting or replacing them with water to avoid being caught, and his cadavers quickly spoiled even during the freezing cold Prague winters. He never let these trials and tribulations derail his goals of learning and teaching.

Bockkalek's desire was to become a very good professor of pathology and anatomy. However, he garnered more prestige a practicing physician than as an academic. He loved teaching medical students, despite not being a good public speaker, often lecturing in a monotone. He would spend hours in the laboratory with his students, going from one dissection table to the next, always taking time to explain the intricacies of the human body to his eager apprentices. In 1853, Bochdalek pressured the faculty's administration into supplying his students with proper dissection tools and microscopes for examining autopsied material, which at the time were considered luxuries. His students loved him, and would faithfully attend his lectures in poorly lit rooms, bringing candles from their homes to illuminate the ends of the tables.

Bochdalek dissected 600–800 cadavers a year. He gained much anatomical knowledge, which his colleagues envied. He had a bright mind and introduced the process of keeping postmortem records and implementing dissection protocols in several dissecting rooms at the General Hospital. In 1850, he returned from a tour of the Scandinavian nations with 1294 anatomical preparations of sea fauna, which he categorized and studied.

For many years, Bochdalek had an assistant/coworker named Gunther, who was a very knowledgeable dissector with no formal education. Gunther was called a "half doctor" because of his medical skills, and was credited with teaching Czech physicians how to pull teeth with pliers.

Several eponyms are associated with Bockdalek, including: Bochdalek's canal (through the tympanic membrane); Bochdalek's tube, connected with Bochdalek's (thyroglossal) duct; Bochdalek's gland, a remnant of a Bochdalek's duct, which may grow into Bochdalek's cyst; Bochdalek's flower basket (the protrusion of the choroid plexus through the foramina of Luschka); Bochdalek's ganglion of the dental nerve plexus at the root of the upper canine; Bochdalek's muscle (triticeoglossus muscle); and Bochdalek's valve (a mucous membrane fold in the lacrimal canaliculus at the lacrimal punctum). The best-remembered is Bochdalek's trigone, the (lumbocostal) triangle of the diaphragm (Figure 3.12).

It was in 1848 that Bochdalek first described the death of a child due to herniation of the diaphragm, which he predicted could be reduced successfully through the advent of surgical treatment. Bochdalek reconstructed the pathological anatomy of this hernia, which he traced back to a congenital malformation occurring *in utero* and resulting in the rupture of an intact membrane of the diaphragm, referred to as the trigone of Bochdalek. He clearly described this posterior defect as the lethal cause of newborn congenital diaphragmatic hernias. His observations laid the foundation for the successful surgical correction of this once-lethal anomaly, although this was not achieved until 1946 (by Robert Gross). It is interesting to note that the naming of this hernia has come under criticism. John J. White suggests that it is inappropriate to name it "Bochdalek's hernia," and that, to be more anatomically accurate, it should be called a "congenital posterolateral diaphragmatic hernia in a newborn infant," as he believes that the hernia does not protrude through the trigone of Bochdalek but through the pleuroperitoneal hiatus. However, J.A. Haller Jr. believes that the Prague pathologist's initial observations should

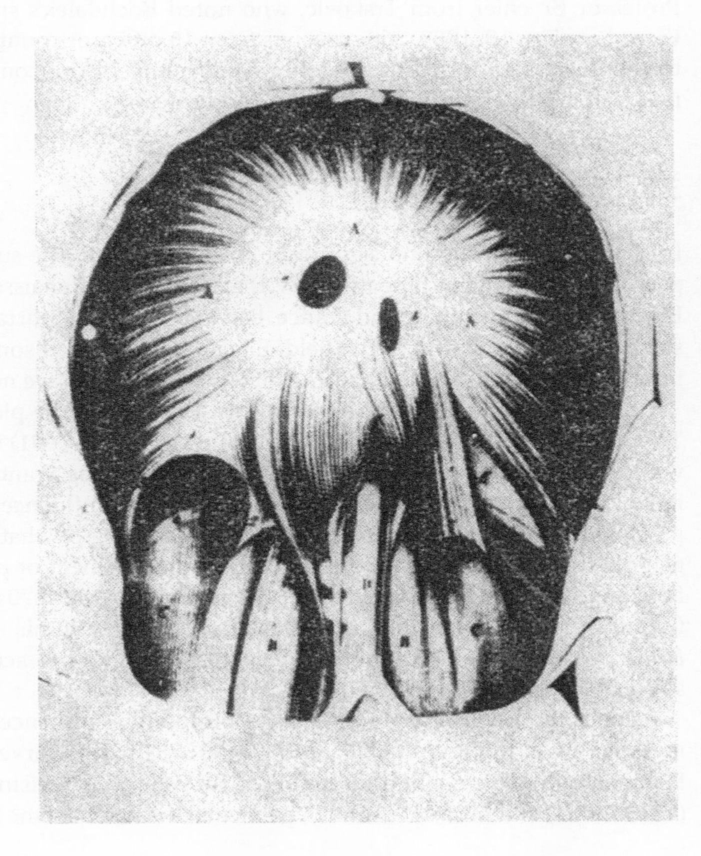

Figure 3.12 Bochdalek's illustration of the posterolateral trigone of the diaphragm, taken from his original thesis in 1848.

not be disregarded because of minor details of the embryological pleuroperitoneal sulcus, since this was not familiar to anatomists of Bochdalek's era.

A man of great determination, Bochdalek was known as a dedicated anatomist and professor who exemplified perseverance. He always faced obstacles while pursuing his dreams, just as much in his personal life as in his career. He had seven children, but only one son and three daughters outlived him. His favorite daughter, Emilie, took care of him after his wife's death, at the age of 58, in 1873. The death of his son, Vincenc, proved to be particularly distressing for Bochdalek, partly because Vincenc was also a prominent anatomist who had published work in 1866 on the anatomy of the tongue.

During his time in Prague, the stigma of anatomy and dissection often prevented Bochdalek's medical research. He was one of the few anatomists who supported the creation of anatomical laboratories in Prague for both research and the education of future physicians. He often felt that pursuing his dreams was pointless. When he wanted to give up, however, his addiction to learning gave him the drive to succeed, and he became Vice-Dean of the Faculty of Medicine in Prague in 1862, and Dean in 1865. He retired in 1872 and moved back to Litoměřice, where he worked in his botanical garden. While retired, Bochdalek was called upon to direct the embalming of Mensdorfa, a noble Czech, and in 1873 he was called upon to embalm His Majesty Ferdinand V, the last crowned Czech king. His devoted study of anatomy and love of teaching was made evident by his students' frequent visits to his home. In 1881, Dean Breisky and Vice-Dean Klebsem conferred an honorary degree on Bochdalek for his 80th birthday.

Vincent Alexander Bochdalek died of pulmonary edema on May 3, 1883. He was dissected by Professor Brechler from Trskovic, who noted Bochdalek's small stature and beautiful teeth. From June 7 to 10, 1983, 260 surgeons from 16 different countries held a conference dedicated to the discussion of Bochdalek's hernia in honor of the contributions made to the modern medical field by this Czech anatomist and professor.

3.3 England

In England, there were no significant contributions to the study of anatomy before the later part of the seventeenth century. By that time, post-Renaissance anatomists in Continental Europe, especially in Italy and France, had made many important advances. However, notwithstanding this late arrival in the field, England contributed something new, not only in factual information, but also in the nature of investigation: what we now call "scientific method."

There were two related reasons for this. First, when the pioneering English anatomists Sir Thomas Willis (1621–1675) and Richard Lower (1631–1691) were working, the Royal Society was a new and active institution, dedicated to pursuing Francis Bacon's recipe for generating reliable knowledge. Willis and Lower were strongly influenced by the Baconian method: dispassionate, meticulous observation; careful recording of what was found; slow progression to general statements; avoidance of any kind of prejudgment or prejudice; and constant vigilance for counter-instances. Concurrently, John Locke (1632–1704) – one of Isaac Newton's few friends – was formalizing the Baconian method in his *Essay Concerning the Human Understanding*. It was a time of intellectual ferment and accompanying social change, and scientific anatomy was one beneficiary.

Second, the English anatomists were profoundly influenced by the discoveries – and the investigative example – of William Harvey (1578–1657). Harvey had studied under Fabricius in Padua, and therefore inherited the skeptical view of Galenism promoted by Andreas Vesalius (1514–1564) and his successors. In particular, he had become fascinated by Fabricius' account of the "little doors of the veins" – the valves – the function of which remained a mystery. On his return to England, Harvey worked for a time as physician and secretary to Francis Bacon, who was then Lord Chancellor. He went on to become the court physician to James I and Charles I of England, a position that gave him great influence, but by then he had absorbed the Baconian method "from the horse's mouth." One illustration of this lies in the definitive experiment that led him to the revolutionary concept of the circulation of the blood: he could pass a solid probe cephalad, but not caudad, through a vein. From this and other evidence, observed many times with no counter-instances, he inferred that blood could flow through the veins in only one direction.

When Harvey's epoch-making book was published in Frankfurt in 1628, his reputation was harmed; he had explicitly rejected Galen's teaching and replaced it with a radical new theory, and the conservative medical profession reacted negatively. In particular, an angry dispute arose between Harvey and the doyen of Parisian anatomists and surgeons, Jean Riolan; much of the correspondence between the two men still survives. Later, Harvey found himself in disagreement with Descartes about the mechanism of heart function. But his experimental studies and observations had convinced him, and he would not recant his views; and within a decade of his death in 1657, his account of the circulation of the blood had beome almost universally believed. Harvey's achievement represents a triumph of the Baconian method before the Royal Society was founded.

In the generation following Harvey's death, a tradition of anatomical investigation began in England. Like similar traditions elsewhere in Europe, it survived through the succeeding centuries, distinctive in character but part of the European whole.

3.3.1 Richard Lower

Richard Lower (1631–1691) (Figure 3.13) was one of three sons of an old Cornish family. He studied at Westminster School in London, and in 1649 was admitted to Christ Church College, Oxford, where he received a BA in 1653, an MA in 1655, and an MD in 1665. In 1666, Lower moved from Oxford to London to establish a medical practice at the request of the Archbishop of Canterbury. He was elected to the Royal Society in 1667, was put up as a candidate for entry into the Royal College of Physicians in 1671, and was elected a fellow of that institution in 1675. In that same year, Sir Thomas Willis died, and Lower became recognized as the leading physician in London and was appointed Court Physician to Charles II. Lower identified with the Whig party, owing to his strong Protestant and anti-Popish position; therefore, in the 1680s, with the fall of the Whigs and the rise of James II, he lost royal favor and his court position. He developed pneumonia after helping to extinguish a fire in London, and died on January 17, 1691, at the age of 60.

During the seventeenth century, most scientific discoveries were inculcated with the religious and political beliefs and practices of the time. The dominant view in medicine in this century strictly followed the traditional teachings of Galen (130–200 CE). However, William Harvey (1578–1657) used evidence to establish his observations, rather than accepting historical principles, and his account of the circulation of blood and the mechanisms of the heart was revolutionary. Lower's first major influence came from Harvey, and he was explicit about his admiration and respect for that pioneer of experimental physiology:

> there are points not mentioned in Harvey's circulation which need consideration ... Harvey himself, indeed, seems to promise further contributions, had age and time allowed ... I have myself tried to fulfil the promises of that excellent man, and to bring nearest to completion than they have hitherto been.

RICHARD LOWER, M.D.
Practicus Londinensis longe
Celeberrimus atq; Felicissimus.
Ætatis Suæ

Figure 3.13 Illustration of Richard Lower (1631–1691), from *Cerebri Anatome* by Sir Thomas Willis, published in 1664. (*Source:* Wellcome Library, London.)

One characteristic of Lower that is commonly documented is his loyalty to his predecessors, including Harvey and Willis. He faithfully defended and expanded on Harvey's basic principle of experimentation leading to scientific discovery, set out by Francis Bacon and accepted as the scientific method by the newly-formed Royal Society. As Lower's career progressed, he became increasingly well known for his dissecting skills. His precision in experimental dissections allowed him to become an eminent contributor to anatomy in England. Donavan wrote of him in 2004:

> Hedid this by precise anatomic dissection and extensive experimentation, utilizing surgery as a technique in his investigations. He deserves recognition as the first surgical experimentalist of the modern medical or post-Harvey era.

Lower's second mentor was Sir Thomas Willis (1621–1675), a scientist at Oxford, whom he served for a period of time as his assistant. Lower proved his interest rested in two main areas: the function of the cardiopulmonary system and neuroanatomy. Already heralded as a dissector and anatomist, Lower brought extraordinary dissection skills to Willis' work on the nervous system and cerebral circulation. This partnership with Willis, and with several others, enabled Lower to make his greatest impact on medicine, so it will provide the focus of this account.

3.3.1.1 Contributions to Understanding of the Nervous System

In an era when experimentation was not required – and not well practiced – to unlock the mysteries of the human body, Lower exemplified skill and intellect more characteristic of later times. He promoted the use of experiments in science, and partly for that reason he became a master of identification and reasoning in previously uncharted areas of human biology. He worked closely with Willis and contributed considerably as Willis completed his *Cerebri Anatome* (Figure 3.14). Over several years, Lower published three distinguished manuscripts that confirmed and defined his contribution to medicine and English society in the seventeenth century and beyond.

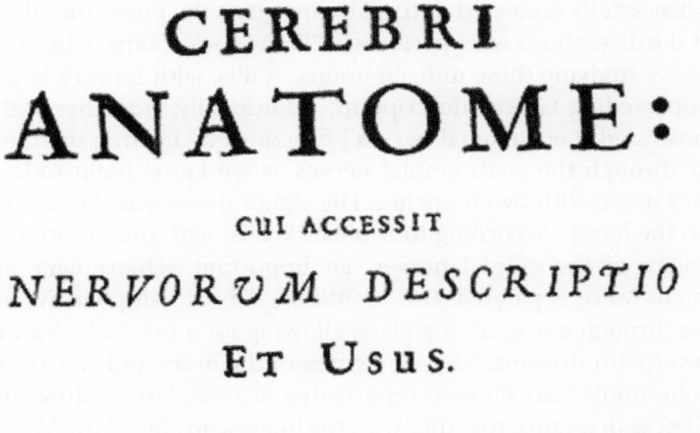

CEREBRI

ANATOME:

CUI ACCESSIT

NERVORUM DESCRIPTIO

ET USUS.

STUDIO

THOMÆ WILLIS, ex *Æde Chriſti*

Oxon. M. D. & in iſta Celeberrima

Academia Naturalis Philoſophiæ Pro-

feſſoris *Sidleiani*.

LONDINI,

Typis *Ja. Fleſher*, Impenſis *Jo. Martyn* & *Ja. Alleſtry*

apud inſigne Campanæ in Cœmeterio

D. Pauli. M DC LXIV.

Figure 3.14 Photograph of the title page of *Cerebri Anatome* by Sir Thomas Willis. This book, in which Willis established the concept of neurology, was published in 1664. It praises Richard Lower in the preface, acknowledging his anatomical skills.

3.3.1.2 Cerebri Anatome

Luckily, Willis recognized the importance of method in studying the brain. There were few accomplished dissectors in the seventeenth century, and in 1665 Willis was so captivated by Lower's expert craftsmanship that he immediately hired him to work with him. Both men had an interest in neuroanatomy, and both determined that in order to outline it in exquisite detail, the brain must be dissected with careful attention to delicate structures, hence the importance of highly experienced hands (Figure 3.15).

The *Cerebri Anatome*, published in 1664, was Willis' work describing the neurological structures and functions of the brain. O'Connor writes, "previous anatomists were let down by 'flawed techniques,' producing artifactual results." Willis knew that dissection was a critical step in producing accurate images and trusted Lower's expertise. In a divergence from traditional methods of *in situ* dissection, Lower approached the brain from below, removed it from the skull, and sliced it from the inferior end upward. It is widely documented that Lower helped extensively with brain and spinal cord dissections for inclusion in *Cerebri Anatome.*

It is extraordinary that Lower successfully dissected unfixed brains given the poor condition of the specimens and the fact that the dissections were performed "in the back rooms of houses and inns," according to O'Connor. By studying these unfixed brains, Willis, with Lower's help, reclassified the cranial nerves, superseding Galen's description. Additionally, "together with Richard Lower, he [Willis] demonstrated their [cranial nerves'] functions by ligating them in dogs." Willis recognized the first through the sixth cranial nerves as we know them today. His seventh nerve was the auditory nerve with two branches. His eighth nerve was the vagus, and he recognized its branches to the heart. According to Molnár, Willis, with the expert aid of Lower, correctly identified many of the cranial nerves, an important achievement in neuroanatomy. Once the specimens were separated and identified, Sir Christopher Wren (1632–1723) would examine them through a magnifying glass, allowing for a precisely drawn image. O'Connor also credits Lower with drawing "elegant figures of the brain and charts of cranial and autonomic nerves." In his book, Carl Zimmer reports that in 1663, Lower "drew up diagrams of the nervous system." These diagrams, together with the images produced by Wren, were used to complete the *Cerebri Anatome.*

Lower and Willis made another significant contribution to neuroscience by proving that the control center of the heart was in the hindbrain. In an experiment carried out on dogs, they tied off the nerves that were transmitted from the hindbrain to the heart, whereupon the heart "quickly engorged with blood and died."

When Willis and Lower were experimenting with seizures, Lower frequently wrote to Robert Boyle (1627–1691), excited that he had illustrated and written about many diseases of the brain. These letters still survive and are enriched with Lower's astonishing observational skills and excellent accounts of case studies.

Willis received credit for the discovery of the arterial circle at the base of the brain, named the circle of Willis. However, Lower was interested in the function of this circle and designed an experiment to determine if there was collateral circulation among the four main arteries. Wren and Lower performed dye injections on humans, as previous injection studies on animals immediately after death had demonstrated that blockage of just one of the cerebral arteries would not lead to sudden death. Lower is credited with establishing that circulation through this arterial circle in the cerebral circulation named for Willis could be maintained even if three of the four contributing arteries were ligated.

3.3.1.3 Vindicatio

Lower wrote *Vindicatio* in 1665 as a harsh retort to Edmund O'Meara (1614–1681), who had attacked Willis for his Harveian approach to research. *Vindicatio* provided a strong statement in favor of the scientific method, anticipating the direction medical research would take in

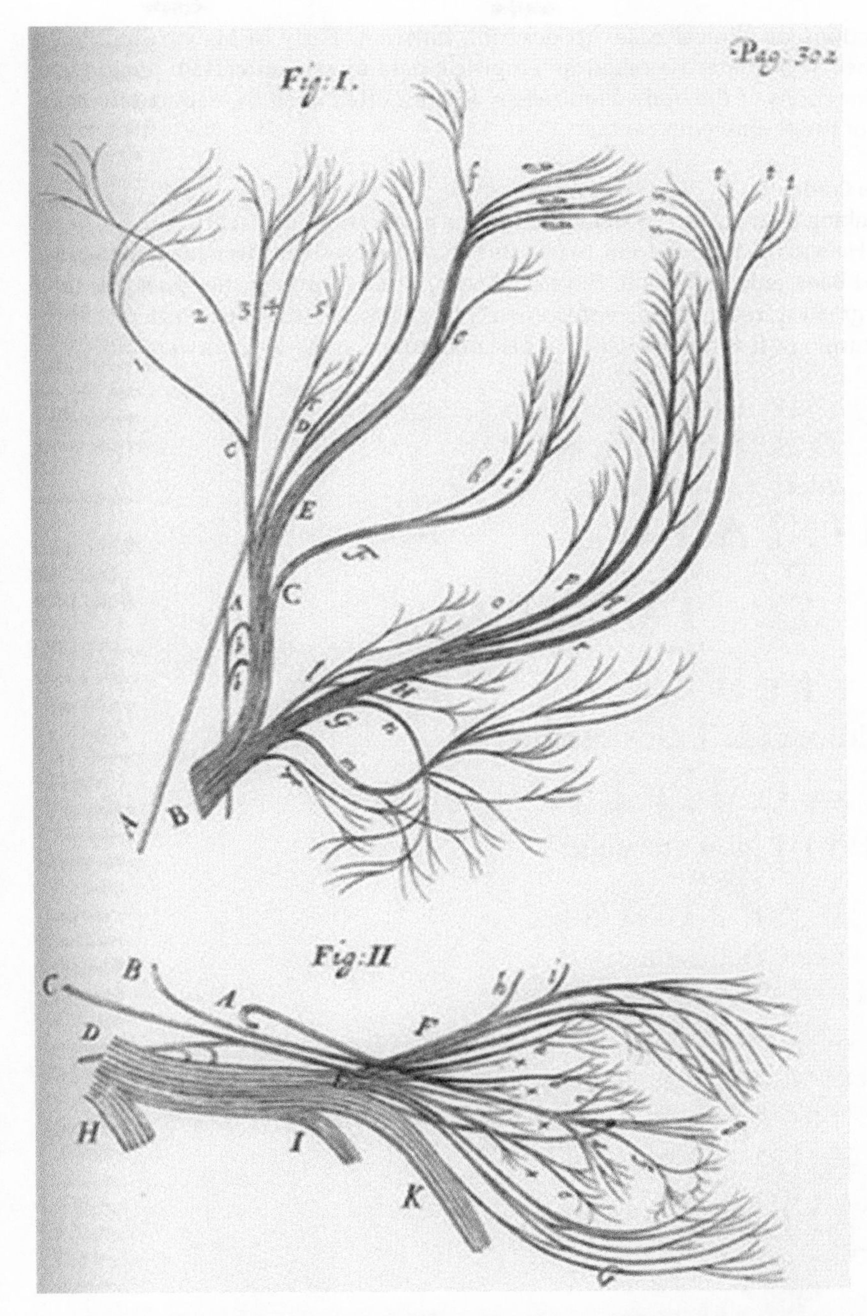

Figure 3.15 Illustration from *Cerebri Anatome* by Sir Thomas Willis. This image, based on a dissection of Richard Lower, illustrates Willis' fifth and sixth cranial nerves and demonstrates Lower's capability for finely detailed dissections.

future centuries. Many renowned scientists of the time agreed that the use of experiments would provide the most powerful and reliable method for making new discoveries. Although Willis occasionally used unsubstantiated conjecture to explain some phenomena, such as "spirits," the majority of his work followed a similar pattern to the modern-day scientific method, using either dissections or clinical cases to describe function. Early in his career, Lower followed Galen's theory, but later he relied on empirical data to explain certain anatomical and physiological functions of the body. Empiricism was not often cited by leading scientists until the later part of the seventeenth century.

3.3.1.4 Tractatus de Corde

Lower's groundbreaking research in physiology led to the publication of *Tractatus de Corde* in 1669 (Figure 3.16). He was the first, in 1666, to transfuse blood successfully from an animal into a human (there had been failed efforts in the early 1600s). It was through the *Tractatus* that Lower gained his greatest recognition; unfortunately, this has led to some neglect of his contributions to neurology. It is noteworthy that his conclusions in the *Tractus* were based on

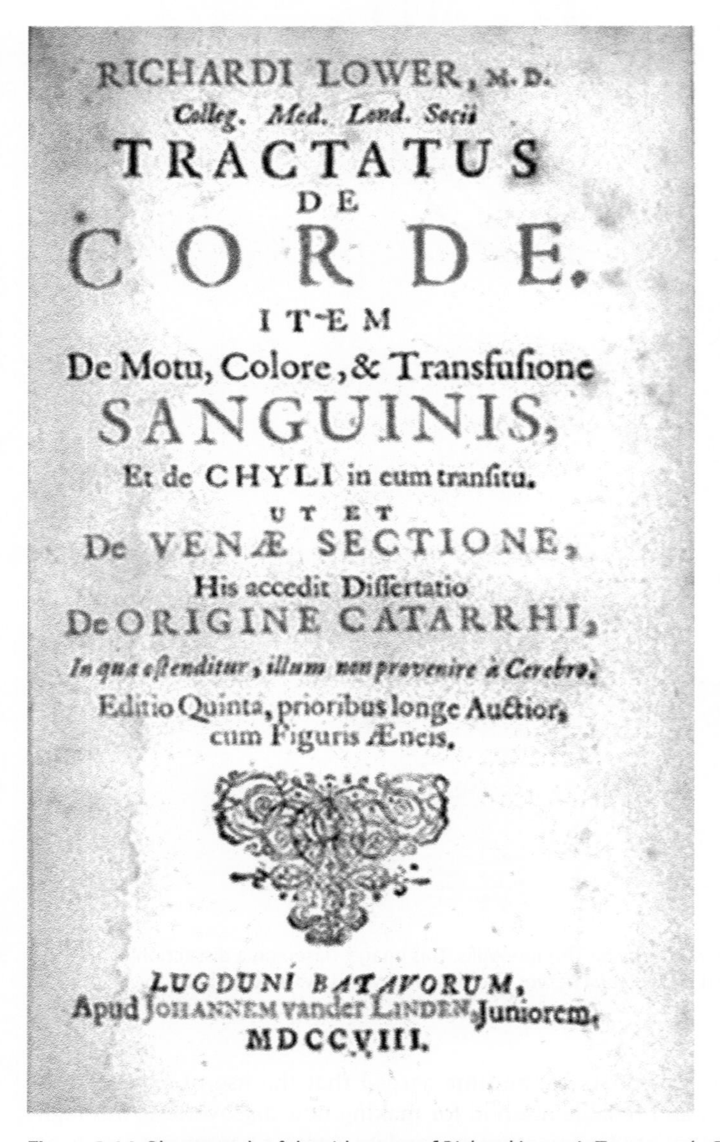

RICHARDI LOWER, M.D.
Colleg. Med. Lond. Socii
TRACTATUS
DE
CORDE.
ITEM
De Motu, Colore, & Transfuſione
SANGUINIS,
Et de CHYLI in eum tranſitu.
UT ET
De VENÆ SECTIONE,
His accedit Diſſertatio
De ORIGINE CATARRHI,
In qua oſtenditur, illum non provenire à Cerebro;
Editio Quinta, prioribus longe Auctior,
cum Figuris Æneis.

LUGDUNI BATAVORUM,
Apud JOHANNEM vander LINDEN, Juniorem,
MDCCVIII.

Figure 3.16 Photograph of the title page of Richard Lower's *Tractatus de Corde*, printed in 1669.

precise anatomical dissections, autopsy studies using Willis' deceased patients, clinical observations, and animal experimentation. Zimmer claims that Lower "clearly established himself as a scientist" as a result of this publication.

3.3.1.5 De Catarrhis

Lower's last major publication was *De Catarrhis*, in which he attempted to challenge a widely held belief regarding cerebrospinal fluid formation and the origins of nasal catarrh. It is believed that this concise piece was published in 1670, complementing the second of several editions of the *Tractatus de Corde*. *De Catarrhis* was eventually published independently in 1672, but it was not until 1963 that an English translation was completed. Lower's investigational experiments into how cerebrospinal fluid was formed and how it circulated led to the discovery of hydrocephalus. Lower's book is of historical significance because it was the first scholarly attempt by an English physician to take a classical doctrine and to disprove it by scientific experiment. The humoral theory held that sweating or micturition would correct an excessive accumulation of fluid in the body. If an individual were not allowed to perform one of these functions, it was believed that catarrh would be released from the overloaded cerebral ventricles from the pituitary stalk and would exit through the cribriform plate into the nasal cavity. Lower, however, suspected that this was not an accurate description, as he had done extensive dissections of the brain. He designed an experiment in which he injected "milk or a black substance" into openings of the base of the brain, but neither flowed through the cribriform plate. Lower established through cerebral vascular infusions and other studies that there was no communication between the subdural space and the nasal passages. He then ascertained that particles passing along the olfactory nerves from the nasal cavity would provide a sense of smell to the brain. "An astute observation indeed," wrote John Felts in 2000.

3.3.1.6 Dissension

Many historical documents regarding scientific findings are infected with dissension in determining the rightful recipient of credit. Willis' *Cerebri Anatome* is among them. Some have attributed the results of Willis' anatomic and physiologic studies to Lower, rather than to Willis. Generously acknowledging Lower's contributions to *Cerebri Anatome*, Willis wrote:

> Richard Lower, a doctor of outstanding learning, and an anatomist of supreme skill. The sharpness of his scalpel and of his intellect, I readily acknowledge, enabled me to investigate better both the structure and functions of bodies whose secrets were previously concealed ... in a short space of time everything about the cerebrum and its appendage within the scull seemed clearly revealed and thoroughly explored by us ... when we were entering upon a much more difficult task, the dissection of the nerves, the really wonderful dexterity of this worker and his untiring perseverance were conspicuous in the extreme

Perhaps it was this recognition that prompted the controversy concerning Willis' role in the research that led to the publication of *Cerebri Anatome*. Anthony Wood (1632–1695), a contemporary Oxford diarist and historian, questioned the intellectual contribution of Willis versus that of Lower and wrote in *Athenae Oxonlenses*, "Whatsoever is Anatomical in that book, the glory of *it* belongs to the said R. Lower, whose indefatigable industry at oxon (Oxford) produced that elaborate piece." It seems, though, that the most forceful criticism came much later, from Sir Michael Foster (1836–1907), the distinguished Cambridge physiologist, who also believed that Lower had been the creative genius behind *Cerebri Anatome*. He declared that Lower was "the henchman of the fashionable Willis whose false name in large measure rested on Lower's careful, unacknowledged work."

However, not all authors agree that Lower was the true mastermind behind *Cerebri Anatome.* Sir Charles Symonds claimed that Wood received his information from a biased source and that there never appeared to be any discord between Willis and Lower. Another dispute over Lower's contributions came from Henry Stubbe (1632–1676), a philosopher and physician in his own right, who published numerous papers on the Royal Society. He discredited Willis for many of his contributions to medicine and acknowledged Lower as their likely originator. However, in a minor section of the same manuscript, Stubbe wrote, "although Lower did the dissections, Willis' was the master mind which inspired and directed the work."

Furthermore, some authors discredit Lower for his explanation of catarrh. In fact, Edwin Clarke explained that Lower was not the first to identify that the nasal passages did not directly connect with the intra-cranial cavity. The original theory, found in Hippocratic writings and further solidified by Galen, was first criticized during the earlier seventeenth century. Jan Baptista van Helmont's *Catarrhi deliramenta* in 1650 denied the existence of catarrh, especially from the brain, in 1650, while Karl V. Schneider refuted the presence of any connecting channels between the brain and nasal cavity in *De catarrhis* in 1664.

Lower stood in Willis' shadow for a decade. However, he never appeared to be resentful, as Willis provided him with a solid foundation in science and a successful medical practice. Indeed, he took full advantage of Willis' teachings, which allowed him to become a well-respected physician in the Royal Society, and led to his being widely regarded as the leading physician of his time. T.M. Brown quotes Anthony Wood as saying Lower was "esteemed the most noted physician in Westminster and London, and no man's name was more cried up at court than his." Perhaps Lower's brilliant career as a scientist was overshadowed by his prominent success as a practicing physician. Fame may have diverted his attention from further research. He did not publish any major works after *De Catarrhis,* although the *Tractatus De Corde* did go through seven editions with significant changes, suggesting that he continued to perform some study.

While there may be disagreement as to the crediting of certain discoveries, it is accepted that Lower worked closely with Willis for several years, assisting with dissections and insight. In addition, history books illuminate Lower's own three major publications and give some insight into the impact of each. As most articles in the literature emphasize, Lower had a marvelous career, making numerous discoveries. This is best exemplified by Woo and Fung:

> Lower undoubtedly belongs in the pantheon of the greatest physicians. His life and work paint a portrait of a passionate and non-conforming mind. These characteristics helped him challenge tradition and make major contributions in anatomy and physiology, bringing great acclaim in his lifetime.

Although this excerpt recognizes Lower in a well-deserved manner, it seems that many authors have omitted a significant portion of his work in neuroanatomy. Despite the strongly held traditions and theories of his day, Lower did not conform, but instead challenged each scientific element that seemed erroneous. This led him to many revolutionary breakthroughs within the disciplines of anatomy and physiology. The refutation of the erroneous account of the origin of catarrh, the dissections and diagrams of the brain (coupled with scientific experiments), and the description of the carotid/vertebral collateral circulation should all shed light on a highly neglected component of Lower's illustrious career: his true legacy was his willingness to challenge widely accepted beliefs in order to uncover unimagined truths, across multiple aspects of medicine. It is a legacy that defines him as an accomplished anatomist and physician.

3.3.2 John Browne

John Browne (1642–1702) (Figure 3.17) figures fairly insignificantly in the annals of history. At first glance, this is surprising since the credits to his name hint at an illustrious career. He was Surgeon in Ordinary to both Charles II and William III, a title that allowed him to superintend and write about "royal touch" ceremonies performed by Charles II to heal scrofula. He is credited as being the first to describe cirrhosis of the liver, in "A remarkable account of a glandulous liver, taken out of a hydropsical person by Mr. John Brown [sic] surgeon of St. Thomas Hospital in Southwark communicated in a letter to one of the secretarys of the Royal Society" (1685). Matthew Baillie's 1793 *Morbid Anatomy* is also credited as being the first description of liver cirrhosis; Baillie might not have known of Browne's publication.

Browne's treatise on muscles utilizes an innovative illustration technique involving labeling each muscle directly. Despite his impressive achievements, his reputation is tempered by his use of plagiarism in this work. The paucity of literature on this seventeenth-century figure indicates the need for further inquiry into who he was and what mark he left on the history of medicine.

In contrast to other prominent physicians of the time, such as Thomas Sydenham (1624–1689) and Richard Wiseman (1625–1686), Browne's history is relatively obscure. He was born in Norwich in the county of Norfolk in 1642, during the reign of Charles I. He studied medicine

Figure 3.17 Portrait of John Browne (1642–1702), from his *Compleat Treatise*.

at Saint Thomas' Hospital in London; then, like many other surgeons and physicians during the seventeenth century, he served in the navy. Following his naval service, he settled for a period in the city of his birth. In 1677, Browne returned to London having been appointed Surgeon in Ordinary to King Charles II. A few years later, in 1683, upon the king's recommendation, he claimed a position on the surgical staff of Saint Thomas' Hospital. This did not last long, however, since in 1691 all the surgical staff, including Browne, were discharged from the hospital for "failure to adhere to the regulations"; they had fallen into irreconcilable disagreement with the hospital's governors.

3.3.2.1 Browne's Publications and Legacy

Browne's publications are found in historical libraries in England, Scotland, the United States, and Australia, but due to their limited print run and their age, they are difficult to access. Fortunately, Kenneth F. Russell – the only scholar to have written extensively on Browne – published an annotated bibliography that enumerates the titles and editions of Browne's work. The body of work covers a variety of topics, including the anatomy of the eye, an account of a hydropsical liver, a treatise on tumors, a discourse on wounds including skull fractures and gunshot wounds, and an early account of the body's defense against disease. His most famous and controversial work, however, is on the subject of the muscles (Figures 3.18 and 3.19).

The muscles of the human body were commonly addressed in publications of the seventeenth century. Browne's contribution was entitled *A Compleat Treatise of the Muscles, as they Appear in the Humane Body, and Arise in Dissection; with Diverse Anatomical Observations Not Yet Discover'd.* The last portion of the title is ironic, seeing as the descriptions of the muscles were taken from a book published in 1648 by William Molins. Russell notes that a manuscript for Browne's version of the publication dated 1675 is nearly a verbatim copy. In this first printed edition, dated 1681, some "padding was done in an attempt to hide the obvious piracy." The plates that accompany the text were copied from the work of Julius Casserius (Figures 3.20–3.23). In keeping with the alterations Browne had made in the text portion of the book, the clothing was changed to seventeenth-century style, and the poses and backgrounds were modified in an attempt to claim them as his own. Russell comments that "it is doubtful if more whimsical or bizarre figures have ever graced a serious treatise on anatomy."

Browne was heavily criticized for this piracy. James Young, his contemporary and a fellow surgeon, wrote a scathing review of his treatise, disclosing Browne's sources and remarking that "he hath not skill enough ... to construe three lines." Similarly, William Cowper (1666–1709) was critical, although his condemnation extended to the work of Molins as well, making a broader comment about the state of anatomical writing in the seventeenth century:

> The many that have lately written on this Subject, especially our English Writers, have rather increas'd than diminished former errors: and particularly that Treatise of Mr. William Molins, and that most erroneous one of John Brown [sic], are chiefly collections of the mistakes of others.

The criticism directed against Browne's piracy was well founded, but it would be wrong to brand him a plagiarist without understanding his actions in the context of the period. The word "plagiarism" (derived from the Latin word meaning "kidnapping") was not coined until much later, and copyright – the law binding the domain of plagiarism – was in its infancy in the late seventeenth century. The 1681 edition of Browne's treatise on muscles ironically includes a Royal Warrant of Copyright prefacing the text, allowing reproduction of neither the plates nor the text for 15 years. Despite ordinances issued by publishing companies and kings, copying

Figure 3.18 Image from the preface of Browne's *Compleat Treatise*. Note the picture of Browne in the background.

was not an unusual practice in Browne's day. Indeed, Cowper's criticism appeared in his *Myotonia Reformata*, which was published while he was compiling the *Anatomy of Humane Bodies*; in the latter, he utilizes – without acknowledgement – plates that had appeared in Godfrey Bidloo's *Anatomia Humani Corporis*.

Browne did not limit himself to plagiarizing others. He also engaged in a form of self-plagiarism in which he reproduced his own work with slight alterations. There were financial incentives for doing so. Russell notes that

> these modifications, coupled with the variations in the titles of the later editions, tend to make them appear to be new works, but such is not the case. Browne must have been an astute business man, and, realizing the value of added material, attempted to give his prospective buyers value for their money.

At least he allowed his buyers to *believe* they were getting value for money. For example, in a subsequent 1684 Latin edition, Browne added a folded sheet entitled *Syllabus Musculorum*

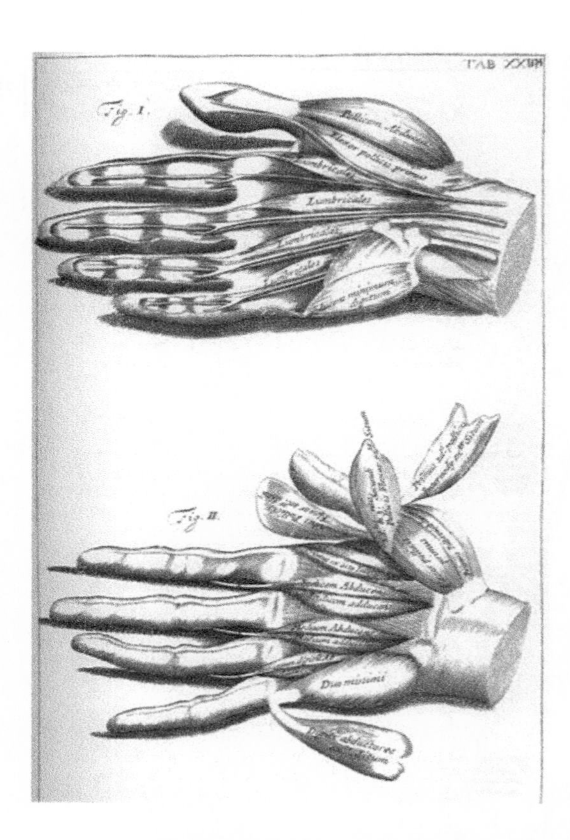

Figure 3.19 Plate from Browne's *Compleat Treatise*, utilizing his direct labeling of anatomical structures of the hand.

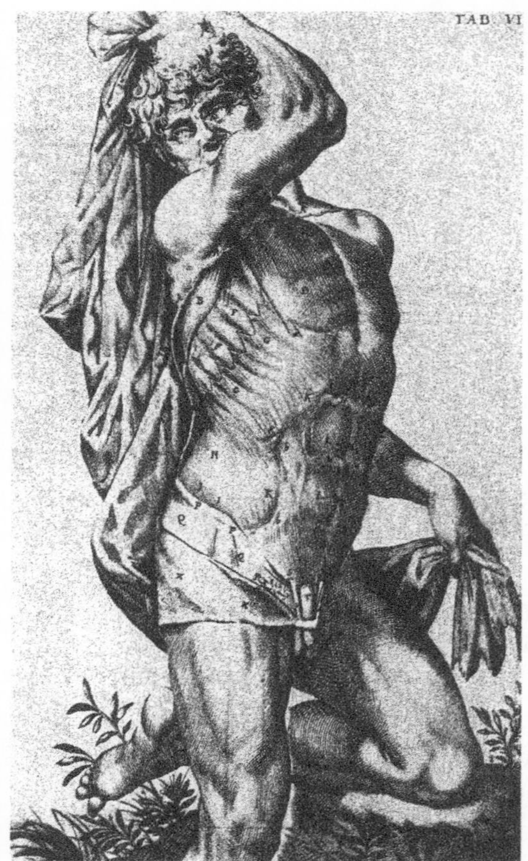

Figure 3.20 Plate VIII from Casserius' text.

Figure 3.21 Image from Browne's text that was obviously copied from Casserius'.

devised by Sir Charles Scarburgh and an appendix on the heart written by Richard Lower. In all such subsequent addenda, however, the authors were credited.

Did anything original come out of Browne's treatise on the muscles? In the same 1684 Latin version, Browne is credited for having pioneered the technique of engraving the name of each muscle on the plate, a practice that is still used in anatomical texts today. This innovation, however, was also discredited by Young as thievery:[Although it]

> looks pretty, and is an ease, and advantage to the Reader … this is not new, nor his own, he stole this also from a muscular scheme, or schemes in Mr. Molines [sic] Parlour, drawn by the accurate Pencil of Mr. Fuller.

3.3.3 James Drake

James Drake (1667–1707) (Figure 3.24) was a renowned political and medical writer of his time. Born in Cambridge, little is known of his family except that his father was a solicitor. Drake was adamant about receiving an education at Cambridge. He began his schooling at Wivelingham and Eton, and was admitted to Caius College, Cambridge in March of 1684, graduating BA and MA. He continued his education there for a few years, earning his MB in 1690 with the encouragement of Sir Thomas Millington (1628–1704), before moving to London to pursue a career in medicine, where he obtained his MD in 1694.

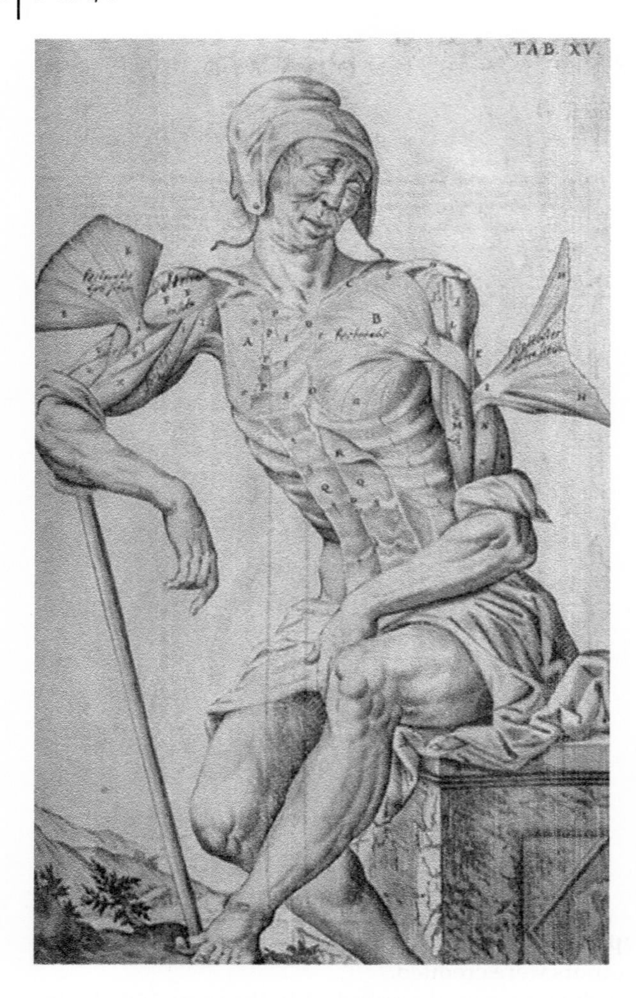

Figure 3.22 Plate XV from Casserius' text.

London opened up many opportunities for Drake, and he took prominent positions in society. He was elected a Fellow of the Royal Society in 1701. Becoming a fellow was a major accomplishment: the Royal Society was made up of the most distinguished scientists of the day and was the oldest scientific academy in existence. After being admitted as a candidate of the College of Physicians in June of 1698, he became a fellow of the college eight years later in September of 1706. James Drake thus became a well-known individual in London, which contributed to the controversy surrounding his political and medical writings.

3.3.3.1 Political Enthusiasm

Drake became enthused with politics and devoted much of his time to writing about his country's political situations. He published a pamphlet in 1702 entitled "The History of the Last Parliament begun at Westminster Feb. 10, in the 12th Year of King William, AD 1700." The pamphlet drew much attention to Drake, especially from the House of Lords. Written in the Tory interest, it accused the Whigs of traducing the princess, soon to be Queen Anne. The pamphlet reported that King William had been plotting to secure the succession of the crown for the elector of Hanover. In their outrage, the members of the House of Lords ordered

Figure 3.23 Figure from Browne's text, showing obvious similarities to Figure 3.22. Note that surroundings have been added to Browne's image but that the muscles are reflected almost identically and that even most of the lettering is the same.

the Attorney General to prosecute Drake. They believed his writing negatively affected the memory of King William. Fortunately for Drake, his trial led to acquittal and he continued to write.

The threat of prosecution did not prevent Drake from continuing to write. Two years later, in 1704, he authored "The Memorial of the Church of England, Humbly Offered to the Consideration of all True Lovers of the Church and Constitution," which he wrote with the help of the Member of Parliament for Ipswich, Mr. Poley. This pamphlet was published anonymously in order to protect the identities of the authors. It reflected harshly upon the Lord Treasurer Godolphin and the Duke of Marlborough, who had started to separate themselves from the Tories, and other officers of the Crown with Whig sympathies. The pamphlet was taken directly to the queen to seek punishment for its authors. In a speech from the throne on October 27, 1705, the queen alluded to the pamphlet for its condescending and insulting content. The House of Commons set out to discover the authors, but although Drake was suspected, no evidence was found against him.

In April of 1706, Drake was prosecuted for writing a newspaper, *Mercurius Politicus*, which attacked the conduct of the government in England. However, a flaw in the information presented in court led to the trial being adjourned. Once again, Drake was acquitted, but the government brought a writ of error. This prosecution took its toll on Drake and has been said to have led to his demise: he developed a fever during proceedings that may ultimately have resulted in his death on March 2, 1707.

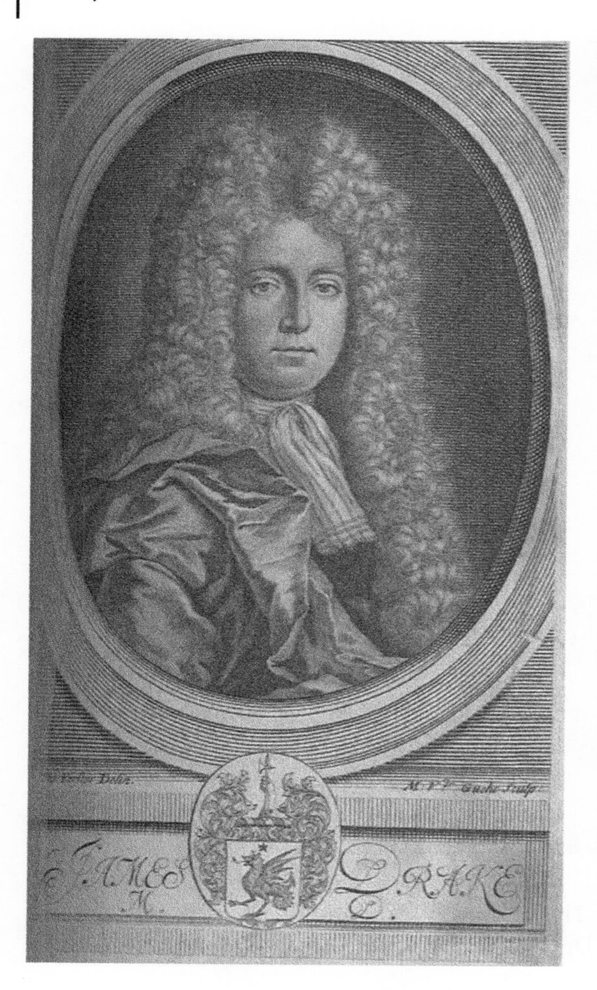

Figure 3.24 James Drake (1667–1707), from the cover page of his *Anthropologia Nova.*

3.3.3.2 Contributions to Medicine

Aside from his notorious political writings, Drake is best known for his greatest medical work, *Anthropologia Nova; or A New System of Anatomy Describing the Animal Economy, and a Short Rationale of Many Distempers Incident to Human Bodies* (Figure 3.25). This work was finished and published shortly before his death in 1707 and contained a commendatory preface written by Dr. Wagstaffe, a physician at St. Bartholomew's Hospital and reader of anatomy at Surgeons' Hall. Drake stated in his preface to the *Anthropologia Nova* that he could not have finished his text without the support of his "learned and ingenious friends" Dr. Branthwait and Mr. Cowper. He especially thanked Cowper for allowing him to use his plates. Unlike some of his contemporaries, such as John Browne, Drake went to great lengths to acknowledge the sources of any unoriginal plates he used. These include figures from the works of Blanchard, Bidloo, Stensen, Wharton, Leuwenhoeck, Bartholin, Vesling, Schneider, Casserius, Malpighi, Du Verney, Willis, Blasius, Vesalius, Ruysch, Kerkring, Lower, Swammerdam, and de Graaf (Figure 3.26). In fact, Drake was a pioneer in orchestrating a collection of anatomical plates from authorities of his day and earlier and not relying exclusively on original works of art. A second edition of *Anthropologia Nova* was published in 1717, and a third in 1727, and it remained popular until Cheselden's *Anatomy*, first published in 1713, replaced it.

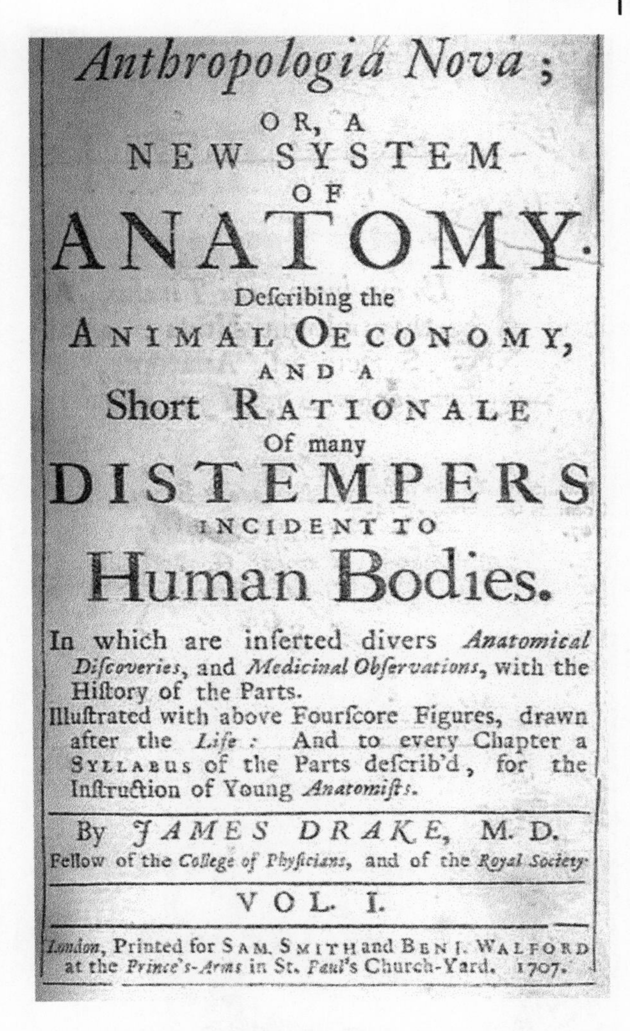
Anthropologia Nova;
OR, A
NEW SYSTEM
OF
ANATOMY.
Deſcribing the
ANIMAL OECONOMY,
AND A
Short RATIONALE
Of many
DISTEMPERS
INCIDENT TO
Human Bodies.
In which are inſerted divers *Anatomical Diſcoveries*, and *Medicinal Obſervations*, with the Hiſtory of the Parts.
Illuſtrated with above Fourſcore Figures, drawn after the *Life*: And to every Chapter a SYLLABUS of the Parts deſcrib'd, for the Inſtruction of Young *Anatomiſts*.
By *JAMES DRAKE*, M. D.
Fellow of the *College of Phyſicians*, and of the *Royal Society*.
VOL. I.
London, Printed for SAM. SMITH and BENJ. WALFORD at the *Prince's-Arms* in St. *Paul's* Church-Yard. 1707.

Figure 3.25 Title page from Drake's *Anthropologia Nova*.

Some scholars have argued that *Anthropologia Nova* was coauthored by Drake's wife, Judith. The work contained detailed information on the reproductive process and stressed the active role of women in conception and heredity, all topics that were apparently very important to Mrs. Drake. She had previous knowledge in the field of medicine that may have aided her in authoring the text. She was an unlicensed practitioner of medicine, especially among women and children. She was a well-known feminist of her day and is remembered for her book, *An Essay in Defence of the Female Sex*. In 1723, she was summoned before the Royal College of Physicians upon accusations of malpractice. Her defense before its president, Hans Sloane, regarding an accuser's claim that she had attempted to poison him, was "the only poison administered was to his ears – in a demand of money."

Aside from his *Anthropologia Nova*, James Drake also contributed to other medical works. His paper on medicine, "Orationes Tres," was published in 1742, many years after his death. He also helped edit the English translation of Le Clerc's *History of Medicine*. This work started with a critique of ancient medicine and continued into the seventeenth century. Drake also became familiar with the physiology of the heart and wrote a paper to the *Philosophical Transactions* entitled "A Discourse Concerning Some Influence of Respiration on the Motion

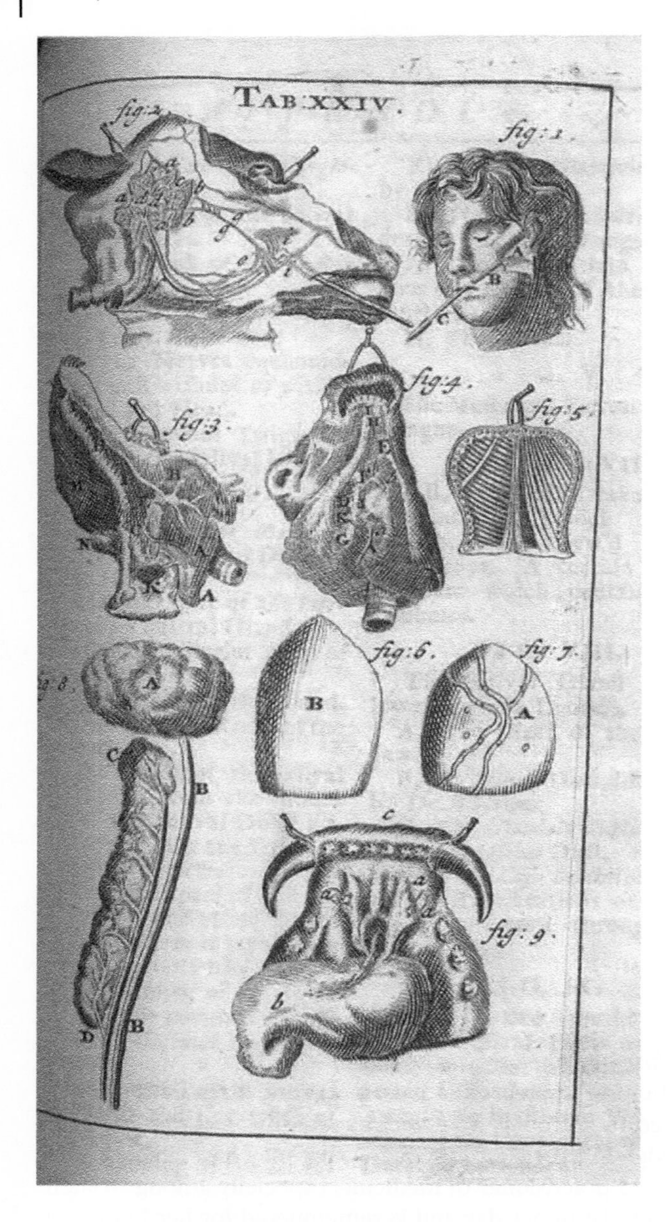

Figure 3.26 Figure from Drake's *Anthropologia Nova*, utilizing a composition of drawings from the works of Blasius, Stensen, Wharton, Lewenhoeck, and Bartholin.

of the Heart, Hitherto Unobserved." He described respiration and its contribution to the circulation, while taking into account the muscles that make up the heart and its overall size and structure; this represented a new discovery.

3.3.3.3 Contribution to Literature

Drake became a distinguished writer during his life. He even wrote comedies, such as *The Sham Lawyer, or the Lucky Extravagant*, which was published in 1697 and "damnably acted at Drury Lane," as stated on its title page. It was borrowed from two comedies by Beaumont and Fletcher, with the Sham Lawyer referencing the plot of *The Spanish Curate* and the Lucky Extravagant referencing the plot of *Wit without Money*.

3.3.3.4 Conclusion

James Drake can best be remembered for his writing talent and the contributions that resulted from it. *Anthropologia Nova* was one of the finest anatomical treatises of its day. Drake mastered the English tongue, and all who knew him or read his work commended his style. He was a multitalented writer with considerable work outside the field of medicine, in politics and literature. Aside from his political dealings, his greatest achievement lies in his careful collection of contemporary and antiquated anatomical drawings, which he used in *Anthropologia Nova*. His talents are made evident in his obituary, which reads:

> The second day of this month [March, 1707] Dr. James Drake, Fellow of this College, died of a fever: a gentleman of very pregnant parts and good learning, as appears by the writings he has left behind him, and deserved a much better treatment from the great world than he met with in it.

3.3.4 Francis Sibson

Francis Sibson (1814–1876) (Figure 3.27), an English physician and anatomist, was a true pioneer of the medical profession. He was an enthusiastic educator at several medical schools and dedicated his life to research. He is most remembered for his description of Sibson's fascia, his experimental use of curare in the treatment of hydrophobia and tetanus, and his detailed description of the positions and movements of internal organs.

Despite his unsettling childhood, Sibson's intelligence and perseverance allowed him to achieve great success in his field. He was born on May 21, 1814, the third of five sons to parents Francis and Jane Sibson in Cross Canonby Township in Cumberland, England. His father was an intelligent though unusual man educated by the Church, but he fell on some troubling times in 1819 when Sibson was still a young boy. As a result of losing his money, he moved his family, now with five sons, to Edinburgh. At the age of five, along with his four brothers, Sibson was

Figure 3.27 Francis Sibson (1814–1876). (*Source:* Wellcome Library, London.)

baptized there, at St. Peter's Episcopal Church. Soon after, his mother and three of his brothers died. Like many children at that time in history, Sibson was forced at a young age to work to support his father and remaining brother, Thomas Sibson, who later became a well-known artist and illustrator for Charles Dickens. So much responsibility at such a young age may have contributed to Sibson's drive and the success he achieved throughout his lifetime.

Sibson became interested in medicine early in his life. He was educated at a school in Edinburgh, and when he was 14 a well-known surgeon-anatomist, Professor John Lizars, who was then Chair of Surgery in Edinburgh, took him as a student. Sibson and Lizers remained friends throughout Sibson's career. Only three years later, on December 21, 1831, at the age of 17, Sibson received his diploma from the Royal College of Surgeons of Edinburgh. At this time, Britain was in the middle of a cholera outbreak and Sibson was eager to give his assistance. He volunteered his time for this purpose immediately after receiving his license in Leith, Newhaven, and Edinburgh hospitals between 1831 and 1832. His hard work and dedication to this cause earned him much respect in his community.

In 1833, Sibson returned to his home town in Cumberland, where he briefly worked with another doctor in general practice in Cockermouth. His next opportunity took him to London, where he studied at Guy's Hospital in the Department of Pathology. In London, he worked with and was mentored by Thomas Hodgkin (1799–1866). Hodgkin was a leader in the field of pathology and is best known for his description of a cancer of the lymphatic system, which was named after him. Then, in 1835, Sibson took on a life-changing role at the Nottingham General Hospital, where he was offered the position of Resident Surgeon and Apothecary. There, Sibson was able to expand his knowledge clinically, while also participating in many research studies on pathological, anatomical, and medical issues. His experience in Nottingham launched his interest in medical research and writing, and provided a solid basis for his future work. His first paper, "A Flexible Stethoscope" (Figure 3.28), was published in the *Medical Gazette* in 1840. He was passionate about his writing and conveying his message to his readers. He gained widespread recognition in the medical community through a paper he wrote in 1844 entitled "On Changes Induced in the Situation and Structure of Internal Organs under Varying Circumstances of Health and Disease." Also in 1844, he wrote an incredibly detailed description of the anatomy, physiology, and pathology of the heart and lungs, including Sibson's fascia, which was named after him. In 1847 and 1848, he published several papers on anesthesia (Figures 3.29–3.31).

One of the more notable experiments in which Sibson participated was performed with the physician Charles Waterton (1782–1865). They studied narcotics, in particular the use of

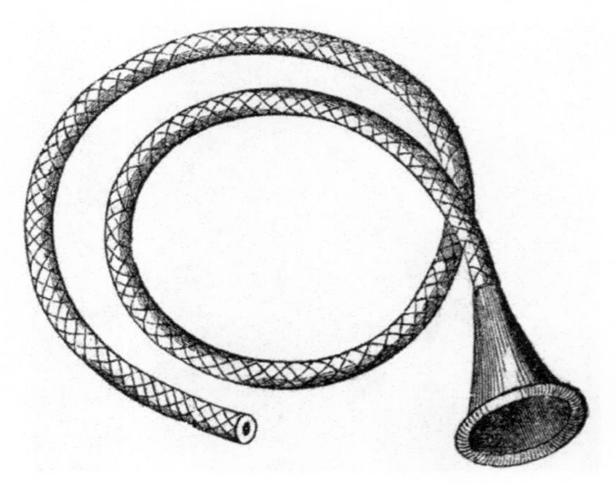

Figure 3.28 Flexible stethoscope introduced by Francis Sibson. (*Source:* From Wikimedia Commons, https://commons.wikimedia.org/wiki/File:Flexible_stethoscope,_F._Sibson_Wellcome_M0019003.jpg.)

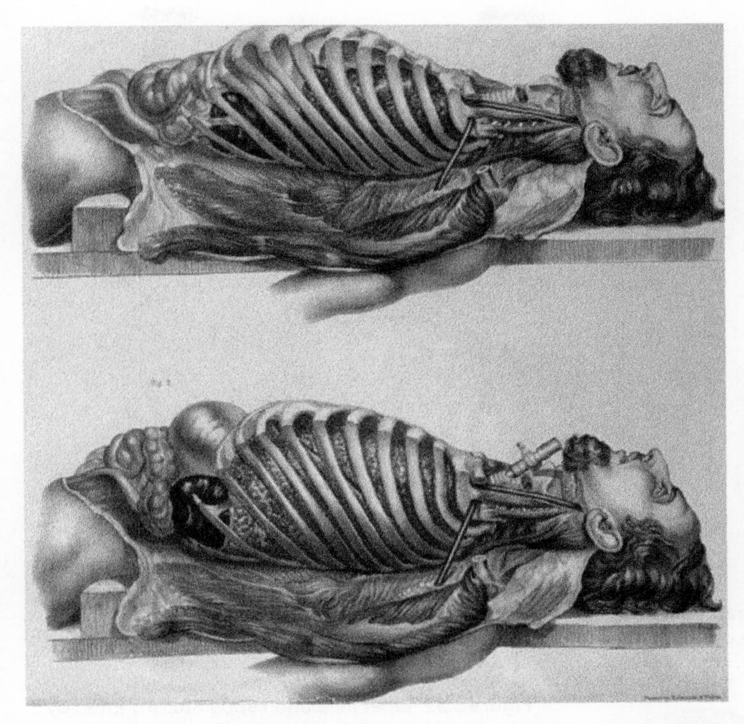

Figure 3.29 Sibson's illustrations of the body of a man lying down with the trunk dissected. Top: The lungs after expiration. Bottom: The lungs after inspiration. (*Source:* Used with permission from Sibson 1869a.)

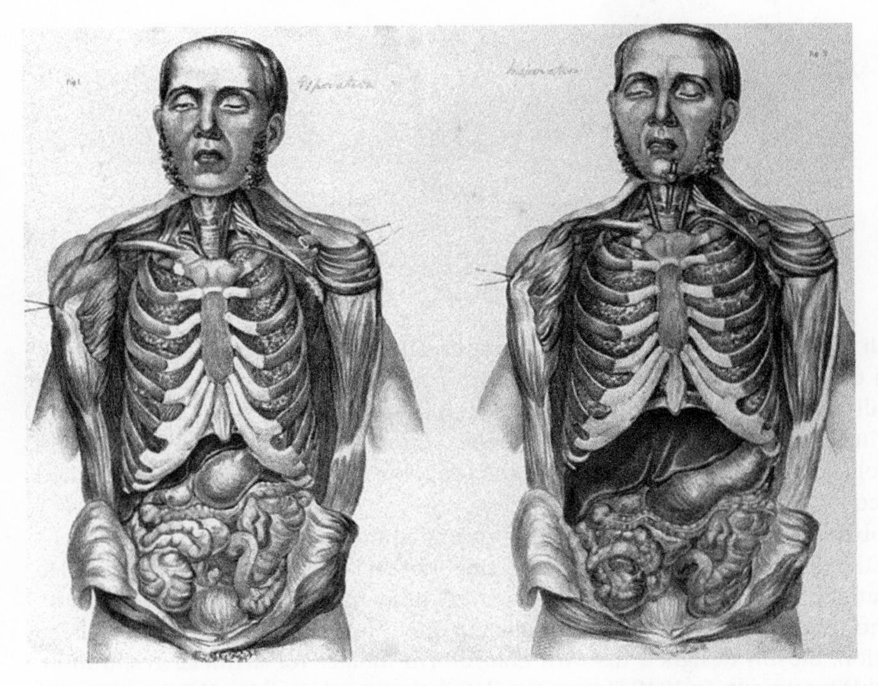

Figure 3.30 Sibson's illustration of the relative positions and movements of internal organs during respiration. (*Source:* With permission from Sibson 1869b.)

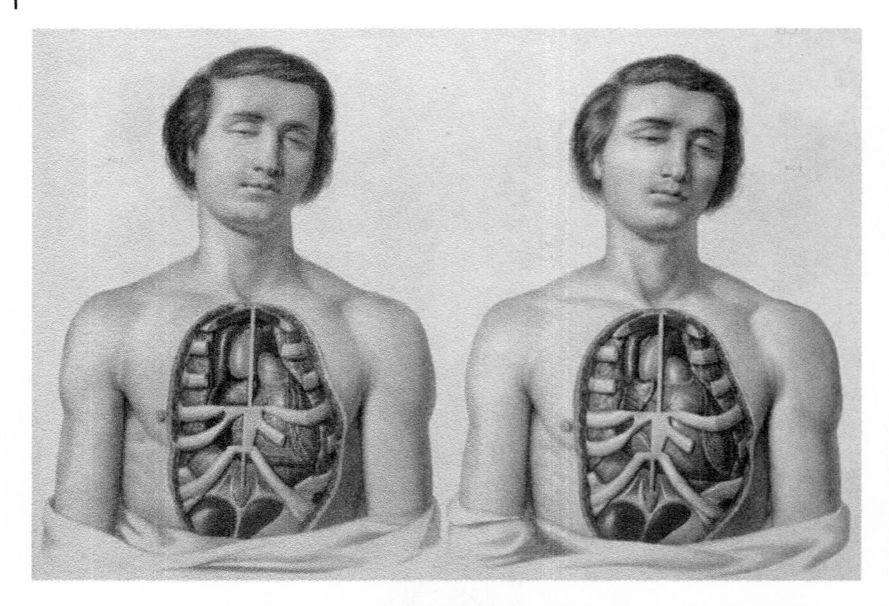

Figure 3.31 Sibson's chromolithograph, showing the relative movements of the heart with inspiration and expiration. (*Source:* National Library of Medicine, http://www.nlm.nih.gov/dreamanatomy/index.html.)

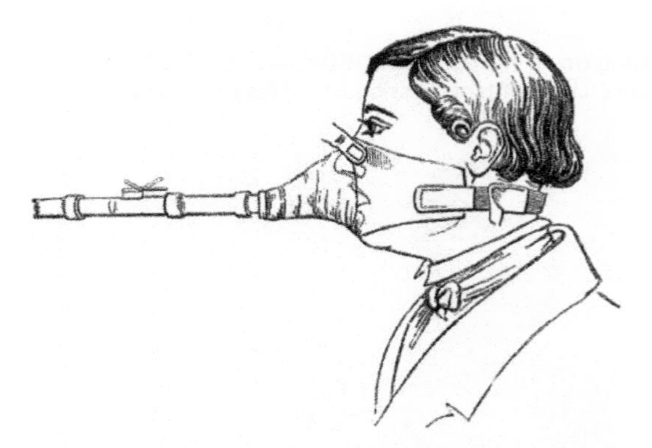

Figure 3.32 Ether inhalation apparatus invented by Sibson. (*Source:* With permission from Ord 1881.)

curare (or "wourali poison" as it was called then), in the treatment of hydrophobia and tetanus. In a letter written to George Ord in May 1838, Waterton stated, "We have already had proof positive that it will cure lock jaw and I think it not improbable that we shall soon have an opportunity of trying its sanatory powers on hydrophobia." Curare was to be used in Nottingham in 1839 when a police inspector was bitten by a rabid dog, but unfortunately the officer died before treatment could be administered.

Sibson was involved in a wide variety of projects throughout his life. For example, he conducted trials with the inhalation of ether for the treatment of facial neuralgia (Figure 3.32). He treated his first patient on January 30, 1847, just a short time after the famous operation by Robert Liston, who used ether as an anesthetic during an amputation at University College Hospital. In addition, Sibson designed his own chloroform inhaler for artificial respiration, which he used in this treatment. In 1849, he wrote a detailed paper describing the position and movements of the heart in both healthy and diseased states. His mission in the study was to determine the position of each part of the heart during systole and diastole in order to understand and describe the friction sounds heard with auscultation. He accomplished this by directly observing the thoracotomy of an unconscious donkey anesthetized using curare. He removed the ribs, while still maintaining respiration, and inserted pins into the heart in various directions to observe their movements. Sibson also published papers on subjects as diverse as the mechanisms of respiration (Figure 3.33), the blowhole of the porpoise, the treatment of gout and rheumatism, the use of ether and chloroform in the treatment of neuralgia, and the 1847 typhoid epidemic in Nottingham.

In 1848, having spent 13 years in Nottingham, he decided to relocate to London to create a practice for himself. This was a very ambitious goal, especially as he had only minimal qualifications at the time, and was met with mixed reactions from his family and friends in Nottingham. Many tried to convince him to stay, while others offered their support and even financial assistance to get his practice started. Before the end of that year, after a 17-year gap since his previous qualifying examination, he obtained a degree of MB with honors in Medicine, and a week later the degree for MD; he was in the top ranking for both. His efforts were praised by a writer in the *British Medical Journal*, who stated:

> Very few men would be equal to the great feat of … passing the whole series of examinations required for the degrees of the University of London in the same year; and to have done so while in practice with such distinction, competing against the pick of the best working men in the regular course in the London schools affords an indication of his remarkable industry and intellectual power.

Sibson thrived in London. He turned a house on Brook Street, in the highly fashionable Grosvenor Square, into a successful practice. He was quick to make new friends and contacts, and by the winter of 1849 he was already hosting private demonstrations on "visceral anatomy"

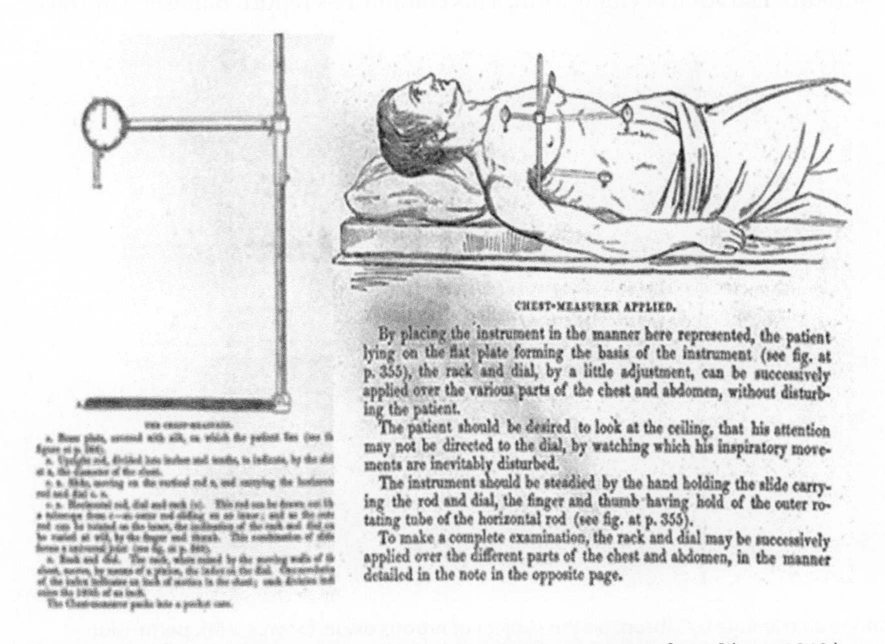

CHEST-MEASURER APPLIED.

By placing the instrument in the manner here represented, the patient lying on the flat plate forming the basis of the instrument (see fig. at p. 355), the rack and dial, by a little adjustment, can be successively applied over the various parts of the chest and abdomen, without disturbing the patient.

The patient should be desired to look at the ceiling, that his attention may not be directed to the dial, by watching which his inspiratory movements are inevitably disturbed.

The instrument should be steadied by the hand holding the slide carrying the rod and dial, the finger and thumb having hold of the outer rotating tube of the horizontal rod (see fig. at p. 355).

To make a complete examination, the rack and dial may be successively applied over the different parts of the chest and abdomen, in the manner detailed in the note in the opposite page.

Figure 3.33 Sibson's chest-measurer. (*Source:* Used with permission from Sibson 1848.)

for a select group of colleagues at his home. In that same year, Sibson became a member of the Royal College of Physicians, and in 1853 he was elected as a fellow. In 1849, he was also elected Fellow of the Royal Society; he was later named to the Joint Treasurers, and he served as a member of its Council from 1872 to 1874. His election into the Royal Society recognized his meticulous work in studying the anatomy of healthy and diseased viscera.

Sibson's next endeavor – and soon the new passion of his life – began with the opening of St. Mary's Hospital in Paddington in 1851. He was one of the first three physicians to work there. When it became a teaching hospital, Sibson enthusiastically took on the role of Chair of Medicine. He soon discovered a passion for teaching, and even lectured at another school, Samuel Armstrong Lane's School of Anatomy in Grosvenor Place. When St. Mary's became a medical school in 1854, Sibson served as a lecturer on medicine. Amidst his challenging schedule, he found time to marry Sarah Mary Ouvry, daughter of Peter Aime Ouvry of East Acton, in July of 1858. They were said to have hosted regular breakfast parties in their home for his medical firms, where topics of conversation varied from anatomy to prayer.

Sibson continued practicing clinical medicine, researching and publishing, and giving lectures. His main interests in clinical practice at St. Mary's were in renal, pulmonary, and cardiac diseases, including aortic aneurysms. He was known for his exceptional clinical teaching skills both in the classroom and on hospital rounds. His teaching style emphasized the differences between the viscera of healthy and diseased individuals. He accomplished this by teaching his students the anatomy of the healthy viscera on cadavers in the lab, so that they would be familiar with the position and movement of the organs when they came to observe pathological states.

One of Sibson's most significant texts, *Medical Anatomy, or, Illustrations of the Relative Position and Movements of the Internal Organs,* was based on his work at St. Mary's Hospital. He began it in 1855, and it was published in seven parts, the last being in 1869. William Ord described it as "the crown of Sibson's labors." Sibson also made many contributions to the field of anesthesia (Figure 3.34). As a member of the Royal Medical and Chirurgical Society of London, he served on the Chloroform Committee, where he further studied the use, therapeutic and toxic effects, and administration of chloroform. This committee's report, published in 1864,

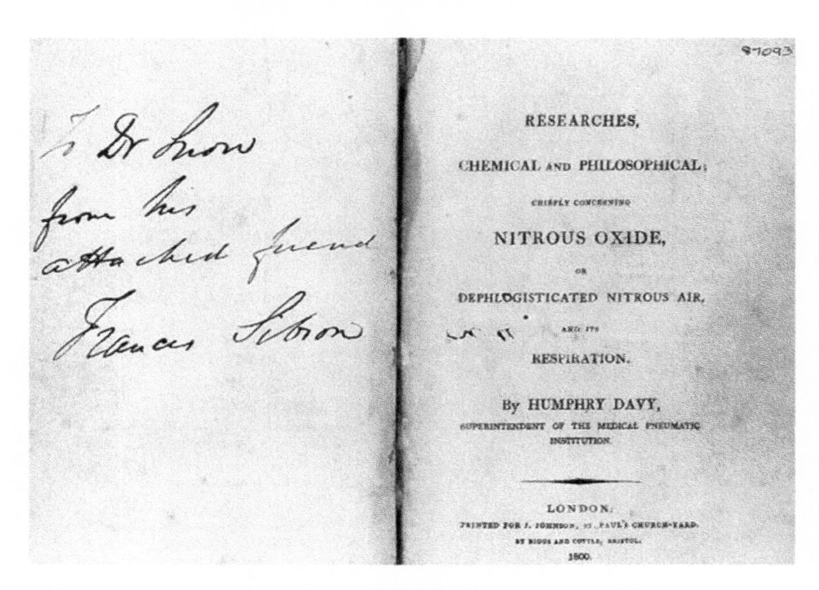

To Dr Snow
from his
attached friend
Francis Sibson

RESEARCHES,
CHEMICAL AND PHILOSOPHICAL;
CHIEFLY CONCERNING
NITROUS OXIDE,
OR
DEPHLOGISTICATED NITROUS AIR,
AND ITS
RESPIRATION.
By HUMPHRY DAVY,
SUPERINTENDENT OF THE MEDICAL PNEUMATIC INSTITUTION.
LONDON:
PRINTED FOR J. JOHNSON, ST. PAUL'S CHURCH-YARD,
BY BIGGS AND COTTLE, BRISTOL.
1800.

Figure 3.34 Inscription to a colleauge by Sibson, on the subject of nitrous oxide. (*Source:* With permission from Davy 1800.)

is regarded as a classic in anesthetic literature. Sibson's descriptions of the thickened supraclavicular pleura, now called Sibson's fascia, proved useful for anesthesia because it was important in the anesthetic blockade of the supraclavicular brachial plexus, which may be punctured, resulting in a pneumothorax. Sibson had described this fascia in his first major paper of 1844.

Sibson held many distinguished roles in the Royal College of Physicians. He had the honor of presenting the Gulstonian, Croonian, and Lumleian lectures. He also served as examiner at the college, curator of the museum, and a censor in 1874. He was chosen to be an examiner and was elected to the convocation and senate at the University of London, where he made many contributions. However, he is also known to have strongly opposed the admission of women to obtain medical degrees.

As a member of the Provincial Medical and Surgical Association, which he joined in 1843, Sibson argued against its proposed change to the British Medical Association. However, he later became the second president of the British Medical Association, succeeding the founder, Sir Charles Hastings, in 1866. He always advocated that a portion of the association's fund should be used for research, as he understood its immense importance. As a member, he attended the annual meetings; in 1867, at a meeting in Dublin, he received an honorary MD from Trinity College, while in 1870, in Newcastle-on-Tyne, he received an honorary LLD from Durham University. Upon stepping down as president after three years, he was named Vice-President for Life. Along with his many passions for medicine, he enjoyed the outdoors, especially mountaineering; he was one of the first to climb the Lyskamm (14 889 ft) in the Swiss Alps. He also took interest in collecting Wedgewood china and architectural drawings.

Sibson was remarkable not only in his work and successes, but also in his character. He was always well received and had the distinct ability to gain people's trust and respect. He was very compassionate, although he could be assertive and forceful when necessary. He was also described as being obsessive about punctuality and intolerant of tardiness. He died suddenly on September 7, 1876 at the age of 62, while on vacation at the Hotel des Bergues in Geneva, Switzerland. He was said to have died from an aortic aneurysm, which was one of his main interests at St. Mary's Hospital. His body was brought back to London and was buried in St. Mary's Churchyard in Acton, Middlesex. His *Collected Works* was published posthumously by William Miller Ord.

3.3.5 William Turner

The English comparative anatomist William Turner (1832–1916) (Figure 3.35) has to be distinguished from the American endocrinologist Henry Hubert Turner, whose name is associated with monosomy X (Turner syndrome). William Turner was born in Lancaster on January 7, 1832; his father, who was a cabinet maker, died when he was five years old. Turner studied medicine under Dr. Christopher Johnson, a local medical practitioner, then went to St. Bartholomew's Hospital in 1850 to study under the supervision of Sir James Paget. After graduation, Paget recommended him to John Goodsir (1814–1867), Professor of Anatomy at the University of Edinburgh, for the post of Senior Demonstrator. In 1857, he graduated with his MB from London University. Turner devoted himself to the study of the human body and comparative anatomy. He wrote a *Handbook to Atlas of Human Anatomy and Physiology*, which contained nine plates selected by Professor Goodsir.

In 1867, Turner became Professor of Anatomy, and following Goodsir's death, he succeeded him as Chair of Anatomy at the School of Edinburgh. From 1873 to 1883, he sat as representative of the universities of Edinburgh and Aberdeen to supervise professional education. In 1877, he was elected a Fellow of the Royal Society of England, and he became President of General Medical Council in 1898, a position he retained until 1905. He received a knighthood in 1886.

Figure 3.35 William Turner (1832–1916).

In 1865, and with the demise of the *Natural History Review*, a quarterly journal of biological science, Turner and Sir George Murray Humphry of Cambridge (1820–1896) came together to establish the new *Journal of Anatomy and Physiology*; the first issue appeared in 1866. Turner published over 200 papers on subjects including neuroanatomy, craniology, comparative anatomy, and anthropology. He also corresponded with Darwin, and supplied him with information on the rudimentary parts of human and animals. His investigations laid the foundations of our current understanding of the nervous system. In the opening address of the 122nd session of the Royal Physical Society of Edinburgh, the retiring Vice-President Johnson Symington said:

> he [Turner] delivered a lecture before the Royal Medical Society ... In this lecture, Sir William Turner not only gave a very valuable summary of the state of our knowledge at that time, but he also added much original matter, and introduced several new terms which have since been generally adopted in the descriptive anatomy of the cerebral cortex.

The 1918 edition of Henry Gray's Anatomy made a note of the intraparietal sulcus of Turner, which clefts the lateral surface of the parietal lobe. Professor Cunningham also specified Sir William Turner as being first to describe the intraparietal sulcus on the brain of the chimpanzee and human, although Cunningham further mentioned that the same feature was independently described by Adolf Pansch of Kiel under the name "sulcus parietalis."

Twenty-three of Turner's students became professors of anatomy throughout Europe. Numbered among his famous students and fellows were Daniel John Cunningham, who succeeded him as Professor of Anatomy at Edinburgh and was editor of *The Textbook of Anatomy*. Notably, Cunningham's *Textbook of Anatomy* was dedicated to Sir William Turner. Turner died on February 15, 1916.

3.3.5.1 Subperitoneal Arterial Plexus

The numerous arterial anastomoses between the visceral and parietal branches of the abdominal aorta, which exist in the retroperitoneal fat and connective tissues, are collectively known as the subperitoneal arterial plexus of Turner. Turner studied four cadavers by injecting stiff colored gelatin into different visceral arteries and making observations of the arteries to which it proceeded. In a classic paper published in *British and Foreign Medico-Chirurgical Review*, he wrote:

> By these injections I have succeeded in demonstrating the existence behind the peritoneum of a well-marked vascular plexus, to which I propose to give the name of subperitoneal arterial plexus. This plexus lies behind the peritoneum, not only in the lumbar and iliac regions, but extends upwards as far as the diaphragm, and downwards into the cavity of the pelvis. It consists of fine elongated arteries, which freely anastomose with each other, so as to form a wide-meshed and somewhat irregular network. It communicates on the one hand with the arteries of the abdominal viscera, and on the other with the arteries of the different parts of the abdominal wall. Through it the visceral and parietal arteries are brought into anatomical communication with each other, so that the vessels in the wall can be injected from many of the visceral branches.

He made only a general account of the subperitoneal arterial plexus and did not provide details or the extents of such anastomoses at a given vascular bed, except in the case of the renal and gonadal arteries. He remarked on the presence of such anastomoses in the transverse mesocolon, mesorectum, mesentary of the small intestine, hilum of the spleen, broad ligament of the uterus, and peri-hepatic region. He suggested that the branches from the subperitoneal arterial plexus supply the retroperitoneal fat, areolar connective tissues, lymphatic glands, and vascular walls. Additionally, his account of renal arterial anastomoses is very interesting:

> in all the cases examined I have traced small arteries passing out of the substance of the kidney, piercing its fibrous coat, and joining the arterial plexus in the fatty capsule. Occasionally these arteries pierced the fibrous coat of the kidney at various points on its surface without following any definite plan, but at other times the arrangement was of a more regular nature. In one case three well-marked arteries pierced the fibrous coat on the posterior surface of the organ – one about the center of this surface, the others close to its extremities. They all passed outwards into the surrounding fat and areolar tissue. Those from the extremities inclined towards each other, joined, and formed a well-marked arterial arch …
>
> an injection may be forced from the renal arteries into those portions of the large intestine which lie in front of the kidneys. Other experiments which I have made show that an injection can travel in exactly the opposite direction from the intestine to the kidney.

Turner also mentioned communications between the spermatic arteries and branches of subperitoneal arterial plexus:

> Not only did the injection pass along these [spermatic] vessels into the substance of the testicles, but it was found that each artery, as it coursed down the lumbar and iliac regions, gave off numerous small branches, which communicated with the subperitoneal arterial plexus

The subperitoneal arterial plexus, as Turner indicated, was forgotten for several decades until the functional significance of such anastomoses became gradually acknowledged in the light of modern radiology and advancing interventional procedures. This arterial plexus is a significant source of collateral circulation to the kidneys and other viscera, and may be involved in the metastatic spread of malignancies.

Turner's major discovery pointed to complexities and subtleties in the circulatory system that had previously gone unnoticed; indeed, the full significance of the subperitoneal plexus for both normal physiology and pathology is only now being recognized.

3.4 France

Although the Renaissance did not lead to revolutionary advances in the study of anatomy in sixteenth-century France as it did in Italy, important contributions were made. Galenism survived in Paris longer than it did in most Italian centers or in England, leading to a protracted and sometimes vitriolic dialog between the Parisian anatomists and others. Nevertheless, there was a strong tradition of anatomical research in France, particularly in Paris, and it flourished during the Enlightenment (the direct descendant of the Renaissance), giving rise to a series of important discoveries. A distinct tradition of anatomical research was pioneered in sixteenth-century France, and that tradition has survived to the present day.

It has been said that developments during the late Renaissance period (the early sixteenth century) served as the stimulus for the scientific revolution that led to spectacular progress afterwards, particularly during the nineteenth and twentieth centuries. Jacobus Sylvius (1478–1555) and others in France devoted themselves to refreshing and preserving the Galenist tradition. This conservatism persisted for more than 100 years, culminating during the second quarter of the seventeenth century, when the Parisian anatomist Jean Riolan (1577–1657), a devoted Galenist, was the most vocal and implacable opponent of William Harvey (1578–1657). After that time, however, French anatomists and (subsequently) physiologists became pioneers of the scientific approach to medicine. In the eighteenth century, France (particularly Paris) was the birthplace of the Enlightenment, the spiritual offspring of the Renaissance.

These developments were especially pertinent to the studies of anatomy and surgery. During the late Renaissance period, the renowned French surgeon Ambroise Paré (1510–1590) played an enormous role in helping surgery regain the status it had lost during the Middle Ages. However, despite their proven skill and expertise, it was not until the end of the eighteenth century that surgeons came to be seen as equal to physicians. Seventeenth-century surgeons struggled to achieve recognition and autonomy in the face of hostility from physicians and barbers. Since the Middle Ages, barbers had not only cut hair but also performed surgery; surgeons of the seventeenth century no longer trimmed beards, but instead focused on academic advancement. Concurrently, the late seventeenth and early eighteenth centuries were characterized by dramatic achievements in anatomy and surgery, and French pioneers were major contributors to these.

France went on to develop a distinctive tradition of research in anatomy and allied disciplines. As the following sections will show, this tradition was notable for its range and variety as well as its importance for the emergence of the knowledge we enjoy today.

3.4.1 Jacobus Sylvius

Jacques Dubois (1478–1555) (Figure 3.36), typically referred to by his Latin cognomen Jacobus Sylvius, had a tremendous impact on the medical field. While many could boast greater anatomical discoveries, perhaps none produced more excellent students than Sylvius. His success

Figure 3.36 Jacobus Sylvius (1478–1555).

in teaching anatomy stemmed from his modernization of Galenic terminology – for example, by replacing Galen's numerical nomenclature with descriptive names. Although the shift from Galenic to post-Galenic thought ultimately left the Frenchman an advocate of a failed ideology, Jacobus Sylvius should still be regarded today as a great medical figure for his contributions as an educator.

Sylvius' father was a struggling weaver in the French town of Louvilly, and the young Sylvius was born into poverty; however, he was able to break the constraints of his economic condition. He enrolled at the College of Tournay, near Paris, where his older brother, Francis Dubois, was a professor. While at Tournay, Sylvius excelled academically as a student of classical languages. It was during this time that he developed a deep personal interest in human anatomy, and his mastery of the classical languages allowed him to delve into the works of Hippocrates and Galen. Sylvius was unable to finish his degree at Tournay owing to a lack of funds, but he trudged onwards despite the setback, following his interest in the medical sciences and apprenticing in anatomy under the Parisian physician and author Jean Fagault. With this training, and with his previous knowledge of Hippocrates and Galen, Sylvius began working as a private anatomy professor in Paris, where he enjoyed a degree of success. However, Paris' Faculty of Medicine eventually forbade him from continuing in this fashion, citing his lack of a medical degree as the reason for their censure, although some historians contend that jealousy of Sylvius' growing popularity was their real motive. By this time, Sylvius' tutorials had generated sufficient funds for him to complete his medical education at the University of Montpellier in the south of France. There, he obtained a bachelor's degree in 1531, and a medical degree in the following year, although there is some uncertainty regarding these dates. He was in his 50s by the time he finally received his medical degree.

By then, Sylvius had mastered the teachings of Galen and Hippocrates and earned a sizable reputation as an anatomy instructor. He returned to Paris armed with his degree and resumed his anatomy tutorials at the College of Treguier. His ability to simplify the medical canon and his sharp focus on practical matters over theoretical flights of fancy characterized his lessons. He was noted as both an excellent speaker and a skilled demonstrator; importantly, he was the first in France to demonstrate from human cadavers. Consequently, his success was enormous, with class sizes reportedly reaching 400–500 students while others "lectured to almost empty benches." In his *Manual of Anatomy*, Sylvius explains the importance of dissection to medical education with the following eloquent introduction:

> I would have you look carefully and recognize by eye when you are attending dissections or when you see anyone else who may be better supplied with instruments than yourself. For my judgment is that it is much better that you should learn the manner of cutting by eye and touch than by reading and listening. For reading alone never taught anyone how to sail a ship, to lead an army, nor to compound a medicine, which is done rather by the use of one's own sight and the training of one's own hands.

Despite Sylvius' exhortation to learn from dissection rather than books or lectures, he referred to the teachings of Galen and Hippocrates with undying devotion, stating:

> After Apollo and Aesculapius they were the greatest powers in medicine, most perfect in every respect, and they had never written anything in physiology or other parts of medicine that was not entirely true.

This belief in the infallibility of Galen formed the core of his ideology, and the dichotomy between these two quotations provides the real meat of his story. On the one hand, we have the Galenic Sylvius, who upholds the Greek master and dissector of apes, pigs, and other lesser animals with "quasi-religious fervor." On the other, we have the post-Galenic Sylvius, who urges his students to value experience and observation over canon; this is the Sylvius who would sneak human limbs under his cloak to the lecture hall, his enthralled students watching as human anatomy was laid bare before them in ways that Galen's works could never provide. Thus, Galenic ideology perhaps survived within the Renaissance period as an aftershock of the Middle Ages – the dogmatism and narrow-mindedness of the Dark Ages reverberating through the rediscovery of the Greek and Roman classics. Likewise, the post-Galenic ideology foreshadows the Renaissance's direct descendant, the Enlightenment. Sylvius seemed able to walk the tightrope between these two camps without experiencing any cognitive dissonance, but the gulf between the dogmatic Galenic and enlightened Renaissance ideologies would prove too much for some of his students.

Sylvius mentored an impressive list of students while at Treguier, many of whom would go on to become well-known anatomists in their own regard. Noted medical figures such as Servetus, Stephanus, Gennarus, Vulpinus, Conrad Gesner, Charles Estienne, and Andreas Vesalius all studied anatomy under him. Sylvius' ability to produce so many top-notch students is perhaps his most lasting legacy, although it should by no means detract from his anatomical credentials. At the time, Sylvius was recognized as the foremost European anatomist. However, his relationship with his former student Vesalius would eventually diminish his professional reputation.

Sylvius' fall from grace was predicated on his unyielding devotion to Galen's teachings, which in some circumstances directly opposed the readily observable anatomical truth. Although Sylvius modernized Galen's work by naming structures previously referred to by numerals (such as the muscles and the vessels), he disliked most criticism of Galen. When confronted with an observable fact that contradicted his idol, he went so far as to claim that the human body had changed in the past millennium, thereby creating the dissimilarity. Such views and attitudes were in direct conflict with the emerging culture of the time. The Age of the Renaissance, beginning its long transformation into the Age of Enlightenment, devalued orthodoxy, replacing it with observable truths; the student Vesalius, a product of this era, pointed out many of Galen's errors in his monumental 1543 text, *De Humani Corporis Fabrica* ("On the Fabric of the Human Body"), creating a rift with Sylvius. Vesalius' attacks on Galenic teaching enraged his professor, who not only demanded that Vesalius recant or renounce their friendship but also attempted to undermine his position in the court of Holy Roman Emperor Charles V. The battle was not one-sided: Sylvius' character was attacked by Vesalius, who resented the Frenchman's well-known miserliness and speculated that his interests in medicine were purely financial. In the year following the publication of the *Fabrica*, many others of the Galenic persuasion attacked Vesalius' work, and in a fit of rage, Vesalius burned his unfinished manuscripts, one of which is said to have been a collection of annotations on Galen's writings. Despite its initially poor reception, the *Fabrica* went on to achieve almost universal acclaim, sounding the death knell of Galenic medicine. The previously unassailable reputation of Galen and those who supported him declined. Vesalius would become regarded as the greatest anatomist of Europe, while Sylvius would at best be remembered as his student's immediate subordinate and at worst as the blind proponent of a flawed, dying ideology.

Despite his anatomical knowledge and teaching ability, Sylvius was equally renowned for his frugality, intolerance, and vindictiveness. His students gossiped that their preceptor fed his servants only bread and water and went so far as to refuse to keep a fire during the cold Parisian winters; likely being of wealthy heritage themselves, and accustomed to the conveniences of life, they seemingly found it difficult to relate to the son of a poor weaver. This class division between Sylvius and the vast majority of his students, and the attendant distance between him and any potential defenders, rendered him more vulnerable to attack and marginalization than he might otherwise have been.

Sylvius' miserliness and blind support of Galenic tenets provided sufficient ammunition for Vesalius to successfully attack his teacher during their embattled years. Consequently, Sylvius' name would never again be as well known as that of his student; while the aqueduct of Sylvius and the Sylvian fissure appear to bear his name, they were probably in fact named in honor of the Dutch physician and anatomist Franciscus Sylvius (1614–1672), although contention exists as to the aqueduct's true attribution. The valve of the inferior vena cava discovered by Sylvius does not bear his name, but is instead referred to as the Eustachian valve in honor of the Italian anatomist and anti-Galenist, Bartolomeo Eustachi (1520–1574). Similarly, none of the muscles or vessels discovered by Sylvius, such as the biceps, triceps, jugular, and subclavian, are named for him.

Sylvius remained a lifelong bachelor, and despite his achievements and fame, he was buried in Paris' Cemetery of Poor Scholars upon his natural death at the age of 77. He administered his tutorials until his final days; the fact that students still came to him even after his professional cachet had declined is perhaps the best testament to the great educator's abilities. But it was a former pupil who mockingly supplied the following epitaph (translated from the Latin) on the wall of the church where Sylvius' last rites were performed:

> Sylvius lies here, who never gave anything for nothing:
>
> Being dead, he even grieves that you read these lines for nothing!

Thus, ignominy haunts Sylvius even in death. The son of a poor weaver who rose above his lot in life to educate and inspire a generation of great anatomists surely deserves a more prominent place in history.

3.4.2 Charles Estienne

Charles Estienne (c. 1504–1564), also known as Carolus Stephanus, studied and earned his degree in medicine in Paris in 1542. This relatively unknown anatomist was the son of Henri Estienne, a famous French painter, and a descendant of the famous family of printers of the same name. He lectured in anatomy at the Faculty of Medicine from 1544 to 1547. His principal works were *De dissectione partium corporis humani libri tres* (1545), an anatomical work illustrated with approximately 60 woodcuts, *Dictionarium historicum ad policum* (1553), said to be one of the first French encyclopedias, *Praedium Rusticum* (1554), a collection of works on ancient agriculture, and *Thesaurus Ciceronianus* (1557). *De dissectione* was followed by a French translation the next year (1546). It was actually begun around 1530, but was interrupted by a dispute in 1539: Estienne had complained of plagiarisms of his published anatomical texts, particularly in Germany (*a cause d'un proces qui survint*). *De dissectione* was so popular that it was published again in 1575.

Estienne was a contemporary of Vesalius, and began his anatomical book prior to the publication of the *Fabrica* in 1543. Like Vesalius, he had been imbued with Galenic tradition by his teacher, Sylvius. Estienne was a humanist, coming from a famous French humanist family, and his interest in anatomy was largely directed at clarifying its nomenclature. In *De dissectione*, he offers one of the first descriptions of venous valves when he comments on the "apophyses membranarum" in the veins of the liver. However, their function was a mystery to him. Estienne's greatest contribution was probably to arthrology. He made several good descriptions of the clavicular joints, the spine and its ligaments, and the temporomandibular articulation. He was the first to trace blood vessels into the substance of bone. Most remarkable of his observations was that of the central canal of the spinal cord. Additionally, his text emphasized the parotid, lacrimal, and thymus glands, the lymphatic glands at the root of the mesentery, the armpit, and the groin.

The images in Estienne's anatomical book were derived from life, following dissections carried out by him and Etienne de la Rivière, a prominent Parisian surgeon. Estienne's anatomical illustrations vary considerably in quality. The anatomy throughout is pre-Vesalian, and the earlier plates are clumsier and perhaps follow the older Venetian-Paduan examples. The later plates approach the bold style of Buonarroti. Small wood blocks depicting the detailed results of actual dissections are inserted carefully into already existing larger blocks that show nudes, both male and female, in heroic poses in a variety of classical landscapes, exposed on marble seats or propped up against trees (Figures 3.37–3.38). It is not clear whether the printer Simone de Colines was using a set of blocks originally prepared for a totally different book, as some have suggested. Others argue that the classical background is deliberate, evoking the antiquity of dissection, and transmuting the gruesome horror of the detail of a corpse to the heroic world of the Greeks and Romans, or of the gods themselves. Nonetheless, his anatomical work offers a great variety of images of parts of the body, most of them new when they were cut, some of which go beyond even what Vesalius would present in the *Fabrica*. For example, Estienne's description and drawings of the human sternum were very detailed and refuted most Galenic teachings. Imbued with the Galenic tradition though he was, he parted company with that tradition in significant respects.

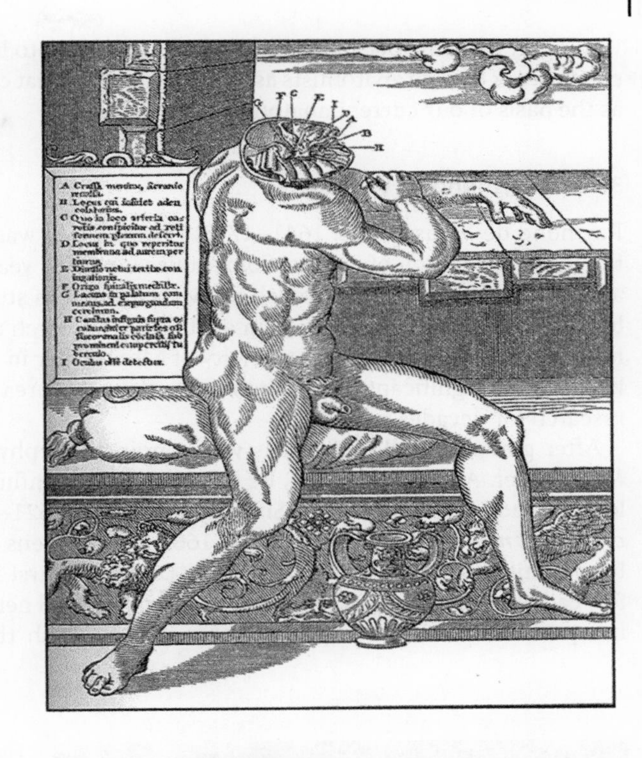

Figure 3.37 An example woodcut from Estienne's *De Dissectione Partium Corporis Humani Libri Tres* (1545) showing an interesting depiction of the intracranial contents from a male specimen. (*Source:* Choulant 1852.)

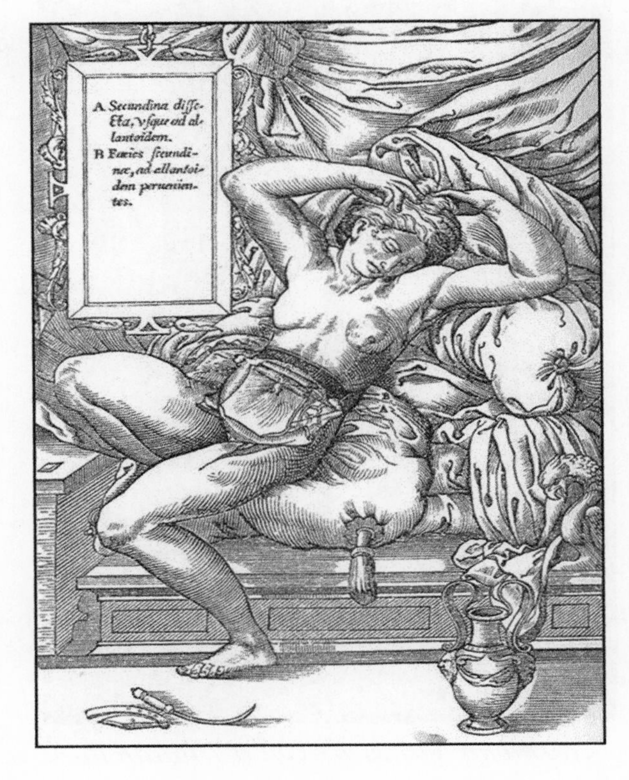

Figure 3.38 A depiction of the anterior abdominal wall of a female from Estienne's *De Dissectione Partium Corporis Humani Libri Tres* (1545). Note the folded knives on the floor. (*Source:* Choulant 1852.)

In 1561, Estienne became bankrupt, and he is said to have died in a debtors' prison. It is the classic work of such anatomists as Charles Estienne that challenged Galenic thought and served as the basis of our current anatomical knowledge.

3.4.3 Raymond de Vieussens

Raymond de Vieussens (c. 1641–1715) (Figure 3.39) was born in Vieussens, a small village in the Rouergue region of France. Some sources put his year of birth at 1633 or 1635. The son of a lieutenant colonel in the French army, de Vieussens studied philosophy at Rhodez in France before embarking on his career in anatomical research and medicine. He graduated with his medical doctorate from the University of Montpellier in 1670. During the course of his career, he provided significant insight into more than one area of medicine and opened up fields of research for decades to come.

After graduation, de Vieussens was appointed as physician to the hospital of St. Eloys in Montpellier. Around this time, he became strongly influenced by the anatomical and physiological studies of the brain of Sir Thomas Willis (1621–1675). Willis published *Pathologicae cerebri, et nervosi generis specimen* in 1667; de Vieussens read it in 1671, and it reputedly led to his pursuit of neuroanatomical research. During his first 10 years at St. Eloys, he dissected over 500 cadavers in his study of the central and peripheral nervous systems. The culmination of his intense work during this period came in 1684 with the publication of his *Neurographica*

Figure 3.39 Raymond de Vieussens (c. 1641–1715). (*Source:* Courtesy of Clendening Library).

Universalis. De Vieussens' work was well received in the European medical community and is heralded as the most complete account of the brain, spinal cord, and peripheral nerves to emerge in the seventeenth century. One of the most important observations in *Neurographica Universalis* was that the spinal cord is a functionally independent structure and not a simple extension of the brain. De Vieussens clarified the relationship between the optic nerve and the lateral geniculate nucleus of the dorsal thalamus. His publication also included the first description of the dentate nuclei, pyramids, and olivary bodies. *Neurographica Universalis* brought de Vieussens membership in the Academy of Sciences in Paris, and he also became a Fellow of the Royal Society of London.

In 1688, de Vieussens was called upon to attend to the royal family and was granted a pension of 1000 livres a year by Louis XIV. A cousin of Louis XIV, Madame De Montpensier, selected de Vieussens as her personal physician until her death in 1693. At that time, de Vieussens returned to his post at St. Eloys in Montpellier to continue his clinical practice and anatomical research.

De Vieussens spent his subsequent years in Montpellier studying the structure and movement of the heart. He particularly focused his research on the clinical and pathological findings of patients suffering from heart disease. In 1706, he published *Nouvelles découvertes sur le Coeur*, in which he presented the detailed anatomy of the lymphatic system and blood vessels of the heart. He improved upon standard fixation techniques and elaborated on their use in his research. He described ducts joining the ventricular cavities to the coronary arteries, the "ducti carnosi." He ligated the superior and inferior venae cavae and pulmonary veins, injected saffron dye into the coronary arteries, and observed that the dye not only coursed through the coronary sinus and the right atrium, as expected, but also flowed into both the right and left ventricles. These data led de Vieussens to the working hypothesis that channels connected the coronary arteries to the chambers of the heart. Adam Christian Thebesius (1686–1732) improved on his techniques two years later, mapping and classifying these vessels according to size and frequency. Ultimately, even though de Vieussens initiated this research, the ducti carnosi became known as the Thebesian vessels.

De Vieussens' observations on the heart were expanded in 1715 with the publication of *Traité nouveau de la structure et des causes du mouvement naturel du coeur* ("Treatise on the Structure of the Heart and the Causes of its Natural Motion"), which described the pericardium, coronary vessels, and muscle fibers in detail. Two excerpts from *Traité nouveau* stand out in particular: the clinical and pathological observations of a patient who was ill with aortic valve insufficiency, and those of another patient who suffered from mitral valve stenosis. These were the first documented clinical presentations and autopsy results of patients with these conditions. *Traité nouveau* was heralded as one of the most important discussions on the correlation between the clinical symptoms of heart disease and the anatomical basis for these findings.

De Vieussens' descriptions of his findings in aortic insufficiency included orthopnea and a bounding and forceful pulse (later named Corrigan's pulse). At autopsy, de Vieussens found that the left ventricle was dilated, the walls of the aortic arch were thick and calcified, and the aortic semilunar valves were "stretched and cut off at their tip." He concluded – reasonably – that every time the aorta contracted, it sent blood back into the left ventricle, forcing it to contract against more pressure, and leading to the patient's forceful pulse.

De Vieussens depicted his patient with mitral valve stenosis as having dyspnea, edema, and a weak, irregular pulse. At post mortem, the patient's right ventricle was hypertrophied and the mitral valve was stenotic. De Vieussens correctly determined that the observed dilation of the pulmonary vein and the inability of the left ventricle to pump blood to the body caused the dyspnea and weak pulse in this patient.

Raymond de Vieussens died on August 16, 1715, almost one year after the death of Louis XIV, his greatest patron. He saw the publication of *Traité nouveau de la structure et des causes du mouvement naturel du coeur* and was privy to a great deal of the acclaim surrounding this work before his death.

De Vieussens' name is still used frequently in cardiovascular medicine today. His namesake valve in the heart was first described in his 1706 publication. Vieussens' arterial ring provides collateral blood flow to the left coronary artery system in cases of proximal left anterior descending coronary artery occlusion. This arterial ring involves the anastomosis between the right and left conal arteries. Several other anatomical structures carry his name. These include the ansa subclavia (Vieussens' annulus), central canal of the cochlear columella (Vieussens' scyphus), limbus fossa ovalis (Vieussens' annulus), superior medullary velum (Vieussens' valve), venae cordae minimae (Vieussens' foramina), and anterior veins of the right ventricle (Vieussens' Veins). Vieussens' legacy persists today via the multiple eponyms attributed to his initial descriptions.

Whether physicians are working with patients who have damage to the nervous system or with those suffering from valvular stenosis or insufficiency, the importance of de Vieussens' work in clinical medicine and anatomy serves as an important basis for the practice of modern medicine. Many physicians leave their mark on the world of medical research by focusing on a single aspect and studying it to the extreme. Others investigate many areas of the body but gain only a superficial knowledge of any of them. Others still provide significant insight into multiple areas of medicine and open up the field of research for decades to come. Raymond de Vieussens was numbered among the latter.

3.4.4 Pierre Dionis

France led Europe in the dramatic achievements in anatomy and surgery that took place during the late seventeenth and early eighteenth centuries, thanks in part to the work of Pierre Dionis (1643–1718) (Figure 3.40). Dionis was the first of a line of eighteenth-century surgeons in Europe who became celebrated for their surgical skills and academic knowledge. Little is known of his early life, although he is believed to have begun practicing medicine around 1661.

Dionis probably received much of his surgical expertise from attending free classes that were held by the Confraternity of Saint-Come. He also attended demonstrations at an institution created by Louis XIII known as the Royal Garden. This institution specialized in the teaching of modern medical sciences and was separate from the Paris University Faculty of Medicine. In his early 20s, Dionis was already considered a master surgeon and a prominent member of the Confraternity of Saint-Come. In 1673, he was assigned by Louis XIV to perform both anatomical and surgical demonstrations at the Royal Garden. During his employment, the Royal Garden remained abreast of the latest scientific advancements and surpassed the Faculty of Medicine in academic excellence. His demonstrations often drew crowds numbering in the hundreds. He gave two 10-day lecture series there, one dedicated to anatomy, the other to surgery. At a time when the majority of physicians still believed and employed the methods of Galen's theory of the circulation, Dionis was instead teaching the widely accepted but relatively new theory of circulation proposed by Harvey (a subject he discussed later in his *Anatomie de l'homme*).

Dionis remained at the Royal Garden until 1680, when he was called upon to be surgeon to the daughter-in-law of Louis XIV, the Dauphine Marie Anna of Bavaria. Following this assignment, he was appointed Surgeon in Ordinary to Queen Maria Theresa of Spain, the wife of Louis XIV, and also to Marie Adelaide of Savoy, who was Duchess of Burgundy and mother of Louis XV. In 1712, Marie Adelaide of Savoy and her husband the Duke of Burgundy died of measles. Despite this tragedy, Louis XIV gave Dionis a reward of 3000 pounds for his good will

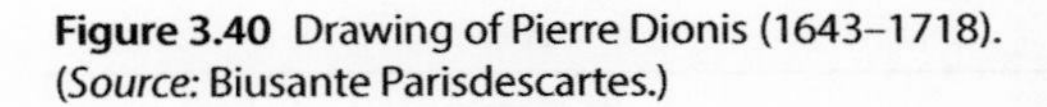

Figure 3.40 Drawing of Pierre Dionis (1643–1718). (*Source:* Biusante Parisdescartes.)

and treatment of Marie Adelaide, and appointed him First Surgeon of the Royal Children's Hospital of France. In 1715, the First Surgeon to Louis XIV, a surgeon named Mareschal, asked Dionis to assist him in treating the king, who was suffering from gangrenous arthritis in the lower limbs. Both Dionis and Mareschal agreed that amputation was the only option, but before the operation could be carried out, Louis XIV died.

During his time as surgeon to the royal family, Dionis had the opportunity to write several books, one of which remained an influential resource in surgical technique for the next century. *Anatomie de l'homme, suivant la circulation du sang, & les derniéres découvertes* ("Anatomy of Man, Following the Blood Circulation and New Discoveries") (Figure 3.41) was published in 1690 and translated into several languages. It includes a discussion of Descartes' circulation hypothesis, which was highly mechanical in conception and in many respects inconsistent with Harvey's discoveries:

> Nonetheless, we must say that this hypothesis is contrary both to reason and to experiment, but at this we should not be astonished. He did not know enough about the structure of the heart, and his meditations took up so much of his time that he was not able to obtain any great knowledge of that structure. All the same we must say that he did all a man could do, who knew nothing of the heart beyond what he knew of it.

Dionis' most famous book, however, was published in 1707 and entitled *Cours d'opérations de chirurgie démontrées au Jardin royal* ("A Course of Surgical Operations Demonstrated at the Royal Garden in Paris"). This was renowned as the first relevant surgical book since the

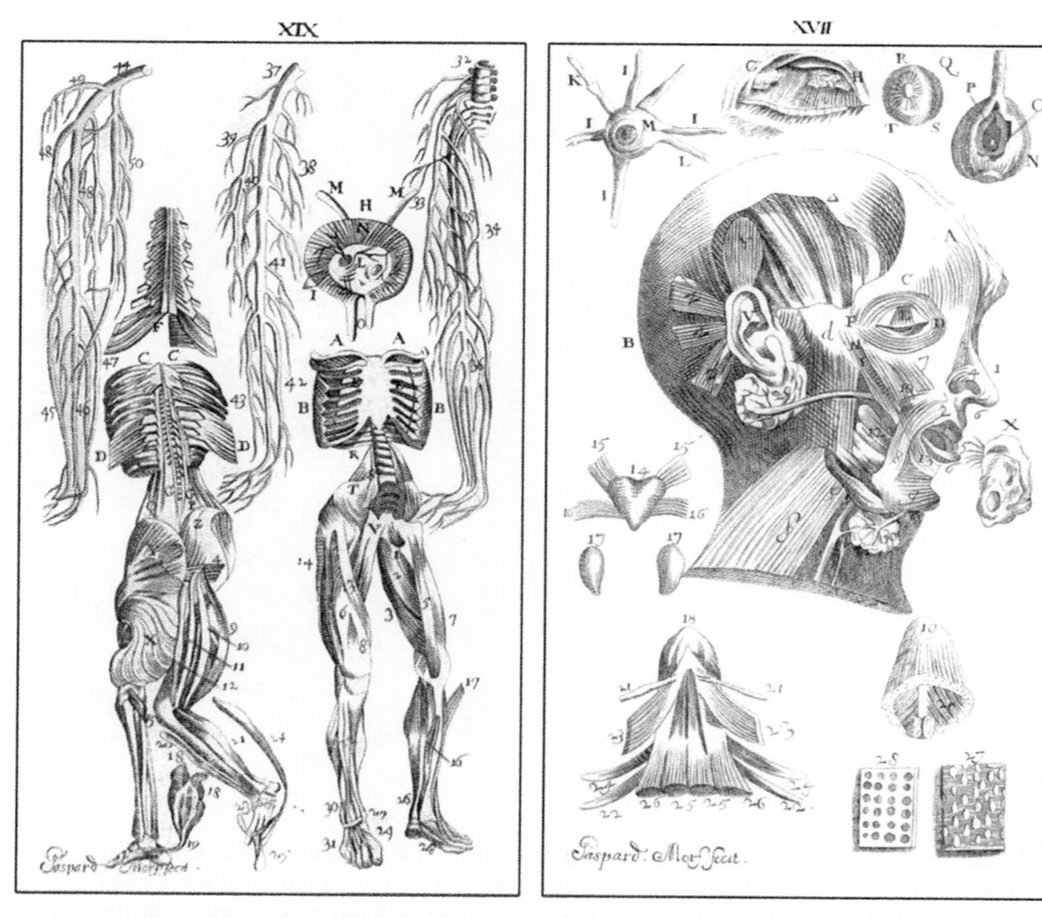

Figure 3.41 Drawings from Dionis'*Anatomie de l'homme, suivant la circulation du sang, & les dérnieres découverts.*

Renaissance, and was translated into many different languages. It was highly influential during the early eighteenth century. Dionis also wrote *Histoire anatomique d'une matrice extraordinaire, cas de rupture au sixième mois de grossesse* (1683), *Dissertation sur la mort subite, avec l'histoire d'une fille cataleptique* (1709), and *Un traité général des accouchements qui instruit de tout ce qu'il faut faire pour être habile accoucheur* (1718). The latter provides the first evidence of a true understanding of the cause and phenomena of extrauterine pregnancy. Concerning the cause of tubal pregnancy, Dionis stated:

> If the egg be too big, or the diameter of the Fallopian tube be too small, the egg stops, can get no farther and shoots forth and takes root: and having the same communication with the blood vessels of the tube, that it would have had with those of the womb, had it fallen into it, is- nourished and grows big to such a degree that the membrane of the tube, capable of no such dilatation as that of the uterus, breaks, and the fetus falls into the cavity of the abdomen, where it sometimes lies dead for many years, and at other times occasions the death of the mother, by breaking open its prison.

This was strikingly original, despite the long history of anatomical research on the reproductive system, and constituted an important contribution to medicine.

Dionis wrote in a simple style reminiscent of Ambroise Paré (1510–1590). His anatomical illustrations were of very high quality and were very informative, especially those showing surgical instruments. The following excerpt from the first chapter of his *Cours d'opérations de chirurgie*, on the role of the surgeon, demonstrates the clarity and elegance of his writing:

> But it must be granted that the Chirurgeon, to whose lot no more than this practical book falls, will frequently run the risque of Killing and Laming his Patients, when without the Direction of a Physician; and, even in the Presence of the Physician himself, will he not be in danger of committing Faults, if his Hand be not guided by his Head? 'Tis certain, that to walk well good eyes and agile and pliant legs are requisite, and that the one without the other is insufficient for that purpose. A blind Man, for instance, provided with good Legs, and led by a quick sighted and faithful Guide, may stumble for want of Light. So, whatever Experience a Chirurgeon may have, if he have not the Knowledge which ought to direct him in his Operations, he will work in the dark; and if he be not a good Theorician, he will never prove an able Practitioner.

Dionis died on December 11, 1718, three years after Louis XIV, and was buried in Saint-Roch Church in Paris. He had two sons, both of whom became prominent surgeons. His grandson, Charles Dionis (1710–1776), became a renowned professor of medicine. Thus, his immediate descendants took up the advances in surgery he had pioneered, and the links he had forged between the skills of the surgeon and those of the physician.

3.4.5 Félix Vicq d'Azyr

Félix Vicq d'Azyr (1746–1794) (Figure 3.42) was a product of the Enlightenment. He was a physician, anatomist, historian, and social reformer. He became the French father of comparative anatomy and the originator of the theory of homology that attempted to link anatomical traits to a common ancestor.

Vicq d'Azyr was born on April 23, 1746, in Valognes in Normandy, the son of a Montpelier-trained physician, Maître Jean-Félix Vicq Sieur de Valemprey. His mother was Catherine Lechevalier, who was descended from nobility. He studied philosophy at the seminary of Valognes, and later at Caen, where he met Pierre-Simon Laplace (1749–1792), who would go on to make a name for himself in physics and chemistry. In 1765, Vicq d'Azyr enrolled in the University of Paris to study medicine. The influential teachers he encountered there included the anatomist and surgeon Antoine Petit (1722–1794) and the comparative anatomist Louis-Jean-Marie Daubenton (1716–1799). In 1772, he defended his thesis on the protective effect of the cranium for the brain and passed his medical certification examinations. Early in his career, he determined that he would be a neuroanatomist, and his treatises on this topic helped lay the foundation for this discipline.

Soon after graduating, he began teaching anatomy; in 1773, in Paris, he taught a popular course on morphology at the Jardin du Roi, currently the Museum of Natural History. Shortly afterwards he fell ill, probably with tuberculosis, and traveled back to Normandy to recuperate. During this time, he began an in-depth study of the anatomy and physiology of fish. On his return to Paris in 1774, with the support of his friend, mathematician and political scientist Marie Jean Antoine Nicolas de Caritat, the Marquis de Condorcet (1743–1794), he was admitted to the Académie des Sciences; he earned his doctorate the same year.

Figure 3.42 Drawing of Félix Vicq d'Azyr (1746–1794).

Vicq d'Azyr continued his study of anatomy in Paris in 1776, following the recommendation of his teacher Petit. In fact, Petit wished Vicq d'Azyr to succeed him as Chair of Anatomy at the Jardin du Roi after his retirement in 1778, but the appointment went to Antoine Portal (1742–1832), who was supported by the Count of Buffon (1707–1788). One year after Petit's retirement, Vicq d'Azyr married, but during the next two years, he lost his wife (the niece of his other celebrated teacher, Daubenton) and child, both to tuberculosis. (Note: some sources state that he was childless.) In 1780, he became Professor and Chair of Comparative Anatomy at Alfort Veterinary School, a post he would hold for the next eight years. Hewas instrumental in establishing the Société Royale de Médecine, and eventually became the permanent secretary of this group. In 1788, he was elected into the Academie Francaise, the highest scientific honor in France. A year later, he succeeded Joseph-Marie-François de Lassone as royal physician to Queen Marie Antoinette (1754–1793), whom he tried to protect during the Revolution. Interestingly, in spite of this, he was in charge of the group that manufactured the saltpeter used in the revolutionareis' gunpowder.

In 1790, he presented his plans for the reformation of the method by which medicine was taught in France to the Constituent Assembly, describing the medical curriculum of his day as *"partout viscieux et nul"* (perverted and worthless). His outline called for the grouping together of the schools of medicine, surgery, veterinary, and pharmacy, with close union to the hospitals, and additionally for the active participation of medical students in clinical work and competitive examinations. This vision resulted in the creation in Paris of three Écoles de Santé (Schools of Health). He spent the last year of his life, 1794, as military physician and superintendent of the anatomical collection that had belonged to the Count of Orléans.

3.4.5.1 Neuroanatomical Contributions

During his eight years in Alfort, Vicq d'Azyr wrote with the aim of presenting a "*grand tableau*" or scheme of all species to illustrate the grand design of the animal kingdom, culminating in humans. However, he was only able to write one of the intended five volumes prior to his death. This work, dedicated to King Louis XVI, described the detailed anatomy of the brain, and contained some of the eighteenth century's most accurate depictions of that organ (Figure 3.43). In the introduction, Vicq d'Azyr emphasized the importance of integrating form and function. The beautifully illustrated plates were drawn by the noted Parisian engraver Alexandre Briceau, whom the author thanked for his skills, stamina, and "endurance of foul odors." All plates were drawn from actual preserved human brains. One was a black-and-white one taken from Soemmering's *De basi encephali*. Preceding each plate, Vicq d'Azyr provided detailed explanations, along with comments by earlier anatomists (e.g., Bidloo, Vieussens, Eustachius, Willis, Monro, von Haller). In his descriptions, Vicq d'Azyr often chided past anatomists, including Vesalius, for inaccuracies in their illustrations and descriptions.

Vicq d'Azyr was one of the first to use horizontal and coronal sections of the brain for the study of its anatomy. He was also the first to use a chemical cocktail (alcohol, saltpeter, and hydrochloric acid) for brain preservation; this would not become a standard method until it was used by Johann Christan Reil at the beginning of the nineteenth century (Figure 3.44). Vicq d'Azyr was the first to describe the substantia nigra, although some have erroneously attributed this discovery to Samuel Thomas Soemmerring (1755–1830). In 1786, he described the mamillothalamic tract (bundle of Vicq d'Azyr) as "the white ribbon that rises from the mamillary

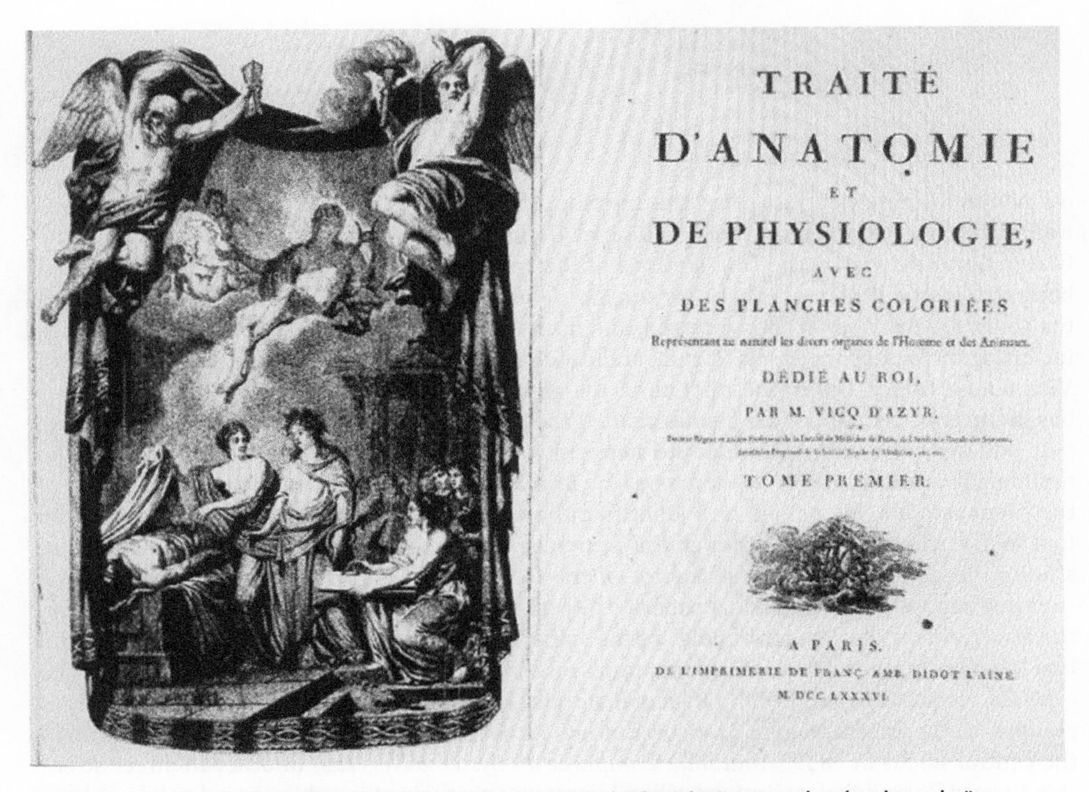

Figure 3.43 Title page from Vicq d'Azyr's *Traité d'anatomie et de physiologie – avec des planches colorës représentant au naturel les divers organs de 'Homme et des Animaux.*

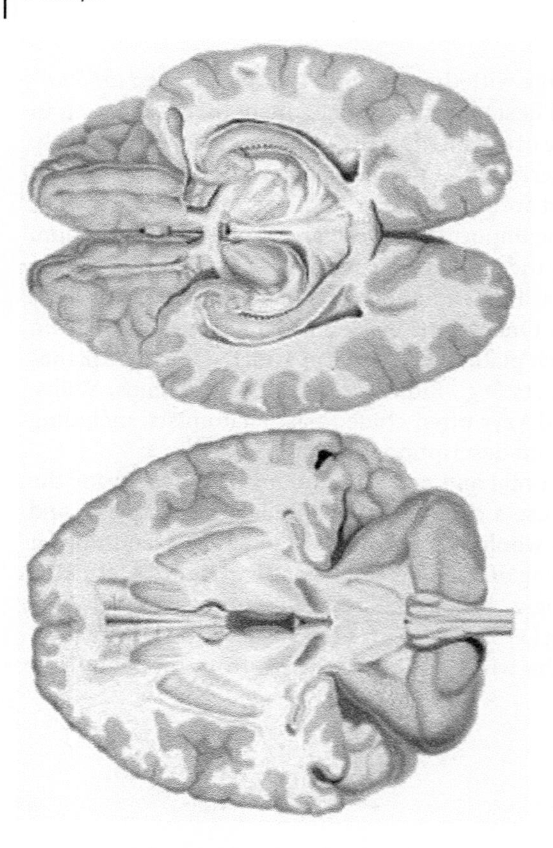

Figure 3.44 Figure from Vicq d'Azyr's opus, illustrating his method of horizontal sectioning of the brain.

eminence, forming a curve toward the anterior tubercle within the optic thalamus." He is credited with rediscovering the white line (line of Gennari) in the calcarine cortex, and he was the first to accurately show the convolutions of the cerebral hemispheres, which historically had been considered little more than *"sur un plat de macaronis"* (macaroni on a dish). He divided the cerebral cortex into frontal, parietal, and occipital lobes. He described the deep gray nuclei, including the basal ganglia, claustrum, and caudate nucleus, as well as the foramen (cecum) of Vicq d'Azyr, which lies in the midline of the anterior pontomedullary junction and receives basilar artery perforators. He also described the locus coeruleus and the centrum semiovale (Vicq d'Azyr's centrum), although the centrum was first portrayed by Vieussens. He gave the first detailed description of the course of the fornix and stria medullaris and of the anatomy of the hippocampus. He named and identified the uncus, cuneus, and precuneus and was the first to describe the insula, although this structure carries the name of Reil. With air-injection studies, he demonstrated the existence of the intraventricular foramen and noted that the superior sagittal sinus drained primarily into the right transverse sinus. He elaborated on the arrangement of the fiber bundles of the spinal cord and their division into a posterior and two lateral columns with an anterior commissure.

It was Vicq d'Azyr, and not the Italian anatomist Rolando, who discovered the central sulcus and pre- and postcentral gyri. François Leuret (1796–1851) first applied the name Rolando to the central sulcus in 1839 (*Anatomie Comparée du Systéme Nerveus*) in order to direct attention toward Rolando's descriptions; when Leuret was preparing his publication, he did not have access to the writings of Vicq d'Azyr. Vicq d'Azyr first mentioned the central sulcus in 1796 in his *Traité d'anatomie et de physiologie*, which was one of the finest publications on the brain prior to the advent of microscopy.

3.4.5.2 Connections to American Forefathers

In 1783, Vicq d'Azyr corresponded with the second President of the United States, John Adams (1735–1826), regarding collaborative efforts between scientific organizations in that country and in France. Later, in 1789, he became friends with American statesman and Founding Father, Gouverneur Morris (1752–1816), during his function as US Minister to France. Parent suggests that Morris may have transmitted letters from the King and Queen of France to London with the aid of Vicq d'Azyr. During Benjamin Franklin's trips to France, he met and also became friends with Vicq d'Azyr. The two discussed various issues. For example, letters of correspondence show they were both intrigued by the potential of infectious agents being transferred from the dead to the living. Franklin brought to Vicq d'Azyr's attention the case of an archeologist who was presumed to have fallen ill as a result of coming into contact with ancient Egyptian mummies. Vicq d'Azyr delivered Franklin's official eulogy to the Academy of Sciences in March 1791, beginning: "A man is dead, and two worlds are in mourning."

Vicq d'Azyr died from uncertain causes (perhaps tuberculosis) on June 20, 1794, during the Reign of Terror. The height of his career marked the beginning of the French Revolution, during which tumultuous time he saw the King and Queen of France beheaded, along with several of his friends and colleagues. His life and career were cut short, but were nevertheless filled with discoveries and descriptions that continue to propel our understanding of the human nervous system today.

3.4.6 Marie-François Xavier Bichat

Marie-François Xavier Bichat (1771–1802) (Figure 3.45) stated a basic principle that laid the foundation of modern clinical anatomy, histology, and pathology: "The more one will observe diseases and opens cadavers, the more one will be convinced of the necessity of considering local diseases not from the aspect of complex organs but from that of the individual tissues." There can be no doubt that Bichat's contributions were immense. To historians and biologists, he is known as a master-genius, the most important medical inventor of the Enlightenment era, the founder of histology and descriptive anatomy, and even the founder of modern medicine.

Bichat was born on November 14, 1771 in Thoirette, a village in the province of Bresse, France. His father, Jean Baptiste Bichat, was a physician trained at Montpellier and the mayor of Poncin in Bugey. Bichat received primary school education at Nantua. He studied the humanities and rhetoric, and gained honorable distinctions in philosophy. In 1791, his father sent him to Lyons, remote from the ongoing French Revolution, where he studied anatomy and surgery under the chief surgeon Marc-Antoine Petit at Hôtel-Dieu, receiving private instruction from him. However, the threat of the encroaching Revolution forced Bichat to leave Lyons. In 1793, he served as a surgeon in the Alps with the army of the Republic.

In July 1794, after the collapse of the Reign of Terror following the execution of Maximilien Robespierre, Bichat moved to Paris and entered the school of Professor Pierre-Joseph Desault (1744–1795), a remarkable French surgeon at the Grand Hospice de l'Humanite (later Hôtel-Dieu of Paris). One day, Bichat volunteered to take over the task of transcribing that day's lecture. His biographer, Monsieur Buisson (one of his pupils), describes Bichat reading the summary of his transcription the following day: "He was listened to with extraordinary silence, and left the theatre loaded with eulogium, and covered with the reiterated applause of his fellow students." The clarity and precision with which he read the abstract struck his professor.

Figure 3.45 Portrait of Marie-François Xavier Bichat (1771–1802) by Pierre Maximilien Delafontaine. (*Source:* From Chateau de Versailles, France/The Bridgeman Art Library, with permission.)

From that time onwards, Bichat received only praise from his mentor, and became his assistant in both school lectures and private practice. Desault treated Bichat as an adopted son, and allowed him to reside in his house as a family member.

The sudden death of Desault on June 1, 1795 was devastating to the young Bichat. Suspicious of murder, he autopsied the body, disproving poisoning as a cause of death. Bichat took care of Desault's widow and assisted in the education of their son. Despite his grief, he began the collection and editing of Desault's unpublished works. Through his efforts, the fourth volume of the *Journal de chirurgerie*, founded by Desault, was published, covering interesting cases of daily practice; so too were the complete *Surgical Works* of Desault, in three volumes.

In 1797, Bichat delivered his first course of anatomical lectures. Being successful in this, he announced other courses of operative surgery and physiology, beginning in the following year. He suffered a life-threatening attack of hemoptysis brought on by the inhalation of noxious fumes from the dissection room, but recovered. In 1798, he established the Société Médicale d'Emulation, a research and educational society. He abandoned surgery in 1799 and devoted his time to experimental physiology, dissection, and autopsies, giving himself over to the exploration of morbid or pathological anatomy. He initially held the informal title of chirurgien-externe, and at the age of 28 (1800) he was appointed physician to the Hôtel-Dieu of Paris. It is said that he dissected around 600 corpses in a six-month period while there, during the winter of 1801–02. He sometimes slept in the morgue so that he could dissect cadavers all day.

In 1799, he published *Traité des membranes général et de diverses membranes en particulier* ("Treatise on Membranes in General and on Various Membranes in Particular"); in 1800, *Recherches physiologiques sur la vie et sur la mort* ("Physiological Researches upon Life and Death"); and in 1801, *Anatomie générale appliquée a la physiologie et a la médecin* ("General Anatomy Applied to Physiology and Medicine"). *Traité des membranes* passed through several editions and was translated into other European languages. The association between diseases and the lesions of tissues was first identified in this book, and the chapter on synovial membranes is still of basic value to the field of orthopedics. Bichat rejected earlier accounts of the synovial glands by Clopton Havers (an English physician and pioneer in osteology), and recognized them as distinct fat pads. The last book Bichat published, although he only managed the first two volumes, was *Traité d'Anatomie Descriptive* ("Treatise on Descriptive Anatomy") in 1801–02. His students, Buisson and Philibert-Josef Roux (1780–1854), posthumously completed this work by adding three volumes based on his ideas and investigations.

Excessive work gradually impaired Bichat's health. He died on July 21, 1802, and was buried in the same vault as Desault. The cause of his death was mysterious. His body was autopsied by Roux (Figure 3.46), who noted some abnormalities of dentition and an occipital skull fracture, which latter he attributed to tuberculous meningitis. In the third volume of *Traité d'Anatomie Descriptive,* Buisson illustrated his mentor's accidental fall down the stairs of the Hôtel-Dieu, followed by transient losses of consciousness, recurrent fainting, and drowsiness, and ultimately by Bichat's death on the 14th day of the illness. This history resembles that of brain concussion and intracranial hematoma. Some have debated whether he died of a febrile illness after handling infectious laboratory specimens. Interestingly, a Persian anatomical textbook of the nineteenth century (by Zeinolabedin Hamadani, a physician who had perfected his knowledge by studying in Paris) made a note of an incredible French anatomist, Bichat, who had "devoted himself to exhaustive anatomical investigations and had died of upper limb infection after wounding himself with instruments during the dissection of a cadaver."

3.4.6.1 Bichat and his Doctrine of Tissues

Bichat coined the term "tissue" in *Traité des membranes,* and suggested that disease results from the lesions of various tissues. He specified that "it is imperiously necessary to separate tissues one by one, if we purpose to go into a minute analysis of their intimate structure." Elsewhere, Bichat argued that "in every organ composed of many tissues, the one may be impaired without disorder of the others; this is indeed what happens in the greatest number of cases." The doctrine of tissues was in contrast to the teaching of Giovanni Battista Morgagni (1682–1771), who some 40 years earlier had declared a predictable association of diseases with pathological changes in *organs.* In *Anatomie générale,* Bichat stated that anatomy:

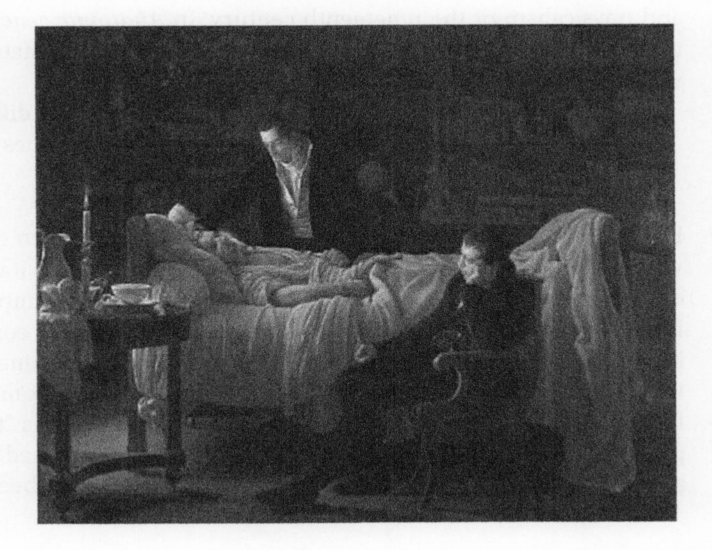

Figure 3.46 "La mort de Bichat" by Louis Hersent. (*Source:* With kind permission from Bibliothèque Interuniversitaire de Médecine (BIUM), Paris.)

> Has its simple tissues, which, by their combinations form the organs, properly so-called. These tissues are 1. The cellular membrane. 2. The nerves of animal life. 3. The nerves of organic life. 4. The arteries. 5. The veins. 6. The exhalants. 7. Absorbents and glands. 8. The bones. 9. The medulla. 10. Cartilage. 11. Muscular fibers. 12. Fibro-cartilaginous tissue. 13. Muscles of organic life. 14. Those of animal life. 15. The mucous membrane. 16. The serous. 17. The synovial. 18. The glands. 19. The dermis. 20. The epidermis. 21. The cutis.

Bichat characterized and distinguished tissues by their changes upon boiling, soaking, or baking, and even based on their reaction to acids and bases, but he never used a microscope. He classified pathological tissues as either inflammatory or scirrhus. As Porter indicates, Bichat's doctrine of tissue linked the classic morbid anatomy of Morgagni to the modern cell pathology of Virchow.

3.4.6.2 Age of Bichat: Moving from "Autopsy of the Dead" to "Autopsy of the Living"

Autopsy of the dead, which was avidly practiced by Bichat, gained legitimacy as a definite tool for the diagnosis of diseases in France. The idea of correlation between autopsy findings and disease symptoms originated from Corvisart of the Paris school. Adherence to these scientific methods gave rise to a kind of "autopsy of the living" through the invention of the stethoscope by Laennec, a pupil of both Bichat and Corvisart. Auscultation became an acceptable mode of examining the patient, leading to the evolution of physical examination. Foucault referred to this as "the age of Bichat," characterized by the vanishing of "medicine of symptoms," during which diagnosis had been based entirely on the symptoms with which the patient presented.

3.4.6.3 "General Anatomy": Physics versus Physiology, and Mechanism versus Vitalism

Bichat argued: "If one attempts to express matters of physiology in terms of the language of physics, rather than clarify matters, one would muddle them beyond all measure." According to him, physical objects were constant, and hence were subject to fixed laws, while living beings were subject to varying laws. Remarking on his rejection of reductionist philosophy, Bichat claimed that body mechanisms cannot be explained by merely physical laws. He believed that life is possible only via invisible vital life forces, and he provided abundant proof of this in his *Anatomie générale*. Such an account of vitalism created an alternative to the pure mechanism and physicalism of the nineteenth century. In *Anatomie générale*, Bichat stepped farther and concluded that vitality must be unique for each tissue. He stated:

> as the greatest part of the organs are formed of very different simple tissues, the idea of peculiar life can only be applied to these single tissues and by no means to the organs themselves.

For this, some have referred to Bichat as a "plurivitaliste," to distinguish his concept from the vitalism of his predecessors. Bichat himself contrasted his vitalist doctrine with Bordieu's idea of "the life of each organ." Although vitalism is deemed untestable or unscientific by most, Bichat tried to institute a methodological approach by conducting experimental studies. No more than half a century later, Claude Bernard, a prominent physiologist, refuted Bichat's view of vital forces by proposing instead "an internal environment [*milieu intérieur*] in which the elements of the tissue live," the constancy of which was "the condition for free and independent life." For a variety of reasons, many of them unrelated to the biology, vitalism declined abruptly in the early twentieth century, and the theory has been abandoned.

Figure 3.47 Statue of Bichat by David d'Angers. (*Source:* Photo courtesy M. Claude BOISSY.)

Bichat's name is remembered via several eponyms, including Bichat's fossa (the pterygopalatine fossa), Bichat's fat pad or protuberance (the buccal fat pad), Bichat's foramen or canal (the cistern of the vena magna of Galen), Bichat's ligament (the lower fasciculus of the posterior sacroiliac ligament), Bichat's fissure (the transverse fissure of the brain), and Bichat's tunic (the tunica intima vasorum). His death was regarded as a national loss. Jean-Nicolas Corvisart (a renowned French physician at the Hôtel-Dieu of Paris and a friend of Bichat) wrote to Napoleon Bonaparte: "no one has done so much, or done it so well, in so short a time." Napoleon ordered a monument of Bichat and Desault to be placed in the Hôtel-Dieu. Under his bust appears "The Napoleon of Medicine." Louis Hersent, a French painter and contemporary of Bichat, commemorated his death with the painting "La Mort de Xavier Bichat," which shows him dying as he is watched over by two of his students, Dr. Esparron and Dr. Roux. Notably, the last work of David d'Angers, the famous French sculptor, was a statue of Bichat, situated in the main courtyard of the Paris Faculty of Medicine, in 1857 (Figure 3.47). At the first International Physiological Congress held in Paris following World War I, a medal was issued in Bichat's honor (Figure 3.48).

3.4.7 François Magendie

François Magendie (1783–1855) (Figure 3.49) was considered a pioneer in experimental physiology. He lived through tumultuous times, from the occupation of Paris to the Hundred Days to Waterloo. Magendie was the son of surgeon Antoine Magendie and Marie Nicole de

Figure 3.48 Medallion of Xavier Bichat, awarded at the 1920 Congress of Physiology, Paris, fashioned by L. Dufour. (*Source:* With kind permission from Rudolph Matas Library, New Orleans, LA, USA.)

Figure 3.49 Painting of a young François Magendie (1783–1855). (*Source:* Biusante Parisdescartes.)

Perey; he, too, became a well trained surgeon. He had a younger brother, Jean-Jacques, named in honor of Jean-Jacques Rousseau (1712–1778). The two Magendie children were raised on the basis of Rousseau's teachings, which entailed fostering the child's personal independence. Consequently, by the age of 10, Magendie had never attended school and was unable to read or write. In 1791, encouraged by the French Revolution, Magendie's family moved to Paris, where his father became more political and focused less on his surgical practice. His mother died in 1792.

When he was 16, too young to be admitted to the École de Santé, Magendie became an apprentice at the Hôtel-Dieu. The surgeon Baron Alexis de Boyer (1757–1833), a friend of his father, took him on as a student and put him to work preparing anatomical dissections. In 1803, Magendie was accepted as a medical student at the Hôpital Saint-Louis. In 1807, he officially became an assistant in anatomy at the École de Médecine and gave courses in anatomy and physiology. He received his medical degree in Paris on March 24, 1808. The Revolution, the succeeding European wars, and the dramatic social changes these entailed made it a chaotic time for the practice of medicine in France. In 1801, Napoleon created a law that stated that no physician could practice medicine without a diploma from a medical school. Even Boyer, Napoleon's personal surgeon, had to be examined.

After his thesis, Magendie's first publication was an article that appeared in the *Bulletin des Sciences Médicales*, published by the Société Médicale d'Émulation, which glorified the memory of Marie François Xavier Bichat (1771–1802). In 1811, Magendie was appointed anatomy demonstrator at the Faculté de Médécine in Paris, and for three years he taught anatomy, physiology, and surgery. He was known as a skilled surgeon during operations at the École Pratique, but his personality seems to have caused ructions. His rude behavior is said to have resulted in a conflict with Professor of Anatomy, François Chaussier (1746–1828). Additionally, it is said that the famous professor of surgery, Guillaume Dupuytren, saw Magendie as a dangerous rival and attempted to obstruct him on multiple fronts at the Faculté de Médécine. Perhaps because of these conflicts, Magendie resigned from his position as anatomy demonstrator in 1813 and began to practice as a private physician; he also began teaching private classes in physiology. This marked the beginning of his transition from surgeon to experimentalist. In 1818, after intense competition, he was appointed to the Bureau Central des Hôpitaux Parisiens, and in 1826 he became Médecin adjoint at the Salpêtrière. In 1821, he founded the *Journal de Physiologie Expérimentale*, and in 1831 he replaced Joseph Claude Anthelme Récamier (1774–1852) as Chair of Medicine at the Collège de France. The reason behind Magendie's shift from surgeon to experimenter has been debated. One suggestion is that political pressure from Dupuytren led him to want to make a name for himself in a field less competitive than that of surgery.

Experimentally, Magendie made multiple physiological and pharmacological discoveries. He disputed the function of the epiglottis during deglutition as necessary in preventing food from entering the trachea, demonstrating in animals that this did not hold true. He showed that pulmonary inflammation was less severe if the vagi were transected, and he made a positive contribution to the study of infection when he demonstrated that the saliva of rabid dogs contained the contagious component.

His first physiological experiments were on the mechanisms of swallowing and vomiting, proving the passive role of the stomach in vomiting. He also demonstrated the hemodynamic importance of elastin in the arteries and provided proof of the role of the liver in detoxification of the blood. Magendie introduced into medical practice a series of recently discovered alkaloids: strychnine, morphine, brucine, codeine, quinine, and veratrine. In 1817, he discovered emetine, the active ingredient of ipecac. He did not hesitate to test on himself all the substances that he found harmless in his animal experiments.

Neuroanatomically, Magendie made important observations of the retina. He showed that the thalamus and cerebral peduncles were concerned with movement and that sectioning of the brain in these areas led to hypertonicity. He was the first to produce decerebrate rigidity with accuracy, and he studied elementary reflex arcs. He observed that puppies could still detect vapors such as tobacco smoke and ammonia after transaction of their olfactory nerves, and concluded that the trigeminal nerve must be involved in conveying such stimuli. Moreover, he ascribed almost all sensory functions of the head to the trigeminal nerve and

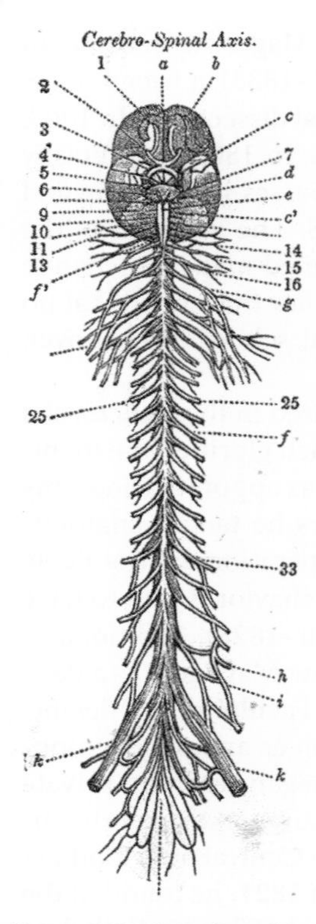

Figure 3.50 Drawing of the brain and spinal cord with associated nerves from Magendie's *An Elementary Treatise on Human Physiology*. (*Source:* Courtesy of the Reynolds Historical Library, The University of Alabama at Birmingham.)

refuted historical ascriptions of sight to this cranial nerve (Figures 3.50 and 3.51). Magendie also demonstrated that transaction of the anterior half of a cerebellar peduncle resulted in rotational movements of the experimented animal.

It has been said that Magendie's most important contribution to science was also his most disputed. Contemporaneous with British anatomist Sir Charles Bell (1774–1842), Magendie conducted a number of experiments on the nervous system, in particular verifying the differences between sensory and motor nerves in the spinal cord. This discovery has been likened to the importance of William Harvey's (1578–1657) description of the circulation of the blood. Moreover, it was essential for Marshall Hall's (1790–1857) discovery of the spinal reflexes. Many British claimed that Bell published his discoveries first and that Magendie purloined his findings. There was an intense dispute over the rightful discoverer, during the course of which Magendie commented:

> In sum, Charles Bell had had, before me, but unknown to me, the idea of separately cutting the spinal roots; he likewise discovered that the anterior influences muscular contractility more than the posterior does. This is a question of priority in which I have, from the beginning, honored him. Now, as for having established that these roots have distinct properties, distinct functions, that the anterior ones control movement, and the posterior ones sensation, this discovery belongs to me ... Why must this scientist (Bell) spoil his work and injure himself by not rendering to his rivals the justice due them? Why must he cling to that barbarous patriotism which rejects everything that does not come from his own country? Why does he persist in his pretensions to discoveries which he has not made?

Olmsted concluded that the designation "Bell–Magendie law" represented a compromise between the points of view of the older anatomist (Bell), who arrived at function by way of observation and inference, and the physiologically-minded scientist (Magendie), who insisted on experimental verification. Magendie also found that electrical stimulation of the dorsal roots in the lumbar region resulted in variations in blood pressure.

Magendie became infamous for his live public vivisections. In fact, a landmark bill banning animal cruelty in Britain described Magendie's public dissection of a dog, in which the animal was nailed down ear and paw, half the nerves of its face dissected, and left overnight for further dissection the following day. During a trip to England in 1824, as a guest of William Hyde Wollaston (1766–1828), Magendie publicly demonstrated his method of vivisection of other cranial nerves in canines. Such experiments were condemned in Britain for their brutality but proved important in determining the multiple functions of such nerves as the optic, facial, and vestibulocochlear. For example, Magendie was the first to ascribe a sensory function to the facial nerve, as sectioning of this structure resulting in the animal crying out in pain.

From 1824 to 1828, Magendie made many important discoveries regarding cerebrospinal fluid. In fact, only three years after the death of its discoverer, the Italian savant Domenico

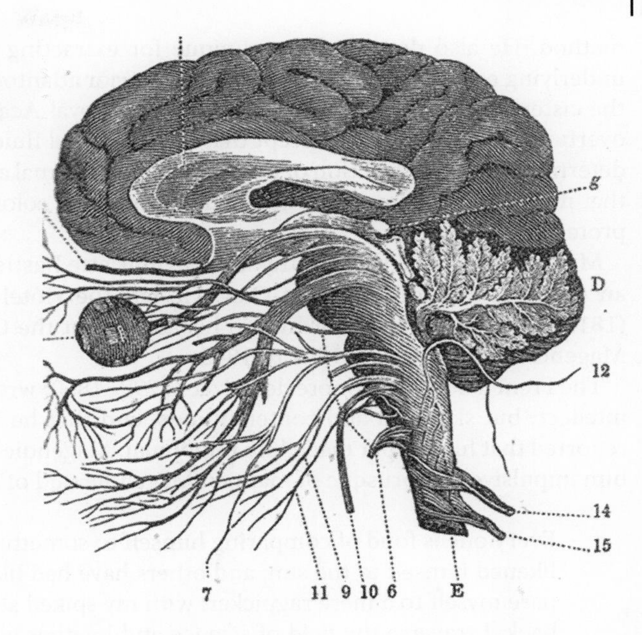

Figure 3.51 Neuroanatomical drawing from Magendie's *An Elementary Treatise on Human Physiology*. (*Source:* Courtesy of the Reynolds Historical Library, The University of Alabama at Birmingham.)

Cotugno (1736–1822), Magendie documented his observations of this fluid and debated whether or not it was derived from the serum or was a fluid *sui generis*. In 1828, he presented these findings in a lecture entitled "Mémoire Physiologique sur le Cerveau" at the Academy of Sciences in Paris, demonstrating the egress of cerebrospinal fluid from the fourth ventricle into the subarachnoid space at the "tip of the calamus scriptorius of the medulla" (foramen of Magendie). Prior to Magendie's description of the median fourth ventricular outlet, it had been realized that there was some form of communication between the fluid contents of the cerebral ventricles and the subarachnoid spaces and cisternae. Magendie compared this pathway of cerebrospinal fluid to a tunnel under the Thames and suggested the pineal gland was a "plug destined to open and shut the aqueduct." He thought this a more practical function than the "seat of the soul," which Descartes had attributed to it. Regarding cerebrospinal fluid flow from the fourth ventricle into the subarachnoid space, Magendie contradicted Bichat's law, which states that fluid secreted by a hollow organ is always secreted toward the interior of the organ and not toward its exterior. When the German physiologist Friedrich Tiedemann (1781–1861) visited Paris, he spoke with Magendie regarding such contradictions. Magendie vivisected an animal to demonstrate his findings to Tiedemann, who then returned to Germany to push for acceptance of this new idea.

Clinically and in the realm of neurology, Magendie described a 34-year-old female with gigantism, which he attributed to a pituitary disturbance, and a child with congenital absence of the cerebellum. Notably, he was approached by a Polish officer who had been knocked unconscious in a battle with the Russians and had lost his sense of hearing and taste. After consulting physicians in Vienna and Trieste, he allowed Magendie to apply an electrical current to his chorda tympani nerve via the external auditory meatus, which resulted in return of his sense of taste. Magendie is said to have also restored this man's hearing; supposedly, the first sound that the man heard following treatment was the beat of the drums at sunset in the Luxembourg Gardens. Magendie also had some success in dealing with various neuralgias by using Galvanic current, and he is said to have amazingly restored sight to a blind man using this method. He also described a technique for extracting cerebrospinal fluid without injuring underlying neural tissue by piercing the posterior atlanto-occipital membrane (i.e., puncture of the cisterna magna). In a memoir read at the Royal Academy of Sciences in 1824, Magendie overturned the historical concept that cerebrospinal fluid was a pathological product. He also determined the composition of this fluid in both normal and pathological conditions. He stated that it took on a yellow color in jaundice, a reddish color in scurvy, and a marked increase in protein in cholera.

Magendie married the wealthy widow Henriette Bastienne de Puisaye in 1830 and acquired an estate in Sannois, Seine-et-Oise. He left the Hôtel-Dieu in 1845, and Claude Bernard (1813–1878) substituted for him in his absence at the Collège de France. Bernard had been Magendie's student from 1841 to 1843.

The French novelist Honoré de Balzac (1799–1850) wrote that Magendie was a distinguished intellect, but skeptical and contemptuous, and that he believed only in the scalpel. He also reported that his temper made him unpopular. Magendie's students are said to have considered him impulsive and brusque of manner. Magendie said of himself:

> Every one is fond of comparing himself to something great and grandiose, as Louis XIV likened himself to the sun, and others have had like similes. I am more humble. I compare myself to a mere ragpicker: with my spiked stick in my hand and my basket on my back, I traverse the field of science and I gather what I find ... Medicine is a science in the making.

François Magendie (Figure 3.52) made epoch-making contributions to experimental physiology, pharmacology, anatomy, and pathology, as well as to the foundations of neurology and neurosurgery. His name lives on today in the eponyms Bell–Magendie law, foramen of Magendie (median aperture of the fourth ventricle), Magendie's space (subarachnoid space), and Magendie's sign (Magendie–Hertwig syndrome, a downward and inward rotation of the eye due to a lesion in the cerebellum, also named for Karl Heinrich Hertwig, 1798–1881).

3.4.8 Jules Germain Cloquet

Jules Germain Cloquet (1790–1883) (Figure 3.53) was born in Paris to a family of four children, with one older brother and two younger sisters. His father, Jean-Baptiste Cloquet, was a draftsman and an engraver with the commercial ports' inspection division in the Levant of the royal French navy under Louis XVIII, but he lost his post during the Revolution. He later earned a living giving private drawing lessons, as well as undertaking journeys to Egypt with the anthropologist Jean-François Champollion, the father of Egyptology. Jean-Baptiste's expertise in the art of drawing was passed on to his son Jules Germain Cloquet, which enabled him to make his mark in the field of anatomy and anatomical illustrations.

Cloquet began his academic career in 1803 at the Lycée Napoleon. He was initially interested in registering at a leading university in France, l'Ecole Polytechnique, to study natural sciences, but his father Jean-Baptiste and Constant André-Marie Duméril, a family friend and professor of anatomy and physiology, persuaded him to continue his education in the field of medicine.

In 1805, Cloquet became an apprentice to André-Marie Duméril in his general practice. Late in 1806, at the age of 16, he moved to Rouen in order to study anatomy at l'École d'Anatomie artificielle de Rouen, where he became known as an excellent student. During the course of his studies, he learned how to prepare the colored wax models that were used as teaching tools in anatomy courses. Wax sculpture training required extensive knowledge in the natural sciences,

Figure 3.52 Drawing of François Magendie in his later years. (*Source:* Biusante Parisdescartes.)

Figure 3.53 Portrait of Jules Germain Cloquet (1790–1883). (*Source:* Biusante Parisdescartes.)

anatomy, physiology, and pathology, which Cloquet acquired as a pupil of Achille Cléophas Flaubert, the father of the famous French novelist Gustave Flaubert. Cloquet became close friends with his mentor and his mentor's son.

Following his studies in Rouen, Cloquet turned toward a career in the military, entering the army medical school at Val-de-Grâce in 1808. This choice was almost certainly influenced by the heavy recruiting efforts made by the army for surgeons and assistant medical officers during the Napoleonic Wars. However, his studies in Val-de-Grâce were cut short when illness forced him to take a leave of absence.

After taking a few months off to rest, Cloquet decided to continue his studies in medicine in Paris. He quickly gained recognition as a talented student with a great ability in dissection and incredible visual and spatial senses that allowed him to draw on the spot and to model anatomical wax specimens. His inherited talent and his learned skills for drawing and modeling proved valuable assets, as he was appointed modeler of anatomical figures at the Faculty of Medicine in Paris in 1809. In 1811, only one year after his older brother Hippolyte, he became an intern and received the same promotion as the famed anatomist Jean Cruveilhier. He would tutor interns from hospitals in Paris in various medical subjects, including anatomy, physiology, and surgery.

In conjunction with his own studies, from 1810 to 1813, Cloquet composed the anatomy, physiology, and surgery courses given by his brother Hippolyte. Jules Cloquet's work was so much appreciated by the faculty that they pushed for his exemption from military service. He made sure that the faculty would not regret its support by becoming one of the two laureate ex æquo prize winners in 1813 for excellence in anatomy and physiology. In addition, he obtained a certificate of merit in chemistry.

In 1814, Cloquet became an assistant of anatomy at the school. Owing to his success as an assistant and to his dissecting and artistic abilities; he was made prosector in 1815. He attracted many pupils with his innovative teaching style, combining the use of anatomical preparations with chalk drawings and sketches on the blackboard.

A brilliant student and innovative teacher, Cloquet was too poor to pay for his tuition. However, following the submission of his thesis on July 17, 1817, he was allowed to obtain his medical degree without paying the customary fees. For this thesis, entitled *Recherches anatomiques sur les hernies de l'abdomen* ("Anatomical Research into Abdominal Hernia"), Cloquet dissected and studied approximately 340 cases of hernia. These cases were derived from 5000 cadavers that he had dissected during a three-year period at the University of Paris, along with his friend and colleague Pierre-Augustin Béclard. Cloquet's thesis was revolutionary in that it was a thorough anatomical study with detailed dissection, drawing, and description. It incorporated four plates, which he drew himself, and which were engraved in lithographs by his father (Figures 3.54–3.57). He described the locations where inguinal and crural herniae are most likely to occur in terms of the cremaster muscle, the peritoneum, and the spermatic vessels.

In 1818, with the help of Béclard, Cloquet translated the third edition of William Lawrence's *A Treatise on Ruptures* from English to French. One year later, he published a second thesis entitled *Recherches pathologiques sur les causes et l'anatomie des hernies abdominales* ("Pathological Research into the Causes and Anatomy of Abdominal Hernia"), which became a valuable reference for the medical profession. This thesis was published in quarto with 10 lithographic plates and included 78 figures. Cloquet eloquently explained that the displacement of the peritoneum does not cause rupturing of the herniated sac. Furthermore, he provided a detailed description of the processus vaginalis and noted that it was not closed at birth. His comprehensive work compared favorably with the studies of Astley Cooper, Franz Hesselbach, and Antonio Scarpa, and represented a new wave of lithography that would become the norm thereafter. Key, the editor of Cooper's 1827 third edition, commented that

Figure 3.54 Plate from Cloquet's 1817 thesis, *Recherches Anatomiques sur les Hernies de L'Abdomen.*

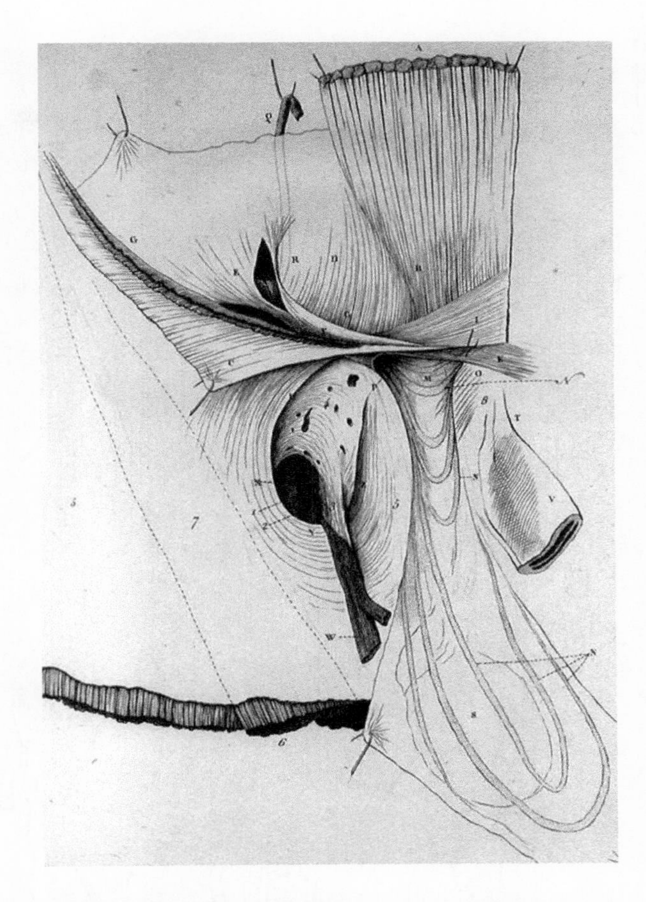

Figure 3.55 Plate from Cloquet's 1817 thesis, *Recherches Anatomiques sur les Hernies de L'Abdomen.*

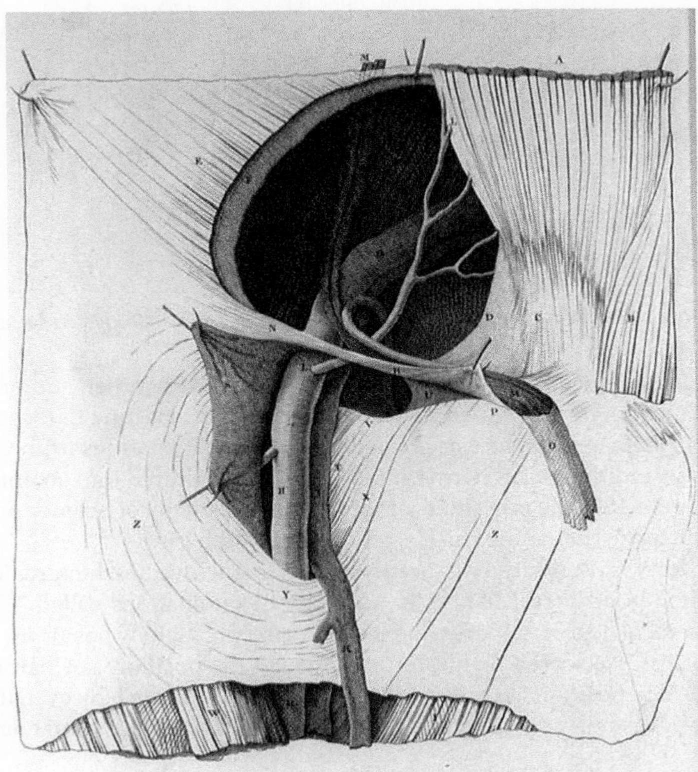

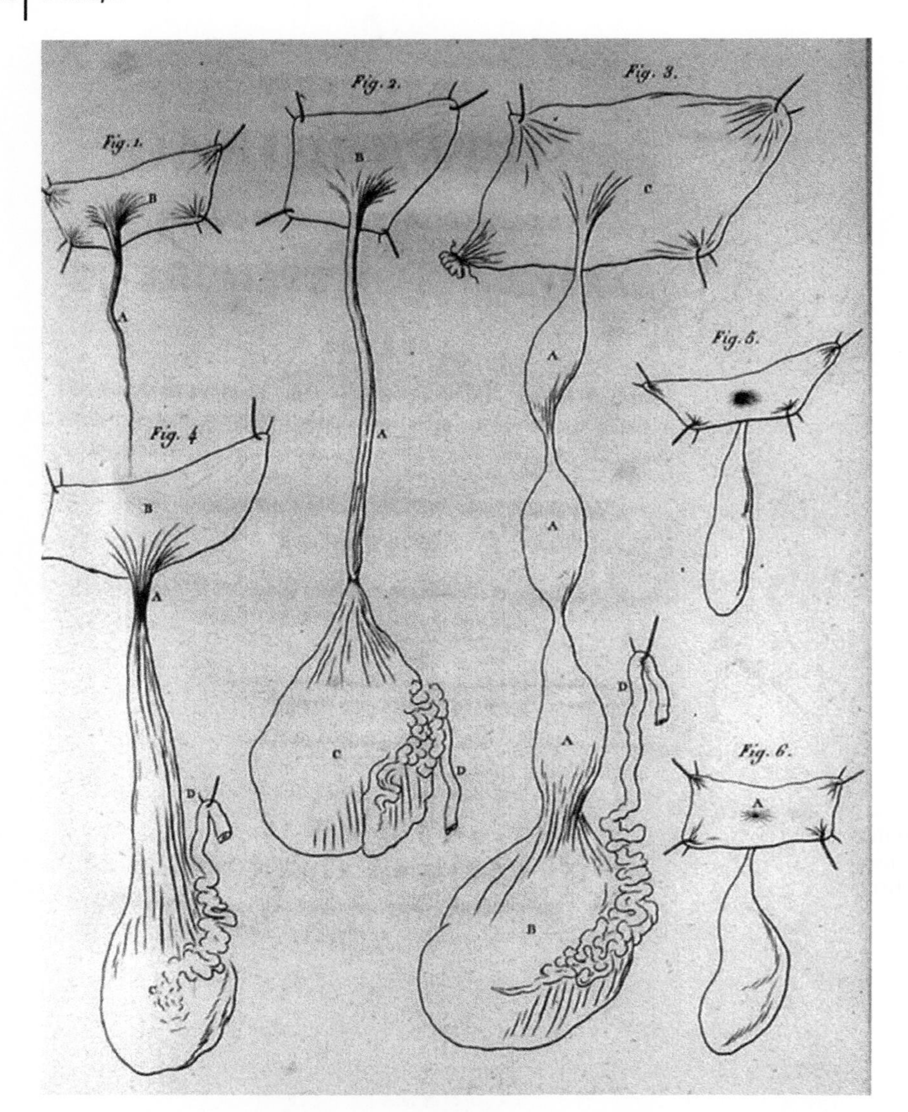

Figure 3.56 Plate from Cloquet's 1817 thesis, *Recherches Anatomiques sur les Hernies de L'Abdomen*.

Cloquet's meticulous work was "on a par" with other authors and anatomists of the era. However, he also stated that it was difficult at that time to locate a copy of Cloquet's thesis in England and that he was obliged to assume that citations from Cloquet's work were correct.

In addition to his hernia studies, Cloquet ventured into anatomical studies pertaining to the eye. In 1818, he presented a report to the Academy of Science on the pupillary membrane and the formation of the small arterial circle of the iris.

In post-Revolutionary France, advancing within the hierarchical system in surgery necessitated good performances in a series of examinations called "Concours." After obtaining his medical degree, Cloquet applied for many prestigious positions, for which he had to pass such exams. He scored highly, but was denied the position as Chair of Pathology at the Faculty of Paris, a position left void by the departure of Béclard. However, this disappointment was short-lived, as he was appointed second surgeon at l'Hôpital Saint-Louis in 1819.

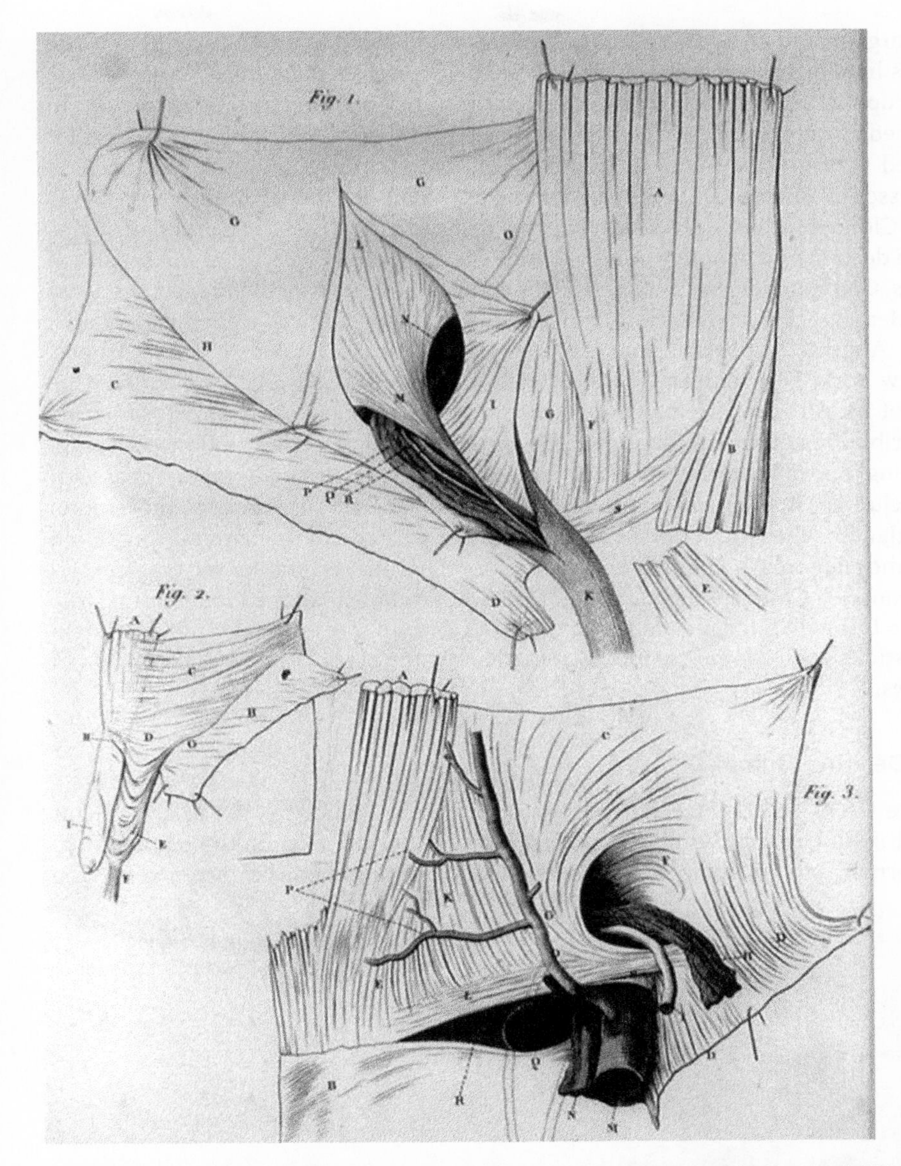

Figure 3.57 Plate from Cloquet's 1817 thesis, *Recherches Anatomiques sur les Hernies de L'Abdomen*.

In 1821, Cloquet was elected a Member of the Royal Academy of Medicine, which had been created a year earlier. It was also in 1821 that he published his final important work on hernia, entitled *Observation sur une hernie vulvaire, suivre de quelques réflexions sur la nature et le traitment de cette maladie* ("Observations on a Vulvar Hernia, with some Reflections on the Nature and Treatment of this Disease"). Between 1821 and 1831, he produced *l'Anatomie de l'homme* ("The Anatomy of Man"). This monumental work was published in five folio volumes and included 1300 figures and 300 plates, illustrated by Cloquet using the new art of lithography. It was a major contribution to the teaching of anatomy because it was so comprehensive and so well illustrated.

A practicing surgeon and anatomist, Cloquet was also interested in alternative therapeutic medicine such as hypnotism and acupuncture. In 1826, he published *Traité de l'acupuncture* ("Treatise on Acupuncture"), detailing the therapeutic use of acupuncture in medicine. In 1829, he performed an amputation of the breast on a patient who was under hypnosis. In 1831, he was appointed Professor of Surgical Pathology at l'Hôpital des Cliniques, and in 1833 he became successor to Antoine Dubois by becoming the Chair of Clinical Surgery.

Jules Germain Cloquet retired in 1858 and became an honorary professor. He was an anatomist and surgeon decorated with many honors worldwide, including Professor at the Faculty of Medicine in Paris, Commandeur de la Légion d'Honneur, and Member of the French Academy of Science and Medicine. He also was a member of medical and surgical academies in Paris, Marseille, Lyon, Angers, Brussels, Rome, Leipzig, Berlin, Bruges, Vilna, Naples, Athens, Philadelphia, New York, New Orleans, Rio de Janeiro, and Persia. He died on February 23, 1883, at the age of 93. Although he may not be remembered as one of the pioneers of hernia surgery, his contributions provided surgeons with detailed anatomical descriptions that were useful in developing innovative techniques. For this reason, Cloquet has many eponyms associated with him, including Cloquet's fascia (areolar membranous tissue superficial to the femoral ring), Cloquet's gland or the lymph node of Cloquet (a lymph node in the femoral ring, often mistaken for a femoral hernia), Cloquet's hernia (a femoral hernia perforating the aponeurosis of the pectineus muscle), Cloquet's ligament (vestigial processus vaginalis), Cloquet's canal (the hyaloid canal, also termed Stilling's canal), and Cloquet's septum (the femoral septum). A man blessed with artistic talents, Cloquet is mostly remembered for his illustrative presentation of anatomical studies.

3.4.9 Augusta Déjerine-Klumpke

Augusta Klumpke (1859–1927) (Figure 3.58) was born in San Francisco, CA to an affluent family. Her father, John Gerard Klumpke, was a well-known businessman in the realm of real estate and was born in Suttrup, England. Augusta had four sisters and one brother, and was the

Figure 3.58 Photograph of Augusta Déjerine-Klumpke (1859–1927). (*Source:* Courtesy of London Library.)

second child in her family. Her mother, Dorothea Mathilda Tolle, an American, went to Europe to seek a cure for Augusta's older sister, who suffered from osteomyelitis following fracture of her femur. During this time, her parents were divorced and her mother became the sole provider for the children. In 1871, the family moved to Germany, and two years later to Geneva, Switzerland.

Klumpke studied chemistry and the natural sciences at an academy in Lausanne. She decided to become a teacher until her mother read in a fashion journal (*La Mode Illustrée*) that a woman, Mrs. Madeleine Brès, had received her doctorate in medicine from a Parisian medical school. This galvanized the young Klumpke, who applied and was admitted to medical school in Paris in 1877. In 1880, she studied at the Charité Hospital. Following a recommendation by her anatomy mentor Professor Fort, she was authorized in 1882 to become a medical extern (a student connected to a hospital but not living there) at L'Hôpital de Lourcine, now Broca Hospital; in 1887, she was promoted to intern, becoming the first female intern in the entire French medical system. Klumpke was one of two female externs admitted to L'Hôpital de Lourcine: the other was Blanche Edwards (1858–1941), an English woman and a famous figure in the French feminist movement. Both of these women prepared their externship projects under the direction of Auguste Broca, the son of the pioneering neurobiologist Paul Broca.

It was during her externship years that Klumpke read a paper by Professor Wilhelm Erb of Heidelberg describing a palsy resulting from injury to the superior part of the brachial plexus. This piqued her interests in the brachial plexus and she went on to develop a series of experiments targeted at this anatomical area, working in the laboratory of Professor Vulpian of the Hôtel-Dieu. Klumpke's paper on injuries to the brachial plexus was published in 1885 in the *Revue de Médécine*. Subsequently, she won the Godard Prize from the Academy of Medicine in 1886. In Klumpke's classic paper *Des polynévrites en general et des paralysies et atrophies saturnines en particulier* ("About Polyneuritis in General and Saturnian Palsies and Atrophies in Particular"), she described a patient with oculopupillary malfunction (Horner's sign) associated with avulsion of the T1 nerve root. Her medical thesis on this topic was approved in 1889, and in 1890 she won the silver medal prize from the Paris Faculty of Medicine and the Lallemand Prize from the Academy of Sciences.

In 1888, Klumpke married the French neurologist Joseph Jules Déjerine, who was at that time the head of the medical facility at the Charité Hospital where Klumpke had initially attended school in 1880. Professor Déjerine was the second successor of Charcot at the Clinique des Maladies du Système Nerveux in Paris, and was a respected figure in his own right, giving his name to a variety of syndromes or conditions. Klumpke contributed to her husband's studies, particularly as a neuroanatomical illustrator, and was a major contributor to the two-volume book *Anatomie des centres nerveux* ("Anatomy of the Central Nervous System") (Figures 3.59 and 3.60), published in 1895 and 1901.

In 1891, Klumpke gave birth to a girl, Jeanne Ivonne. Jeanne also studied medicine, and in fact became one of her mother's pupils. After Déjerine's death in 1917, Klumpke continued his research in addition to her own. During her career, she published over 56 papers on neurology and neuroanatomy. She also wrote a detailed description of her own life, translated into English by Bogousslavsky. She was a pioneer in rehabilitation therapy of soldiers with injuries to the nervous system, particularly spinal cord injuries, during the First World War. She contributed much to the present knowledge of heterotopic ossification following spinal cord injury. In 1914, she was elected the first female president of the French Society of Neurology.

Klumpke died on November 5, 1927, and was buried next to her husband in Paris at the Père Lachaise churchyard. She is remembered as the first female writer of a neuroanatomy textbook and for her description of the palsy resulting from injury to the inferior parts of the brachial plexus.

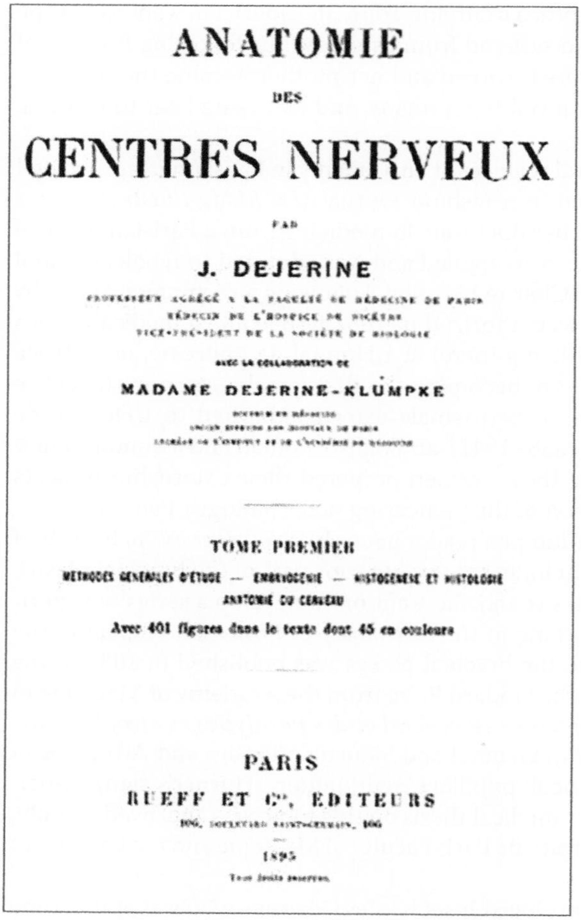

ANATOMIE

DES

CENTRES NERVEUX

PAR

J. DEJERINE

PROFESSEUR AGRÉGÉ À LA FACULTÉ DE MÉDECINE DE PARIS
MÉDECIN DE L'HOSPICE DE BICÊTRE
VICE-PRÉSIDENT DE LA SOCIÉTÉ DE BIOLOGIE

AVEC LA COLLABORATION DE

MADAME DEJERINE-KLUMPKE

TOME PREMIER

MÉTHODES GÉNÉRALES D'ÉTUDE — EMBRYOGÉNIE — HISTOGENÈSE ET HISTOLOGIE
ANATOMIE DU CERVEAU

Avec 401 figures dans le texte dont 45 en couleurs

PARIS

RUEFF ET Cie, EDITEURS

1895

Figure 3.59 Title page of *Anatomie des Centres Nerveux* by J. Déjerine, published in 1895 with collaboration from his wife, Augusta Déjerine-Klumpke.

3.5 Germany

3.5.1 Adam Christian Thebesius

The Silesian physician Adam Christian Thebesius (1686–1732) (Figure 3.61) was one of the most renowned scientists studying the coronary circulation and its pathologies in the eighteenth century. His groundbreaking work opened the field of coronary microcirculation for generations to come. Knowledge of the anatomy and the pathophysiology of the coronary circulation is essential for the physicians and medical researchers of today.

Descended from a long line of priests, lawyers, and physicians, Thebesius was born in Silesia in modern-day Germany in 1686 and steered into medicine by his uncle, a physician. Trained at some of the most prominent institutions of his time, including Leipzig, Halle, Utrecht, and Leiden, Adam Christian divided his education between practical skills and research. His research experience had a profound impact on his interest in medicine: in 1708, at the age of 22, he published his anatomical thesis *Disputatio medica inauguralis de circulo sanguinis in corde*, describing the cardiac veins that emptied into chambers of the heart, the *vasa cordis minima*. He is also known for his studies of the coronary sinus; the valve of the coronary sinus is named the Thebesian valve.

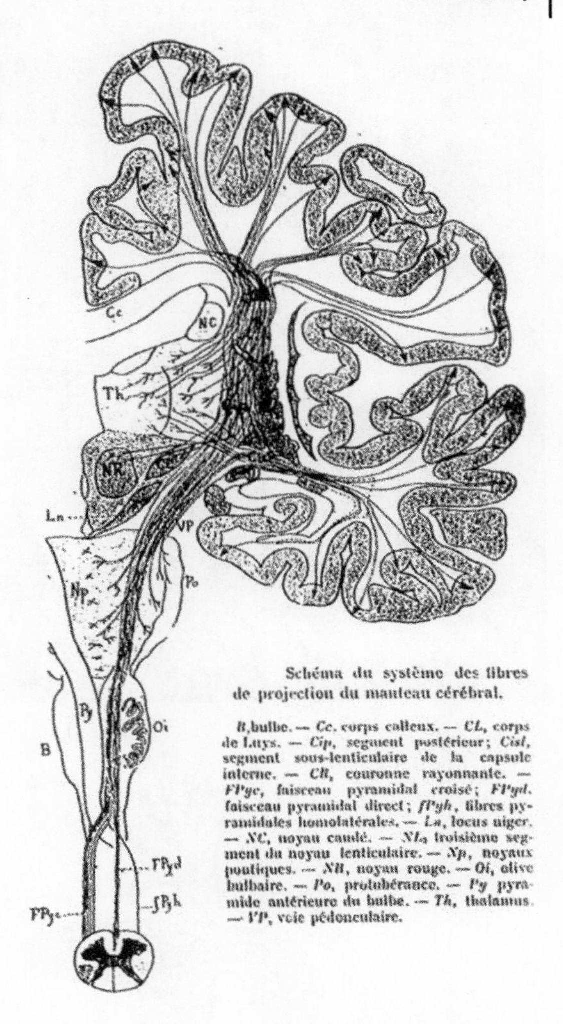

Cc
NC
Th
NR
Ln
VP
Po
Np
Py
Oi
B
FPyd
fPyh
FPyc
Schéma du système des fibres de projection du manteau cérébral.
B, bulbe. — Cc, corps calleux. — CL, corps de Luys. — Cip, segment postérieur; Cisl, segment sous-lenticulaire de la capsule interne. — CR, couronne rayonnante. — FPyc, faisceau pyramidal croisé; FPyd, faisceau pyramidal direct; fPyh, fibres pyramidales homolatérales. — Ln, locus niger. — NC, noyau caudé. — NL3 troisième segment du noyau lenticulaire. — Np, noyaux pontiques. — NR, noyau rouge. — Oi, olive bulbaire. — Po, protubérance. — Py pyramide antérieure du bulbe. — Th, thalamus. — VP, voie pédonculaire.

Figure 3.60 The first schematic drawn by Klumpke for *Anatomie des Centres Nerveux*, demonstrating the anatomy of the projection fibers of the cerebrum. The pyramidal tract is outlined.

Thebesius' research career ended when his father died and he returned home to support his family in Silesia. He established his practice in Hirschberg, a town in the mountainous region of Jelenia Gora in Poland. Prior to 1945, it was one of the largest cities in the former state of Silesia. Thebesius remained there for the rest of his life.

His work as a provincial doctor was described as exemplary. He treated all patients regardless of income and status, and often provided care to the poor without fee. But Thebesius had other interests as well, to which he devoted much of his free time. He built a small observatory and studied astronomy, and acquired a good reputation as a poet.

In 1715, Thebesius became *Stadtphysikus* for Hirschberg and surrounding villages, including the health resort of Warmbrunn. He was made a medical advisor to the Hapsburg court and Fellow of the Leopoldian-Carolian Academy of Nature Researchers. He continued to pursue his other interests. His son, Johann Gottfried, followed him into the medical profession. In fact, the Thebesius family provided the area of Silesia with physicians for six generations.

Thebesius' most notable mark in research came while he was still a student at Leiden, when he documented the small veins connecting the chambers of the heart with the coronary vessels (Figure 3.62). Two years prior to this publication, one of the most celebrated

Figure 3.61 Portrait of Adam Christian Thebesius (1686–1732), with his diagrams of the cardiac veins in the background.

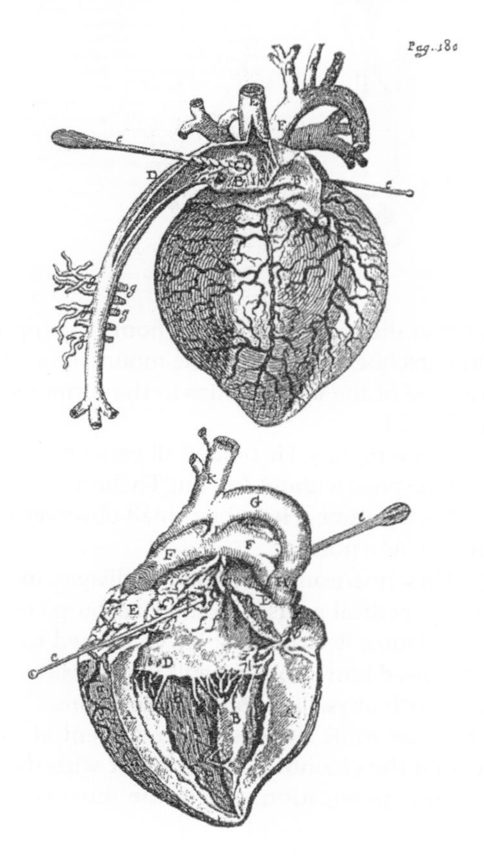

Figure 3.62 Two diagrams by Thebesius, showing the cardiac veins and the internal anatomy of the heart.

men in French medicine, the physician Raymond Vieussens (1635–1713), had published *Nouvelles découvertes sur le coeur* ("New Discoveries on the Heart"). Vieussens described ducts joining the ventricular cavities to the coronary arteries, the *ducti carnosi.* He had ligated the vena cava and pulmonary veins, injected saffron dye into the coronary arteries, and observed that the dye not only coursed through the coronary sinus and the right atrium, as expected, but also flowed into both the right and the left ventricles. This led Vieussens to the working hypothesis that ducts connected the coronary arteries to the chambers of the heart. Thebesius improved on Vieussens' techniques by reversing the direction of the injections; dye and wax were forced through the coronary sinus, allowing him to map and classify the vessels according to size and frequency.

Thebesius raised the question as to why the Creator had placed these small veins in the walls of the heart, and then very obligingly answered it by saying that they made continuous blood flow in the heart possible by serving as an exit at the beginning of systole. After answering the question in a manner so satisfactory to himself, he gave due credit and much praise to the Creator for being so foresightful as to anticipate the need and usefulness of these little vessels. Although the research was initiated by Vieussens, Thebesius' more detailed experiments on these vessels overshadowed the former's work. A group of French physicians charged Thebesius with plagiarism, but he prevailed; his thesis went through several editions, and the *vasa cordis minima* became known as the Thebesian vessels.

Contemporaries of Thebesius and Vieussens disagreed on the significance and ramifications of their observations. So too do members of the modern scientific community: the importance of the Thebesian vessels and their role in the pathology and microcirculation of the heart is still under investigation. Thebesius' work on the valve of the coronary sinus today informs research into the role of microcirculation in ischemic coronary artery disease, almost 400 years after it was first described.

3.5.2 Franz Kaspar Hesselbach

The surgeon and anatomist Franz Kaspar Hesselbach (1759–1816) was a native of Hammelburg, a district of Bad Kissingen in Lower Franconia, Bavaria. He quit grammar school in Fulda without graduating and obtained no further academic education. He became an apprentice, then a private pupil, and eventually an anatomical prosector under Karl Kaspar von Siebold (1736–1807) at the charitable foundation, the Juliusspital at Würzburg.

Hesselbach worked as an unpaid prosector for six years until 1789, when he was granted a salary. Owing to Hesselbach's activity, the anatomical specimen collection at Würzburg was substantially increased, reaching over 1000 prosections. Later, Hesselbach became a lecturer at Würzburg. His interest in anatomy occupied most of his time, and he gave many tutorials on the subject. Subsequently, he was appointed Professor of Surgery at Würzburg.

As a surgeon, Hesselbach is best known for his work with hernia operations, including his anatomical and surgical treatise *Anatomisch-chirurgische Abhandlung über den Urspurng der Leistenbrüche* (Figure 3.63), published in 1806, which describes a number of structures relating to hernias. The following still bear his name: Hesselbach's fascia (the cribriform fascia, the part of the superficial fascia of the thigh that covers the fossa ovale, which is a defect in the anterior proximal thigh that allows the greater saphenous vein to enter the femoral vein), Hesselbach's hernia (a loop of bowel that herniates through the cribriform fascia), and Hesselbach's ligament (the interfoveolar ligament, a thickening in the transversalis fascia on the medial side of the deep inguinal ring, i.e. along the inferior epigastric vessels, connected superiorly to the transverses abdominis muscle and inferiorly to the inguinal ligament). The work also included comments regarding what would come to be known as the inguinal triangle. In addition,

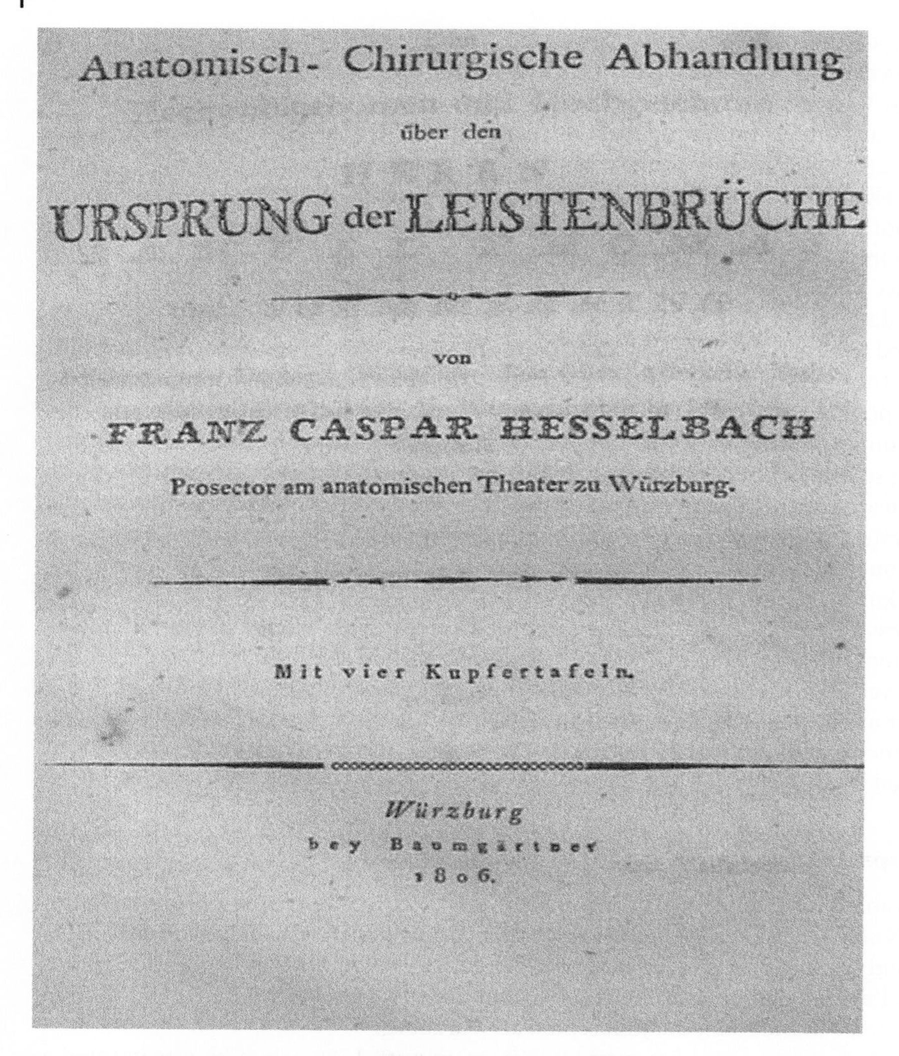
Anatomisch- Chirurgische Abhandlung

über den

URSPRUNG der LEISTENBRÜCHE

von

FRANZ CASPAR HESSELBACH

Prosector am anatomischen Theater zu Würzburg.

Mit vier Kupfertafeln.

Würzburg

bey Baumgärtner

1806.

Figure 3.63 Title page from Franz Kaspar Hesselbach's 1806 anatomic and surgical treatise regarding the origin of groin hernias (*Anatomisch-chirurgische Abhandlung über den Urspurng der Leistenbrüche*).

Hesselbach described the femoral hernia in 1798 and distinguished internal (direct) and external (indirect) inguinal hernias in 1810 (Figure 3.64).

The triangular area of the anterior abdominal wall that bears Hesselbach's name is usually described as being bordered by the deep epigastric vessels laterally, the lateral margin of the rectus abdominis medially, and the inguinal ligament. However, Hesselbach's definition differed somewhat from this one. The discrepancy apparently occurred in Quain's 1828 text of anatomy, where Hesselbach's name was first attached to the description given here. Hesselbach's original depiction of this geometric area in *Anatomisch-chirurgische Abhandlung* used the superior ramus of the pubis (not the inguinal ligament) inferiorly, the lateral edge of rectus abdominis medially, and the femoral vein and inferior epigastric artery laterally. Later, however, the inguinal ligament replaced the pubic bone as a component of the triangle. Hesselbach pointed out that the two lower corners were important. The medial corner, between the margin

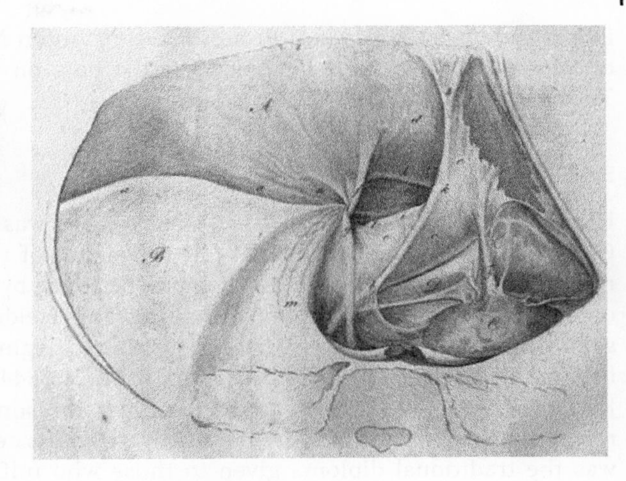

Figure 3.64 Drawing from Hesselbach's *Anatomisch-chirurgische Abhandlung*, depicting the triangular region that today bears his name. The illustration is titled by Hesselbach an "Internal view of the groin region of the adult, left side." Note the inferior epigastric artery (f), the anterior abdominal wall (A), the internal ring (n), the ilium (B), and the vas deferens (J).

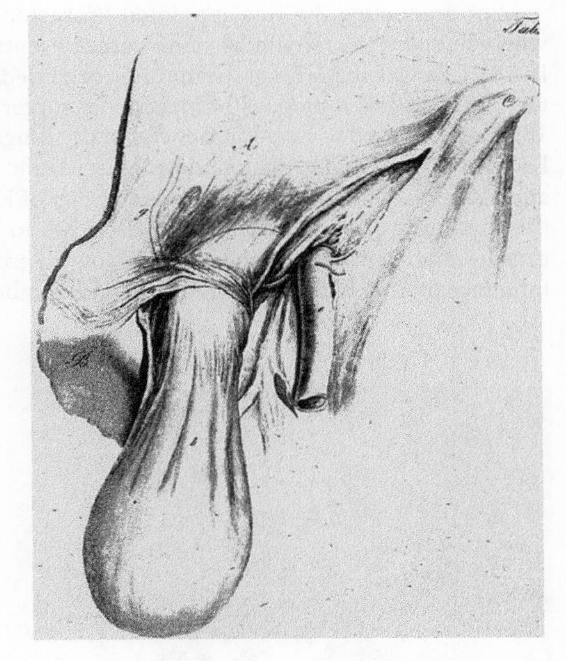

Figure 3.65 Drawing from Hesselbach's *Anatomisch-chirurgische* Abhandlung, demonstrating a large indirect hernia sac (b), entitled "External Groin Hernia in an 88-Year-Old male on the Left Side." Note the inferior epigastric artery (g) in the background. Also note the anterior superior iliac spine (a), pubic bone (B), and external oblique aponeurosis (A).

of the rectus abdominis and the pubic bone, he described as the site of direct inguinal hernias, while the lateral corner, between the pubic ramus and the femoral vein, is the site of the neck of a femoral hernia (Figure 3.65). This area of anatomy is of considerable interest and importance to the surgeon. Hesselbach is often given the distinction of being the first to define the difference between direct and indirect inguinal herniae. However, in 1804, two years before Hesselbach's publication on the subject, Cooper stated that Henry Cline, a London surgeon, was making these observations in his lectures at St. Thomas' and Guy's Hospitals in 1777.

After the takeover of Würzburg by Bavaria in the early 1800s, the title Doctor of Medicine was conferred on Hesselbach by the Würzburg medical faculty in recognition of his support of and contributions to the anatomical institute. He taught surgery at the Juliusspital during the illness and after the death of Professor Georg Anton Markard (1775–1816), who was Chair of Surgery and Oberwundarzt. He left the post on the arrival of Professor Cajetan von Textor (1782–1860).

3.5.3 Hubert von Luschka

Hubert von Luschka (1820–1875) (Figure 3.66) was born Hubert Luschka in Konstanz, Germany, the son of a foreseter and the eighth of twelve brothers. He wanted to study medicine, but due to poor family finances he began by studying pharmacology at two of the oldest universities in Germany, Freiburg and Heidelberg, in 1841. Later, his brothers supported him financially so that he was able to begin his study of medicine. He continued his studies in Heidelberg from 1843 to 1844. In 1844, he passed the state examination in Karlsruhe, practicing medicine and surgery for some time in Meersburg. In 1845, he received his doctorate from Freiburg University. His degree was that of *privat docent*, which was the traditional diploma given to those who fulfilled the criteria for becoming both teachers and researchers.

After graduating, he was appointed Prosector in Anatomy at the University of Tübingen, where he spent the remainder of his career, eventually becoming chairman of the department. In 1845, he was made an assistant to Georg Friedrich Ludwig Stromeyer (1804–1876), a military surgeon and a pioneer of orthopedic surgery during the First War of Schleswig (1848–1851), who served as Surgeon General of the Kingdom of Hanover in the Austro-Prussian War. Early in his career, Luschka undertook scientific travels to Paris, Vienna, and northern Italy, and also practiced medicine in his native city of Konstanz. He followed his renowned teacher, the anatomist Friedrich Arnold (1803–1890), to Tübingen from 1848 to 1849, during which time he studied osteology and anthropology in great detail. It was Arnold who had the greatest influence on Luschka's career. (Arnold is remembered for Arnold's nerve, the auricular branch

Figure 3.66 Hubert von Luschka (1820–1875). (*Source:* Courtesy of University of Tubingen.)

Figure 3.67 The Anatomical Institute at Tübingen during von Luschka's day. (*Source:* Courtesy of the University of Tübingen.)

of the vagus nerve; his son Julius [1835–1915] was a pathologist, remembered for his descriptions of hindbrain hernias [Arnold–Chiari malformations].) In 1849, Luschka was invited to Tübingen as extraordinary professor, and following Arnold's retirement, he was appointed *ordinaries* (1853) and director (1855) of the Anatomical Institute (Figure 3.67). Today, he is considered among the leading anatomists of the nineteenth century.

3.5.3.1 Teaching and Anatomical Research

Luschka's lively lectures made him a favorite among his students, and his three-volume textbook of clinical anatomy was considered the gold standard for medical students of his day. His work focused on the aspects of anatomy pertinent to medicine and surgery, and he coined the term "clinical anatomy." His *Anatomie des Menschen in Rücksicht auf das Bedürfnis der praktischen Heilkunde* ("Human Anatomy in Consideration of the Needs of Practical Medicine," 1862–1869) aimed to provide such a link. He was one of the first to conduct detailed research on normal cadavers rather than on pathological specimens, and he developed the process of making frozen sections of anatomical specimens for study. Luschka had a special interest in the pelvic girdle and, specifically, the sacroiliac joints; he described the anomalous coracoclavicular joint. He went against the prevailing opinion of his day when he suggested that the sacroiliac joint was a true joint on the basis of his observations that it contained a synovial membrane and was lined with cartilage. Also remarkable for the time were his criticisms of Darwin's theory of evolution. Luschka investigated the larynx and its contribution to voice production and described the musculature of cleft palates. He identified the coccygeal body in 1860. He made detailed histological observations of the carotid body and, in 1862, was the first to describe a carotid body tumor.

3.5.3.2 Contributions to Neuroanatomy

Luschka's best known contribution to neurosurgery is his discovery of the foramen of Luschka, or lateral aperture of the fourth ventricle (Figure 3.68). In 1855, in his *Die Adergeflechte des menschlichen Gehirns* ("The Choroid Plexus of the Brain"), he confirmed that the foramen of Magendie existed and described a lateral outlet of the fourth ventricle, stating:

> On either side, the outer angles of the fourth ventricle assume the form of a gutter leading outside, whereby the latter portion of the choroid plexus passes outside of the fourth ventricle, while the arachnoidea stretches freely over the place in question. The fourth ventricle therefore by its exterior angles, has an open communication with the subarachnoid space. The hiatus, where the pia mater passes over into the ependyma, meanwhile, through the lateral portion of the choroid plexus of the fourth ventricle, is so much contracted, that only a narrow slit remains, which is, however, sufficiently ample to furnish an entrance here, under the arachnoidea, to a liquid which is injected from below with a tubulus, the tela choroidea inferior still being fully preserved. This anatomical fact is of considerable importance, because in some animals, for instance the horse, the lower extremity of the fourth ventricle is completely closed up, in which case the exterior angles of this cavity are the only means whereby a communication may be effected between the fourth ventricle and the subarachnoid space.

Interestingly, Fisher points out that the existence of the foramina of Luschka was doubted by many anatomists even after its confirmation by Key and Retzius in 1872.

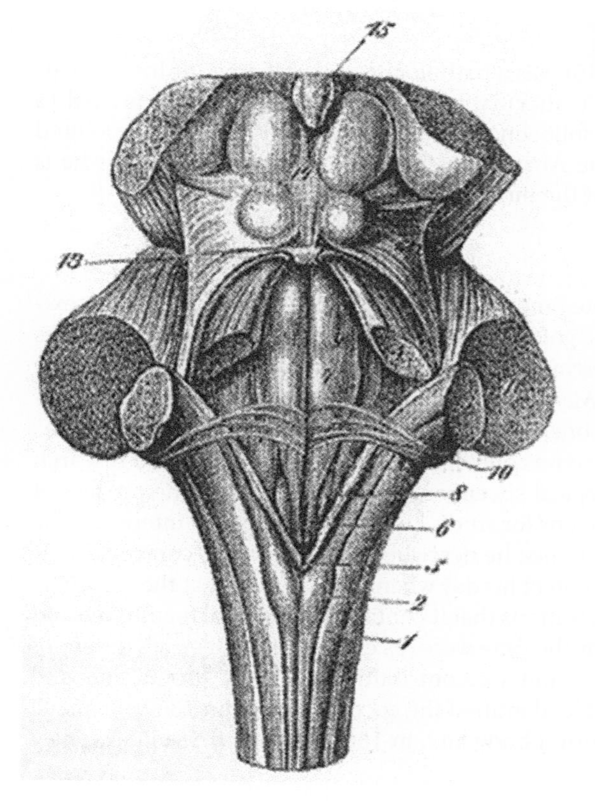

Figure 3.68 Illustration demonstrating the lateral apertures (foramina of Luschka) of the fourth ventricle.

Luschka had a keen interest in hydrocephalus and contributed to our understanding of the source of cerebrospinal fluid. He stated:

> it cannot escape unbiased observation that the most important sources for the formation of the cerebrospinal fluid are the choroid plexuses; however, as may be concluded from the metamorphosis of its epithelium, the ependyma, and outer vascular membrane (pia mater) also participate in it.

Luschka's landmark publication on this topic, *Die Adergeflechte*, yielded many new discoveries, several of which were based on research performed on the bodies of healthy criminals. He found communication between the subarachnoid spaces of the head and both the ventricles of the brain and the subarachnoid space of the spinal cord, although such connections had been denied by Virchow (1821–1902) and Kölliker (1817–1905) at the University of Würzburg. He demonstrated them by suspending corpses by the feet for several hours and showing that the subarachoid space was full and protuberant, and very much enlarged by the fluid collected under it. A tube was introduced into the cervical subarachnoid space of other specimens and ink mixed with water was injected. These experiments illustrated the retrograde flow of fluid and air from the subarachnoid space into the fourth ventricle, Sylvian aqueduct, third ventricle, and, finally, lateral ventricle. Air was also injected into the spinal subarachnoid space to illustrate its passage up into the ventricular system in specimens held upside down. Basilar skull fractures were imitated by trephining the sphenoid through the nostrils, mimicking the egress of cerebrospinal fluid from the nose following trauma. Luschka found that the central canal of the spinal cord may exist in the adult but is obstructed with cellular debris and amyloid corpuscles.

Luschka identified connections between the glossopharyngeal and vagus nerves and hypothesized that pressure on the hypoglossal nerve from a dilated vertebral artery could result in the loss of function of the tongue. He wrote on the phrenic, dural, and cardiac nerves. He first described the intracranial venous sinus lodged in the petro-squamous suture (Luschka's sinus) in his *Anatomie des Menschen* ("Human Anatomy"), as well as describing the drainage of the cavernous sinus, the anatomy of the inferior petrosal sinus, and the portal system and blood supply of the pituitary gland. On the latter, he stated:

> a very large supply of blood is dispatched towards the anterior lobe of the gland through slender twigs of the internal carotid, which originate from it chiefly during its passage through the cavernous sinuses; some blood also is derived to the gland from the delicate network which belongs to the pia mater of the infundibulum.

In 1860, Luschka built on Rathke's 1834 descriptions of the development of the anterior lobe of the pituitary gland, observing that remnants of this part of the sella could be found anywhere along its pathway from the pharynx to the hypophysis. He made detailed observations and discovered the sinuvertebral or recurrent meningeal nerves (recurrent nerve of Luschka) that innervate the spinal dura, posterior longitudinal ligament, and posterior annulus. He also discovered the uncovertebral joints (Luschka's joints) (Figure 3.69); he called them the "half joints" and described them as follows:

> The cavity is not closed by the synovial membrane, there are not even blood vessels. Within the covering of the cavity can be found no epithelium, which is usually found as a cartilaginous mass. However, it is possible for abnormal development to occur, creating a true synovial membrane with epithelium on the inner side, resulting in a transformation from half to complete joints.

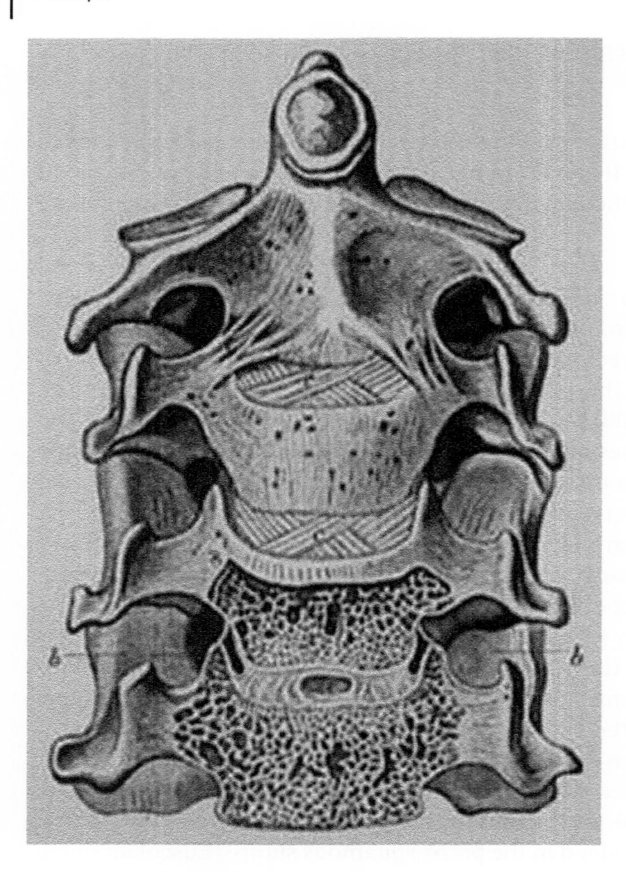

Figure 3.69 Illustration of the uncovertebral joints (b), as discovered by Luschka.

Additionally, he described what are now known as tears of the annulus. Also in regard to the spine, he found that individuals lose height when they stand and that the height of an individual decreases with age. Interestingly, Luschka disputed the findings of the noted neuroanatomist Louis Pierre Gratiolet (1815–1865) that the Khoikhoi, a native people of southwestern Africa, had lesser development of their frontal lobes, and particularly the superior frontal convolutions, as compared to Europeans.

In 1865, Luschka was ennobled for his works, and from that time onwards he used *von* in his name. On July 11, 1874, he received the degree of doctor *honoris causa* for his significant contributions to the anatomical sciences. He died less than a year later (March 1, 1875), at the age of 55, in Tübingen.

3.5.4 Adolf Wallenberg

Adolf Wallenberg (1862–1949) (Figure 3.70) was born in Stargard, outside the West Prussian city of Danzig (modern-day Gdansk, in Poland). Wallenberg's father Samuel was a physician, and his grandfather a rabbi. His father died when he was only six. At the age of 20, Wallenberg went to Heidelberg and subsequently to Leipzig to study medicine under Wilhelm Erb (1840–1921), Adolf Strumpell (1853–1925), and Carl Weigert (1845–1904). He graduated with a thesis on poliomyelitis and received his doctorate from Leipzig University in 1886. Returning to his homeland, Wallenberg spent most of his working life as an internist in the Department of Medicine at the general hospital (Städtisches Krankenhaus) of Danzig.

Figure 3.70 Adolf Wallenberg (1862–1949).

From 1886 to 1888, he was an assistant there. Refusing offers of several chairs, Wallenberg remained at Danzig and even worked as a general practitioner in the city. Becoming a reputed physician, he was appointed chief physician to the city hospital in 1907, and became a professor in 1910. Later in life, he was involved in a serious accident when the horse harnessed to the carriage in which he was riding bolted. It is said that he suffered a skull base fracture, diplopia, and anosmia. Wallenberg believed that this traumatic event caused his personality to become more compulsive.

Wallenberg began his neuroanatomical studies in a primitive laboratory in his own apartment, which he later moved to his hospital. He studied both humans and animals, wrote on the comparative anatomy of the brainstem, outlined the anatomy of the trigeminal lemniscus, and described the olfactory system and its role in the assessment, recognition, and taste of foods. He described the connections between the basal forebrain and intermediate brain centers (tractus fronto-quintalis, tractus diencephalo-quintalis, and pars baso-quintalis fasciculi) in birds and fish. Using a clinico-anatomical methodology based on his previous research findings, he made precise notes of his patients' symptoms and performed careful neurological examinations. He published on the alterations of the central nervous system in children with cerebral palsy and on diseases of the brainstem.

In 1895, Wallenberg reported his observations on infarction of the lateral medulla in a 37-page paper titled "Akute Bulbaraffektion (Embolie der Art. cerebellar post. Inf. sinistr?)." Based on the anatomical studies of a contemporary French physician, Henri Duret (1849–1921), he hypothesized that the lesion might be in the lateral fossette of the medulla oblongata supplied by the posterior inferior cerebellar artery, although he was unable to provide sufficient evidence. In a second paper, Wallenberg demonstrated at autopsy that there was complete occlusion of the left posterior inferior cerebellar artery in a patient with symptoms of lateral medullary syndrome. Wallenberg's classic report of his first patient with lateral medullary syndrome was as follows:

> A 38-year-old man, with poor vision caused by a pre-existing ocular condition suffered an attack of vertigo without loss of consciousness. At the same time, he developed pain and hyperesthesia on the left side of the face and body, hypoesthesia of the right half of the face, and loss of pain and temperature sensitivity in the right extremities and the right half of the torso, with retention of the sense of touch.

Contrary to the thoughts of Charles Foix (1882–1927), an eminent vascular neurologist, Wallenberg considered vascularization of the lateral fossette to be multi-arterial. Ludwig Edinger (1855–1918), the famous German neuroanatomist, became aware of Wallenberg's works and invited him to attend meetings of the German Neurological Society. From that point on, Wallenberg and Edinger were friends and scientific collaborators. Wallenberg published 67 papers, four in collaboration with Edinger. Most of his papers were published in the *Archiv für Psychiatrie und Nervenkrankheiten* (currently *European Archives of Psychiatry and Clinical Neuroscience*) and *Deutsche Gesellschaft für Neurochirurgie* (currently *Journal of Neurology*). Together, Wallenberg and Edinger described the avian brain, and *Jahresberichte über die Leistungen auf dem Gebiete der Anatomie des Zentralnervensystems* ("Annual Reports on the Achievements in the Areas of the Anatomy of the Central Nervous System") was published following their collaboration. In 1929, Wallenberg received the Erb award for his achievements in the field of the anatomy, physiology, and pathology of the nervous system.

Unfortunately, Wallenberg's life ran concurrent with World Wars I and II. During World War I, he served as a military physician, which temporarily stopped his research. He began again at the end of the war, and was one of the organizers of the German Neurological Society Meeting in Danzig. Even after his retirement, Wallenberg continued his investigations in the basement of his hospital. In 1938, he ended his research because of the intolerable pressure of the Nazi regime. As a Jewish scholar, he was banned from working and from publishing in journals. Although hesitant to leave his homeland, he fled with his wife to England, via Holland and Belgium. The neuroanatomist Wilfrid Le Gros Clark (1895–1971) arranged for Wallenberg to go to Oxford, where he stayed for nearly four years. There, he worked with Le Gros Clark on histological specimens at the Anatomical Institute. In 1943, after receiving an American visa, he immigrated to the United States and settled 50 miles from Chicago.

Wallenberg spent the rest of his life in an apartment in the Manteno State Hospital in Illinois. He learned English, lectured in the hospital, and was even consulted in difficult clinical cases. On April 10, 1949, he died at the age of 86 from ischemic heart disease. Wallenberg's life and career were overshadowed by war and the unstable political situation. However, this early neuroscientist maintained his mental and physical integrity and continued his investigations and mentorship abroad. He was known as being an unusually modest, warm-hearted man who worked with endless patience. His daughter, Marianne Wallenberg-Chermak, became a physician and wrote about her father's life in *Grosse Nervenärzte* (1963). Scientifically, Wallenberg represented a triumph for the clinico-anatomical methodology in the early twentieth century, at a time when scientists looked for strict anatomical and pathological correlations of disease entities. The Adolf Wallenberg Prize has been offered by the German Neurological Society since 1980 and is awarded to outstanding researchers in the fields of cerebrovascular disease, cerebral blood flow, and cerebral metabolism.

3.5.5 Korbinian Brodmann

Korbinian Brodmann (1868–1918) (Figure 3.71) was born to Josef and Sophie Brodmann in Liggersdorf, Hohenzollern. He was educated in his home town until the age of 12, when he studied the humanities in Überlingen. He attended Gymnasium in Sigmaringen and ultimately graduated from Konstanz. He began his medical career in 1889 by studying medicine at the Universities of Munich, Wurzburg, Berlin, and Freiburg. On February 21, 1895, he received his Approbation (medical degree), allowing him to practice across Germany. His initial intention was to establish a general practice in the Black Forest, but in the year following his licensing, he decided to specialize by studying psychiatry at the Universities of Lausanne (in Switzerland) and Munich. While in Munich, he also worked in psychiatry at the University Pediatric Clinic and Polyclinic under Dr. Hubert von Grashey.

Brodmann spent the summer of 1896 recuperating from a bout of diphtheria and working as an assistant at the private Neurological Clinic, focused on nervous diseases, in Alexanderbad im Fichtelgebirge, northern Bavaria. The clinic was directed by Oskar Vogt, who described Brodmann as having "broad scientific interests, a good gift of observation, and great diligence in widening his knowledge." Following his recovery, Brodmann set out to prepare himself for a career in research by studying pathology in Leipzig. He received his Promotion in 1898 following

Figure 3.71 Signed photograph of Korbinian Brodmann (1868–1918).

the successful defense of his thesis regarding chronic ependymal sclerosis. After receiving this degree, he worked under Otto Binswager in Jena at the Univerity Psychiatric Clinic.

In 1900, Brodmann transferred to Frankfurt to work at the Municipal Mental Asylum. During his brief stint there, he met Alois Alzheimer (1864–1915), and this relationship proved crucial in sparking his interest in the neuroanatomical foundations of neurology and psychiatry. He was also influenced by such notable figures as Karl Weigert (1845–1904) and Franz Nissl (1860–1919), who were recognized for their cellular staining techniques, as well as Ludwig Edinger (1855–1918), who identified the oculomotor parasympathetic nucleus. He only remained in Frankfurt until 1901, but this period proved essential in shaping his future successes.

In the same year that Brodmann had received his Promotion in Leipzig (1898), Oskar Vogt (1870–1950) had begun creating his multidisciplinary brain research institute, the Kaiser-Wilhelm-Institut für Hirnforschung in Berlin-Buch (Neurobiologisches Institut), with divisions for neuroanatomy, neurohistology, neurophysiology, neurochemistry, and genetics. Remembering Brodmann's gift for keen observation from their relationship in Alexanderbad, Vogt invited him to work with him and his wife Cécile Vogt-Mugnier (1875–1962) in Berlin. When Brodmann arrived, Vogt suggested he undertake a methodical comparative study of the cells of the cerebral cortex. The relationships he formed in Frankfurt proved useful at the Neurobiological Institute, as he used new histological staining techniques developed by Wiegert and Nissl to organize the cerebral cortex topographically.

While at the Neurobiological Institute, Brodmann began editing the *Journal für Psychologie und Neurologie* (later the *Journal für Hirnforschung*, and in 1994 the *Journal of Brain Research*). As Vogt and Vogt-Mugnier performed research in myeloarchitectonics, a field focused on the parcellation and layering of nerve fibers, their work complemented Brodmann's (Figure 3.72). In April 1903, Brodmann made a joint presentation with them at the meeting of the German Psychiatric Society in Jena. With his presentation, Brodmann stated "functional localization without the lead of anatomy is utterly impossible."

Figure 3.72 Photograph of Brodmann and several colleagues. From left to right: Brodmann, Cécile Vogt, Louise Bosse, Oskar Vogt, Max Lewandowski, and Max Borchert.

Between 1903 and 1908, Brodmann published a series of seven communications on comparative mammalian cytoarchitectonics (studying 64 different species). He defined cytoarchitectonics as "the localization of the individual histological elements, their layering, and their parcellation in the adult brain." The sixth and best-known of these was published in 1908, and contained the famous map in which he organized the unique histological regions of the human cortex. These works formed the basis for the publication of his 1909 monograph *Vergleichende Lokalisationslehre der Grosshirnrinde in ihren Prinzipien dargestellt auf Grund des Zellenbaues* ("The Principles of Comparative Localization in the Cerebral Cortex Based on Cytoarchitectonics") (Figures 3.73 and 3.74).

In his map of the human cortex, Brodmann identified 47 histologically distinct regions using novel staining techniques introduced by Franz Nissl, and in his study of primates he described 52 regions. At the time, there were three schools of thought on the mapping of the brain, to which such men as Meynert, Betz, Ferrier, Kaes, Bechterew, Edinger, Flechsig, Lewis, Clarke, and Hammarberg had contributed. Some sought to localize function on the basis of the presence of individual histological elements. Others believed that each cortical layer was associated

Figure 3.73 Title page of Brodmann's 1909 *Vergleichende Lokalisationslehre der Grosshirnrinde in ihren Prinzipien dargestellt auf Grund des Zellenbaues.*

Vergleichende Lokalisationslehre der Großhirnrinde

in ihren Prinzipien dargestellt auf Grund des Zellenbaues

Von

Dr. K. Brodmann

Assistenten am neurobiologischen Laboratorium der Universität zu Berlin.

Mit 150 Abbildungen im Text.

Leipzig
Verlag von Johann Ambrosius Barth
1909

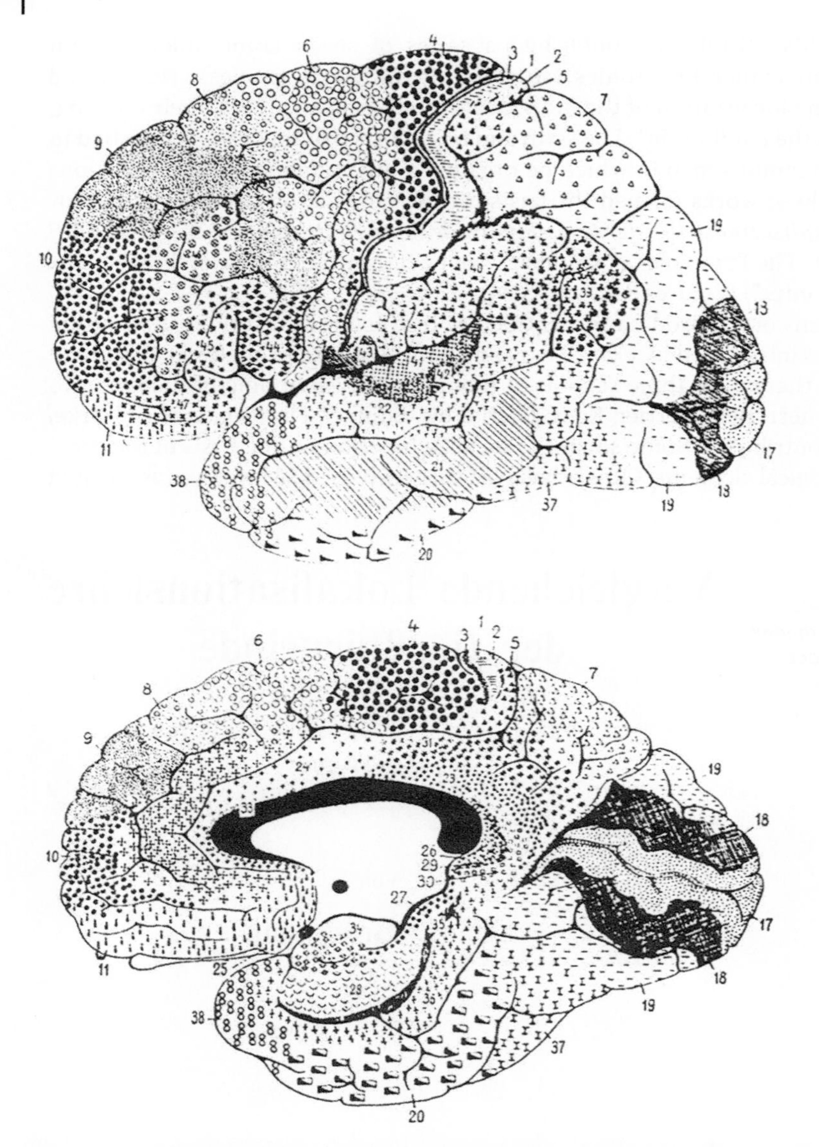

Figure 3.74 Illustration from Brodmann's *Vergleichende Lokalisationslehre* showing the cerebral hemispheres of the human, with Brodmann's "areas" applied.

with a specific function. The third school of thought, to which Brodmann belonged, was something of a hybrid between the two. It claimed that regions of the brain that contained similar structures in terms of both layering and cell type could produce specific functions. Brodmann believed there was little evidence that cell type determined function, and was adamantly opposed to assigning functions to specific layers. He said:

> These [associations of individual layers with specific functions], and all similar expressions that one encounters repeatedly today, especially in the psychiatric and neurological literature, are utterly devoid of any factual basis; they are purely arbitrary fictions and only destined to cause confusion in uncertain minds.

Brodmann's approach to functionality has affected some scholars' assessments of his work. In 1905, three years before Brodmann presented his main paper on the human cortex, Alfred Walter Campbell (1868–1937) published a monograph titled *Histological Studies on the Localisation of Cerebral Function,* in which he described 17 regions of the cortex, each tied to specific functions such as vision, sensation, and olfaction. Campbell's works are little known to most physicians because Brodmann's all but erased them from history. However, some believe that *Histological Studies,* which emphasized function far more heavily than did Brodmann, was the pioneering work on cytoarchitectonics.

While working with the Vogts, Brodmann submitted a habilitation regarding the cytoarchitectural division of the prosimian cortex to the medical faculty in Berlin, at the suggestion of Emil Kraepelin (1856–1926). Habilitation was the academic qualification conferred on an individual who had received a doctoral degree and submitted an additional thesis that allowed him or her to become a Privatdozent, which was requisite for becoming a tenured university professor. Brodmann's habilitation was inexplicably rejected by Berlin's faculty, which prevented him from acquiring a secure professorship.

In 1910, frustrated by the rejection of his habilitation thesis and in search of more stable work, Brodmann traveled to Tübingen, where he worked as a physician. In his free time, he set up his own brain anatomy research laboratory (Figures 3.75 and 3.76). In 1913, he was appointed as a professor to the Faculty of Medicine in Tübingen on the recommendation of Robert Gaupp (1870–1953) and was warmly welcomed by its faculty, including famed anatomist August von Froriep (1849–1917). The outbreak of World War I interrupted Brodmann's research, and he volunteered in a mental hospital in Tübingen. In February 1916, he was awarded the William Cross with swords by the King of Wurttemberg, in recognition of his service.

Despite his welcome in Tübingen by Gaupp and others, Brodmann was still unable to obtain a position with the economic security that would allow him the opportunity for research. It was not until May 1, 1916 that he was able to do so, when Berthold Pfeiffer offered him the Prosectorship at the Mental Asylum near Halle an der Saale. Walther Spielmeyer, who worked with Brodmann at the Psychiatric Research Institute in Munich, applauded Pfieffer for recognizing Brodmann's abilities and need for security. Perhaps encouraged by his newfound security, Brodmann married Margarete Francke on April 3, 1917. Their first child, Isle, was born in 1918.

Brodmann moved to Munich in that same year to take charge of the Department of Topographical Anatomy at the prestigious German Psychiatric Research Institute, run by Emil Kraepelin. Kraepelin had formed a powerful collaboration by inviting both Brodmann and Nissl, because he saw the importance of neurohistology and cytoarchitectonics in the future of

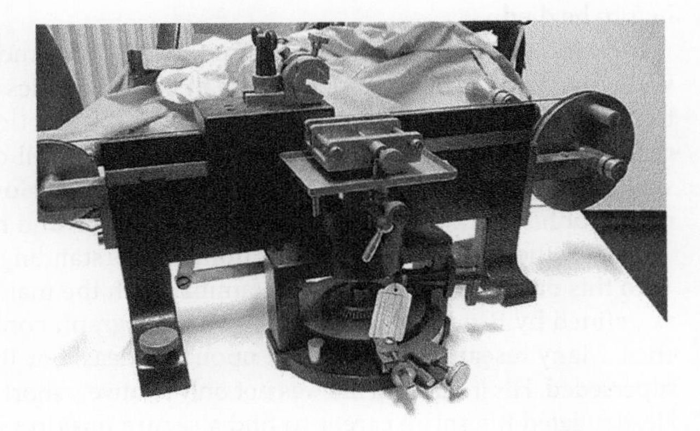

Figure 3.75 Microtome used by Brodmann in making sections through the cerebral cortex of animals and humans.

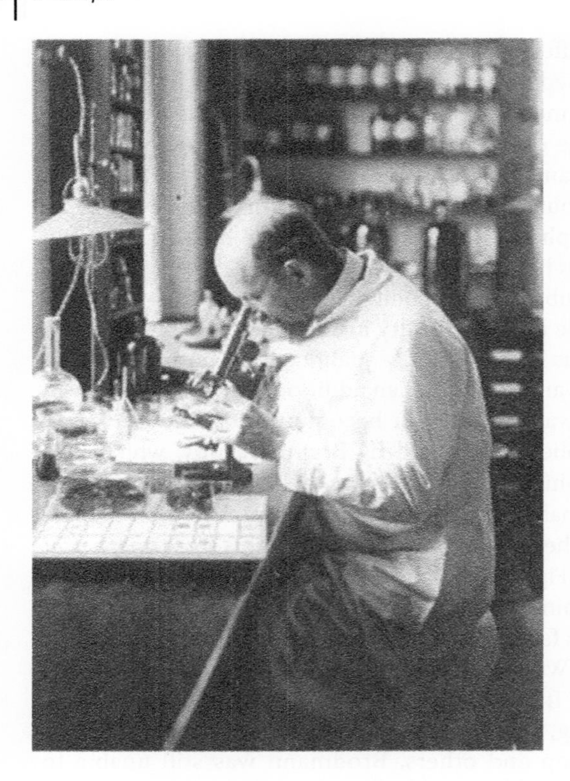

Figure 3.76 Brodmann at work at his microscope.

neuroanatomical research. Unfortunately, just as this group came together, Brodmann died of sepsis on August 22, 1918, only five days after falling ill. Oskar Vogt said of Brodmann:

> just at the moment when he had begun to live a very happy family life and when, after years of interruption because of war work, he was able to take up his research activities again in independent and distinguished circumstances, just at the moment when his friends were looking forward to a new era of successful research from him, a devastating infection snatched him away.

An intense worker, Brodmann was noted to be making writing motions with his finger just before he died.

Even though Brodmann died before the age of 50, most physicians are still familiar with his work, particularly his map of the human cortex. Critics have claimed too much credit is given to Brodmann, saying his work lacked a focus on functionality. Consequently, it has been suggested that the lesser-known Alfred Walter Campbell deserves credit for producing the first significant work in the field of cytoarchitectonics. However, despite the strictly anatomical nature of his work, Brodmann did value function, and he hoped a topographical map like his would lay the foundation for a functional understanding of the brain.

To this day, most physicians are familiar with the majority of the areas of the cerebral cortex as defined by Brodmann, and his 1909 monograph continues to guide the study of neuroscience. Many researchers have built upon his ideas, but Brodmann's original work has yet to be superseded. His influential life was not only relatively short but also marred by under-appreciation. He struggled his entire career to find a secure position suitable for a researcher of his stature.

Despite these obstacles, he managed to contribute work so significant that it has become an integral part of today's medical education and neuroscience research. However, for some reason, there is very little written in English regarding the life of this man whose expertise spanned neurology, psychiatry, physiology, zoology, and anthropology.

3.6 Greece

3.6.1 Aristotle

Aristotle (384–322 BCE) (Figures 3.77 and 3.78), a pupil of Plato (427–347 BCE), the tutor of Alexander the Great, the predecessor of Peripatetics, and the father of comparative anatomy, was an absolute cardiocentrist, believing that the heart was the most important organ in the body and the first to form in the embryo. He also was the first to make clear his scientific observations of the cardiovascular system in animals and in human corpses. In combination with his metaphysical approach, his studies were used to explain the functionality of each part of the heart for at least half a millennium. During this time, the teachings of Hippocrates (460–370 BCE) and Galen (103–201 CE) were widespread. According to Aristotle, the emotions of pleasure and pain, and all other sensations, had their origins and terminations in the heart, which is centrally located and thus reaches every other body part. He also believed the heart was the origin of the blood.

Descriptions of the cardiovascular system are found throughout Aristotle's writings, but are mostly concentrated in the two treatises *Historia Animalium* ("The History of Animals") and *De Partibus Animalium* ("The Parts of Animals"). During the past two millennia, many efforts

Figure 3.77 The School of Athens: Plato (left) and Aristotle (right). A fresco by Raphael Sanzio, an Italian painter and architect.

Figure 3.78 Marble bust of Aristotle. Roman copy after a Greek bronze original by Lysippus c. 330 BCE. The alabaster mantle is modern. (*Source:* From Wikimedia Commons, https://commons.wikimedia.org/wiki/File:Aristotle_Altemps_Inv8575.jpg.)

have been made to understand Aristotle's model of the heart and veins, and conflicting interpretations have been put forth. Some have thought his model was simply an error (e.g., the embryologist Georges Cuvier, 1769–1832), that it was numerological mysticism or metaphysical doctrine (e.g., the biologist D'Arcy Wentworth Thompson, 1860–1948), or that it was based upon the principle of the mean; alternatively, others have beleived it an excellent and accurate early description of the cardiovascular system.

Aristotle was born in 384 BCE in Stagira, a Greek colony and seaport on the coast of Thrace, to a family of physicians. His father Nichomachus was a physician to the court of King Amyntas II of Macedonia. He died when Aristotle was still a boy. At the age of 17, and through the efforts of his guardian, Proxenus, Aristotle moved to Athens, the intellectual center of the time, to study under Plato. For nearly 20 years, he attended Plato's lectures at his academy, and eventually began to lecture himself, while gradually expressing his divergence from Plato's views. Whereas Plato believed that reality existed in ideas, knowable only through reflection and inspiration, Aristotle saw ultimate reality in physical objects, knowable through the experience of the five senses.

With Plato's death in 347 BCE, Aristotle accepted the invitation of Hermeas, the ruler of Atarneus and Assos in Mysia, to attend his court. He stayed there three years, founded a school, and married the niece and adopted daughter of the king, Pythias, who bore him a daughter. When Hermeas was overtaken by the Persians in 345 BCE, Aristotle went to Mytilene on the island of Lesbos, where he studied biology and other natural sciences. In 343 BCE, he was invited by King Philip II (King Amyntas' son) of Macedonia to become the tutor to his 13-year-old son, Alexander. He was deeply appreaciated at the Macedonian court, and reportedly was supplied with thousands of slaves to collect specimens for his investigations in the natural sciences. In 338 BCE, King Philip defeated the league of Athenians and Thebans and conquered Greece. With his assassination in 336 BCE, Alexander succeeded to the throne.

Following Alexander's departure for the Asiatic campaign of 335 BCE, Aristotle returned to Athens. There, together with his student Theophrastus of Efesos (370–287 BCE), he set up a school known as the Lyceum or the Peripatetic School (Aristotle had a habit of walking around the school garden while discoursing, so his followers were later referred to as Peripatetics; that is, "to walk about"). His morning lectures were for advanced students and his evening discourses for the interested general public. While in Athens, his wife Pythias died. Aristotle then married Herpyllis of Stagira, who bore him a son named Nicomachus, after his father.

When Alexander the Great died in 323 BCE, the pro-Macedonian government in Athens was overthrown and a charge of impiety was levelled against Aristotle by the hierophant, Eurymedon. Fleeing to his mother's homeland, Chalcis, on the island of Euboea, Aristotle died of a stomach illness in 322 BCE.

Aristotle was a one-man university, not only to his fellow Athenians, but also to Arabs, Jews, and medieval Eurpeans. According to Will Durant, no other philosopher has contributed so much to the enlightenment of the world. Although he was not a physician, his contributions to medicine were immense. It is said that he dissected over 50 different species, founding the science of comparative anatomy. Interestingly, he arranged various zoological forms on the basis of their increasing perfection, extending from lower to higher animals. Aristotle's life reflects his philosophy that "The ideal man bears the accidents of life with dignity and grace, making the best of circumstances."

3.6.1.1 Aristotle's Scientific Works

After Aristotle's death, his writings were kept by Theophrastus, who had led the Peripatetic School in Athens. Following the capture of Athens by Sulla in 86 BCE, these treatises were brought to Rome, where they gained the attention of scholars. A new edition was prepared, the *Corpus Aristotelicum* ("Aristotelian Corpus"); this is the basis of our present knowledge of Aristotle's works.

The works on the natural sciences were divided into eight major sections: (i) on youth, old age, life and death, and respiration; (ii) on the breath; (iii) on the history of animals; (iv) on the parts of animals; (v) on the movements of animals; (vi) on the progression of animals; (vii) on the generations of animals; and (viii) problems. The first English translation of the *Corpus Aristotelicum*, known as "The Oxford Translation of Aristotle," which took almost half a century to finalize, was began by Benjamin Jowett in 1885 and completed by W.D. Ross of Oriel College in association with Balliol College. "The Revised Oxford Translation of the Complete Works of Aristotle" was later published by Jonathan Barnes in 1984, and included a more consistent translation, minimizing paraphrases and following the style of Aristotle's Greek, which according to the editor was tense, compact, and abrupt, with condensed arguments and dense thoughts.

3.6.1.2 Aristotle's Model of the Cardiovascular System

The middle cavity of Aristotle resembles in its character the left cardiac chamber of the Hippocratic *De Corde*. Galen considered these right and middle cavities as being parts of a single right ventricle. Contrarily, Vesalius regarded the left and middle cavities of Aristotle as a left ventricle subdivided by the anterior leaflet of the mitral valve.

With interpretation of the Aristotelian cardiovascular system, one must consider four elements. First, although an analogy has been perceived between different animals, Aristotle claimed that some degree of inconsistency or variation may be seen on dissection because all species are not necessarily equal in terms of their vasculature. Interestingly, he noted the size of body as being the main determinant for such variability. Second, in *Historia Animalium* and *De Partibus Animalium*, Aristotle applied two different means to delineate the cardiovascular system. While the dissection and experiment were basic for the former, a metaphysical account was predominant in the latter. For this reason, some have distinguished between Aristotelian cardiovascular morphology and cardiopulmonary physiology, noting that the functional integrity of the heart and lungs have been tightened in this model. Although in the Aristotelian legacy, all three cavities of the heart have passages to the lung, this acts only as a coolant for the heart (Aristotle did not mention that Plato had also said that the lung was for cooling the heart). Such a functional explanation of the structures, as Lonie declared, might be compatible with or become an inevitable consequence of the mechanistic approach to nature that passed from Plato to Aristotle and from the Stoics to Galen. Respiration, together with pulsation (a normal event) and palpitation (an abnormal phenomenon), was related to cardiac function in the Aristotelian doctrine. Third, the dates of *Historia Animalium* and *De Partibus Animalium* are generally obscure to us. If, in fact, *De Partibus Animalium* was composed before *Historia Animalium*, one may conclude that the Aristotelian experimentation was intended to prove or preserve a pre-set regimen of metaphysical thought. Although the subtle discrepancies between the two may validate his methodology by showing its liberty, Aristotle attempted to reach a coherent reconciliation between nature and logic by constructing a kind of metaphysical surveillance. Lastly, terminology should be taken as seriously as methodology, for the former has evolved over a period of three millennia. For instance, some have used the term "three-chambered" as an alternative for "triventricular" in designating the Aristotelian heart model, which would aid the layperson in realizing that the cavities of Aristotle are not necessarily cardiac ventricles.

Aristotle's methodology must be borne in mind in interpreting his cardiovascular model. For example, in the *Historia Animalium*, he noted that he had gained his information from dissections of emaciated animals. The method used to kill such animals may account for some differences observed between cavities. If they were strangled, then the vena cava and the right ventricle might have been dilated, being full of dark blood, while the left atrium and ventricle were drained into the distensible veins. This could have created the impression of a great vein as well as a large right cavity, in contrast to a medium-sized middle chamber and small left cavity. Also the designation of artery (meaning "air-containing" in Greek) by the post-Aristotelian observers may stem from the fact that arteries at autopsy are often empty of blood. Notably, it was presumably Alcmaeon (500–450 BCE) who noticed that in slaughtered animals, some vessels are empty of blood and others are not, and thus differentiated arteries from veins.

The middle cavity of Aristotle, for its connection with the aorta, is certainly the left ventricle. The left cavity most likely represents the left atrium or auricle, as it is connected to the lungs. The right cavity is morphologically the right ventricle, for its communication with the vena cavae (great vein). Additionally, Aristotle made note of a venous lake at the junction of the vena cavae and the right cavity, suggesting that the cavity was part of the vein. Here, he hesitated to morphologically attribute this venous lake to either the great vein or the right cavity, for it held the features of both, being both thin and a receptacle. Did Aristotle recognize the right atrium? If so, why did he not count it as a distinct cavity, and enumerate only three? In fact, the morphological account of the right atrium as a venous lake is explicit. Elsewhere in the *De Partibus Animalium*, for philosophical reasons, Aristotle acknowledges that the advantage of an odd cavity is that it provides a common source of pure and fine blood to both sides of the body (the other two cavities mainly supply the ipsilateral sides with either warm or cold blood). This, according to Thompson (a biologist and translator of *Historia Animalium* in Barnes' edition), represents a tradition or mysticism attached to the number three, which is probably in accordance with Plato's notion of the three corporeal faculties. Moreover, Harris (a medical historian) notes that, being based on the principle of the mean, Aristotle's odd cavity was seen as providing a blood intermediate to that of the other two. Thus, from a morphological viewpoint, Aristotle had successfully distinguished four chambers of the heart, but for metaphysical reasons, and perhaps to follow previous teachings, he decided to describe only three.

Interestingly, Aristotle criticized the teachers Diogenes of Apollonia, Synnesis, and Polybus for engaging in speculation and not making observations or conducting experiments. In Aristotle's cardiovascular model, there are only two blood vessels: the great and the lesser veins, corresponding to the inferior vena cava and aorta, respectively, both of which bifurcate in the abdomen after passing the kidneys. He noted the different natures of these vessels – the former were larger and membranous, while the latter were narrower and sinewy – and of the blood contained within. However, both sets of vessels are titled "phelbs" or "veins." Branches from the great vein and aorta accompany one another throughout the body. In addition, a designation of "sinew" is given to the branches from the aorta – ligaments inserting around the joints and nerves – the transection of which causes paresthesia. In this doctrine, the parts of the heart cannot be considered in isolation.

Aristotle had no notion of blood circulation in the modern sense. On his model, the heart only generates blood, which is conveyed to the periphery by two blood vessels in a manner analogous to that of irrigation channels. His notion of the rostral part of the great vein, which is above the heart, fits with the characteristics of the superior vena cava. Being divided into two major trunks (brachiocephalic veins) for the upper limbs and neck, this rostral segment also gives rise to the azygos vein and, mistakenly, the pulmonary arteries.

Knowledge of the cardiovascular system was expanded in the post-Aristotelian era by Hippocrates, Herophilus of Chalcedon (335–280 BCE), Eristratus of Chios (310–250 BCE), and Galen of Pergamum, and in medieval times by Avicenna (980–1037 CE) and Ibn Nafis (1213–1288 CE), who observed blood circulation in the body. Many of them made observations that predated William Harvey's demonstration of circulation in the seventeenth century.

In conclusion, the Aristotelian cardiovascular model consists of two related but slightly dissimilar passages based on experimentation and tradition, which can be perceived as the morphological and metaphysical accounts of physiology, respectively. Restricted by his own methodology of dissecting dead animals, Aristotle described the anatomy of the heart and blood vessels in a way that was accurate and excellent for his time, but incomplete.

3.6.2 Galen

Claudius Clarissimus Galen (129–200 CE) is one of the most renowned figures in the history of science. He was honored by Marcus Aurelius as the "first among doctors, but the one and only philosopher." Galen was born in Pergamum (now Bergama, Turkey), the only child of Aelius Nicon, a wealthy architect and scholar with interest in mathematics, astronomy, philosophy, and Greek literature. As was the norm for the educated governing class, Nicon had his son spend three years studying philosophy, beginning at the age of 14. In 145 CE, as tradition has it, Asclepius, the god of medicine and healing, conveyed a dream to Nicon that his son was to learn medicine. Accordingly, Galen studied medicine at the Pergamum sanctuary of Asclepius (Aesculapion of Pergamon) for four years. Following his beloved father's death in 149–150 CE, Galen decided to continue his medical studies for the next nine years in Smyrna (now Izmir, Turkey), Corinth, Crete, Cilicia, Cyprus, and Alexandria. In 157 CE, at the age of 28, he returned to Pergamon, where he practiced medicine as a surgeon for gladiators (vulnerarius). There, with the severe injuries of the gladiators, Galen enhanced his knowledge of anatomy, physiology, trauma, and sports medicine.

In 162 CE, Galen moved to Rome, where much of the medical community was led by charlatans. Soon, he gained a great reputation for his substantial knowledge as a physician, but the criticisms he leveled at the Roman physicians caused friction, and he was forced to abandon the city for Pergamon in 166 CE. However, Galen's renown led the emperors Marcus

Aurelius and Lucius Verus to summon him to northern Italy in 168 CE to serve the army in its wars against the German tribes. Rather than involving himself in warfare, Galen devoted his time to treating the imperial court, and in particular Commodus, the son of Marcus Aurelius. In 169 CE, while Rome battled the plague, Galen was appointed as personal physician to Marcus Aurelius and Commodus, then to Commodus alone following his father's death, and finally to Septimus Severus. Galen continued to study, write, and teach while under the shelter of the Roman Empire. Although he wrote over 500 papers, books, and treatises, totaling more than four million words, only a portion (~120 works) could be saved from a fire in the Temple of Peace in 191 CE; many of his treatises on philosophy were destroyed. Despite this loss, Galen spent the next few years of his life continuing his medical writing and producing further major works. Carolus G. Kühn has compiled the complete Greek text of these works, along with Latin translations, in a 22-volume collection. Galen died in 200 CE, at the age of 70 years.

3.6.2.1 Galen's Works and Publications

While known as the founder of experimental physiology and embryology, Galen paid tribute to his forerunners – in particular Hippocrates, whom he referred to as "divine." Galen was the first to confute Aristotle's and Erasistratus' belief that what flows within the arteries is merely an undefined spirit called *pneuma*, not blood. Instead, Galen believed that *pneuma* was a physical substance streaming through the nerves and governing sensation and motion. While human dissection was not allowed, Galen performed dissections and vivisections on animals, assuming that human organs were identical. As he declared in *De Anatomicis Administrationibus* ("Anatomical Procedures"), Galen was the first to conduct a vivisection on a pig for a large audience, in order to demonstrate the activity of the organs pertinent to voice and respiration. He recommended that his students practice their skills dissecting dead animals before attempting vivisection. In a traditional medical competition, and under tutorship of Satyrus, Galen adroitly presented the structure and function of the exposed muscles in an anthrax victim. This event, as well as his practice of deliberation, equipped him to become unrivaled in anatomical competitions and demonstrations.

Among Galen's renowned patients were Eudemus (philosopher and his earlier teacher), Flavius Boethus (consul and later governor of Palestine), and several Roman emperors (as already mentioned). Galen successfully cured Eudemus of quartan fever, elevating his status in Roman society. Moreover, in Pergamon, Galen served as physician and surgeon to wealthy gladiators. Interestingly, and thanks to his skillful trauma management, the number of casualties decreased dramatically to only five while Galen held this position.

Galen's name lives on eponomously in several terms: the veins of Galen (two internal cerebral veins, which join to form the great vein of Galen), galenicals (numerous medicinal plants and other ingredients that Galen gathered in a single preparation), Galen's nerve or Galen's anastomosis (a branch of the internal laryngeal nerve that communicates with the recurrent laryngeal nerve), *Galenism* (a system of medicine based on Galen's works and words that remained the unique authority for over 1000 years), Galen's ampulla (the posterior portion of the great cerebral vein), Galen's bandage (a type of bandage wrapped around the head; the middle of the undivided part is applied to the dressings, and the two posterior tails are tied behind the head), Galen's cerate (a cream made of melted white wax, almond oil, distilled water, and rose water), Galen's dislocation (luxation; a supra-acromial dislocation of the lateral end of the clavicle), the Galen foramen (the opening of the superior vena cava into the right atrium), and the Galen ventricle (the laryngeal ventricle; a lateral evagination of mucous membrane between the false and true vocal folds, reaching nearly to the angle of the thyroid cartilage).

3.6.3 Polybus

Little is available in the English literature on the life of Polybus, a teacher at the Aesculapian School on the island of Cos in the fourth century BCE. According to Galen, Polybus was a pupil and the son-in-law of Hippocrates of Cos. He was a contemporary to Archelaus, the king of Macedonia. Hippocrates' two sons, Thessalus and Draco, were also his students, but Polybus was his successor. When Hippocrates traveled to other cities for medical purposes, the directorship of the Cos School was given to Polybus. Being the physician of Cos, he was distinguished from the King Polybus of Corinth (from the story of Oedipus). Polybus was so skillful in medicine that Hippocrates sent him abroad during the time of the plague to assist other cities. Following Hippocrates' death, he continued his education and, together with his brothers-in-law, founded the school of Dogmatism. This school believed that before treating a disease, physicians must be acquainted with the nature and function of the body and its changes under the influence of pathology.

Although controversial, several treatises from the Hippocratic collection (*De Natura Hominis, De Salubri Victus Ratione, De Natura Pueri*, and *De Affectionibus*) have been attributed to Polybus. *De Natura Hominis* ("On the Nature of Man") reads as a synthesis of the medical thoughts of the time, mixing the Hippocratic method with naturalism. Aristotle respected Polybus as the author of *De Natura Hominis*. However, Galen described it as reflecting the unmodified doctrines of Hippocrates, supplemented by the writings of other ancient philosophers such as Empedocles, Parmenides, Mellisus, Alcmaeon, and Heraclitus.

3.6.3.1 De Natura Hominis

De Natura Hominis is a philosophical/physiological treatise containing anatomical sketches and discourses on the pathophysiology of diseases. Galen, in his commentary on the work, credited Polybus as the writer of a brief section, the "Regimen of Health." However, others have doubted the accuracy of this assertion. In fact, some scholars have treated Polybus, not Hippocrates, as the founder of the theory of the four elements of the universe (dry, wet, hot, and cold). If *De Natura Hominis* was written by Polybus, then he participated in the generalization of these elements into medical practice through the so-called "theory of humoralism," which states that every person is composed of blood, phlegm, and yellow and black biles, and that the preponderance of one relative to the others brings on illness. Interestingly, Aristotle ascribed the theory of the four humors to Polybus and a different theory to Hippocrates.

3.6.3.2 Polybus' Theory of the Vasculature

Polybus' vascular model was cited in *Historia Animalium* ("History of the Animals"; Immanuel Becker's version, 512b8) and probably taken from *De Natura Hominis*. It states that:

> There are four pairs of veins. The first extends from the back of the head, through the neck, past the backbone on either side and reaches the loins where it passes on to the legs, and goes on through the shins to the outer side of the ankles and on to the feet. Another pair of veins runs from the head, past the ears and goes inside along the backbone, past the muscles of the loins, on to the testicles, and onwards to the thighs, and through the inside of the hams and through the shins down to the inside of the ankles and to the feet. The third pair extends from the temples, through the neck, underneath the shoulder-blades, into the lung; the one running from right to left travels underneath the breast and on to the spleen and the kidney; the other from left to right traveling from the lung underneath the breast and into the liver and the kidney; and both terminate in

> the rectum. The fourth pair extend from the front part of the head and the eyes underneath the collar-bones; they stretch on through the upper part of the upper arms to the elbows and then through the forearms on to the wrists and the joints of the fingers, and also through the lower part of the upper arms to the armpits, and keeping above the ribs, until one of the pair reaches the spleen and the other reaches the liver; and after this they both pass over the stomach and terminate at the penis. (From Barnes' edition of Aristotle's work; translated by D'Arcy Wentworth Thompson)

In ancient Greece, several authorities assigned the head and brain as the origin of the blood vessels. Alcmaeon (500–450 BCE) was perhaps the first to write in this regard, and also to attribute all senses and consciousness to the brain. In describing his experimental model of the cardiovascular system, Aristotle first mentioned the speculations of Syennesis of Cyprus, Diogenes of Apollonia, and Polybus of Cos. As a cardiocentrist, he rejected the previous cardiovascular theories, including that of Polybus, for being non-experimental. Galen also made his most serious criticism against Polybus' theory, saying:

> Who, having seen the dissection of animals, would accept this theory for even one day? No creature possesses four vessels as he [Polybus] has written. As there is one single large vessel leading into each of the legs, it is not the case that the vessels of the inside of the ankles come from one vessel, and those of the outside of the ankle from another; but rather they are all branches of a single vessel. How could he have overlooked such an important organ as the heart? Indeed, he made no mention of the brains. Is this less noble than the ankles? He has imagined that some vessels are carried from the lungs to the kidneys! Therefore, it is clear from all these things that he has not merely mis-observed, as some dissectors might have mis-observed some small blood vessels, but that he has observed nothing at all. For someone who does not observe the most important things is truly not described as mis-observing, but as not observing at all. No one who has observed something at dissections would be unaware that his theory about vessels being carried down from the head into the whole body is like a drunken hallucination. He had said nothing true in the anatomy even in one passage, but Hippocrates says nothing untrue at all in the second book of the Epidemics. (From Galen's commentary on *De Natura Hominis*)

Polybus' model of the vasculature is, in a number of its components, comparable to the theory of the four principle elements of the universe and to that of the four body humors. Such an analogy gives the impression that Polybus applied these theories in order to derive his anatomical concepts. Hence, Polybus' model represents a mystical approach to the understanding of anatomy and physiology, and is probably based on the numerology associated with the number four, in much the same way as the Hippocratic assertion of the four body humors.

3.7 Holland

During the late sixteenth century, Protestants in the Netherlands were engaged in a long struggle, the Eighty Years' War, against their Spanish Catholic overlords. Thanks to Prince William, founder of the House of Orange, the Protestant heartlands of the north were freed of Spanish rule by the 1570s, although Spain continued to dominate the south. The Protestant North had no university, and William was determined to make good this deficiency; he wanted his citizens to be educated enough to read the Bible, and he needed more broadly educated men to participate in government. Moreover, the Netherlands were by no means immune to the intellectual furor of the Renaissance, and William recognized the need for a university that would take up the new style of learning rather than the medieval scholasticism that still dominated Europe's established seats of education.

William founded the University of Leiden in 1575, and it became a haven for Protestants from other parts of Europe who wished to pursue Renaissance-style learning. Among its other features, Leiden encouraged the practice of anatomy. Within a generation, it had become one of Europe's leading universities, and it remained at the forefront of academic activity during the Enlightenment of the eighteenth century. It is therefore not surprising that Holland's contributions to medical science – to anatomy in particular – were centered on Leiden for at least two centuries following its foundation. It remains an internationally renowned place of learning today.

3.7.1 Franciscus Sylvius

Dutchman François de le Böe (1614–1672), referred to by his Latin cognomen Franciscus Sylvius, is celebrated in two commonly used eponyms, the aqueduct of Sylvius and the Sylvian fissure. Although he is less well known than the French anatomist with the same surname, Jacobus Sylvius (1478–1555), Franciscus Sylvius was a monumental figure in his time.

Born in Germany, Sylvius spent the majority of his career in the Netherlands; his Dutch heritage can be traced back to his paternal grandfather, a wealthy merchant originally from the town of Cambray, in the Southern Netherlands. The older de le Böe was forced to emigrate from his native homeland owing to the prevailing political climate. It was the time of the Reformation, and much tension existed between Catholics and Protestants. The Holy Roman Emperor Charles V and his successor, Phillip II, considered the growing Protestant movement in the Netherlands to be worrying, and employed harsh tactics to suppress the religion. Consequently, in 1568, the Dutch revolted against Imperial rule: the beginning of the Eighty Years' War. The Southern Netherlands remained under Catholic control, forcing the Protestant de le Böe family to leave Cambray for Germany. The German count, Philipp Ludwig II of Hanau-Muenzenberg, allowed Dutch Protestants fleeing Catholic rule, including the de le Böes, to settle near the town of Hanau, east of Frankfurt, and it was there that on March 15, 1614, Sylvius was born.

Sylvius was educated at various universities across Europe. His primary schooling was completed at the French Protestant Academy of Sedan, while his medical education began at the Dutch University of Leiden in 1633. Sylvius left the University of Leiden in 1635, making brief visits to the German universities of Wittenberg and Jena, before finishing his education at the Swiss University of Basle in 1637. His access to Dutch universities was fortunate, as anatomy was well regarded in this region, and the University of Leiden is noted for being the first in the Netherlands to offer anatomy as an undergraduate course. There is little information regarding Sylvius' performance as a medical student, possibly because he frequently moved between institutions.

Despite his lack of academic renown, Sylvius was successful in his endeavors. He worked for a short time as a physician in Hanau after graduating from medical school, before returning to Leiden in 1638 and establishing a successful career as a private anatomy tutor; he is most noted for having instructed the Danish physician Thomas Bartholin (1616–1680) in this capacity. Bartholin published Sylvius' neuroanatomical findings regarding the structure referred to as the Sylvian fissure while he was in the process of updating a medical text written by his father, Caspar.

The relgiopolitical events occurring during Sylvius' life may have prompted him to avoid conflict in medical and academic fields, as his doctoral thesis was unnoticed by academia, and he was passed over for a position at the University of Leiden. This prompted the young doctor to move to Amsterdam in 1641. There, he worked as a physician, and his private practice grew. More significantly, he founded a school of medicine based on the teachings of Jan Baptista van Helmont and others. This theoretical framework, called iatrochemistry, modernized the Galenic humoral theory by integrating it with chemical information being discovered at the time. It appears that he took no part in the fashionable, raging debates between the Galenic and the post-Galenic Enlightenment camps, but instead bridged the two via iatrochemistry.

With these successes, Sylvius secured a position as Professor of Practical Medicine at the University of Leiden in 1658, demanding an unusually high salary of 1800 guilders; in 1669, he was named Rector Magnificus of the University. Sylvius was adept at describing neuroanatomy, particularly the cranial venous sinuses. Consequently, his students often attached his name to neural structures as a mark of respect. This probably explains the use of Sylvius' name to describe the cerebral aqueduct, a structure described by others prior to him, although lively debate concerning this topic is still ongoing in historical circles. The term "aqueduct" was first given to this structure by Giulio Cezzari Aranzio (1530–1589).

Recognition and fame are often achieved through notoriety, yet the life of Franciscus Sylvius proves that dedication and skill can suffice.

3.7.2 Niels Stensen

The Dane Niels Stensen (Latin Nicolaus Stenonis; 1638–1686) (Figure 3.79), a pioneer in anatomy and geology, was born during the Thirty Years' War to Sten Pedersen, a goldsmith who worked regularly for King Christian IV of Denmark, and Anne Nielsdatter, who came

Figure 3.79 Niels Stensen (1638–1686).

from the Danish province of Fyn. Living in Copenhagen, Stensen studied at the country's most exclusive school, the grammar school Vor Frue (1647–1656). He was very astute, and considered by some to be a savant, speaking fluent German, Dutch, French, Italian, Latin, Greek, Hebrew, and Arabic. He attended university in Copenhagen (Københavns Universitet, 1658–1660), studying medicine, mathematics, and philosophy. A fellow Dane, the teacher Thomas Bartholin, persuaded Stensen to study in Holland, where he excelled in anatomy under the tutelage of the anatomist Gerard Bläes. Serendipitously, Stensen discovered the parotid duct (Stensen's duct) in sheep, dog, and rabbit heads; he is most remembered for this observation. However, although it was discovered by Stensen, his teacher Bläes laid claim to the discovery of this structure in 1661. The ensuing dispute resulted in Stensen's moving to the University of Leiden, where he began compiling his *Anatomical Observations*, which he published in 1662. He dedicated this work to his professors in Copenhagen and Leiden, among them the mathematician Jakob Golius (1596–1667), who was a student of Galileo – as was the brother to Stensen's Medici benefactor, Ferdinando II and Vincenzio Viviani (1622–1703), a future collaborator. It is recorded that a physician visiting Paris in 1665 stated, "he [Stensen] could count the bones of a flea, if fleas have bones." In Leiden, he met Jan Swammerdam, Frederik Ruysch, Reinier de Graaf, Franciscus Sylvius, and the famous Jewish philosopher Baruch Spinoa.

Stensen moved to Florence, Italy between 1665 and 1666. There, Grand Duke Ferdinand II of the Medici Palace appointed him to a hospital post that allowed time for research. Later, he was elected to the Academia del Cimento (Experimental Academy), a body of researchers inspired by Galileo's experimental and mathematical approaches to science. In Montpellier, he met Royal Society members Martin Lister and William Croone, who introduced Stensen's work to the group. While in Rome, he met Pope Alexander VII and Marcello Malpighi, the founder of microscopic anatomy.

As an anatomist, he made multiple discoveries and described muscles of the tongue and the esophagus, and he named the levator costae muscles. In addition, his name is associated with the discovery of the lateral foramina (incisive foramina) of the anterior hard palate, which carries branches of the descending palatine vessels (Stensen's foramina) and the vorticose veins of the eye (Stensen's veins). He discovered the tarsal glands of the eyelids and understood that the lacrymal glands, not the brain, produced tears. He rediscovered the vitelline duct in 1664. In 1667, he discovered the follicles of the ovary, although he did not publish this discovery until 1675, three years after his friend de Graaf. He verified the existence of Peyer's patches in 1673 and demonstrated that placing a ligature around the descending aorta resulted in paralysis of the lower limbs, and that when the ligature was removed, function was restored (Stensen's experiment). Furthermore, he discovered that glands produced the cerumen of the ear canal. His best-known anatomical works include *Anatomical Observations* and *Elementary Mylogical Specimens*, published in 1662 and 1669, respectively.

Early in his career, Stensen's former professor, Thomas Bartholin, referred to him as the "royal anatomist" and the century's "new Democritus," but later Bartholin became envious of Stensen and tried to block his ascent in academia. As an anatomist, Stensen became well known across Europe through his public demonstrations of anatomical dissections in the *theatrum anatomicum*. In 1664 came the news that in view of Stensen's "uncommon learning," he had been made a doctor of medicine in absentia at the University of Leiden.

3.7.2.1 Contributions to Anatomy and Physiology of the Heart

In March 1664, Stensen returned to Copenhagen, where he published *De Musculis et Glandulis Observationum Specimen* ("Specimen of Observations on Muscles and Glands") (Figure 3.80), in which he made observations that advanced understanding of the heart: he described the

Figure 3.80 Cover page from Stensen's *De Musculis et Glandulis Observationum Specimen*. Note the depictions of the heart on the left and right sides.

nature of muscle contractions and reported that the heart pumped blood and was not a generator of heat, as others thought. Stensen stated:

> The heart cannot, therefore, be a special substance, not the seat of heat, of innate warmth, of the soul. It cannot even produce its own fluid as blood, nor can it bring forth spirits, as for example the vital spirit.

In a letter of August 26, 1662, Stensen told Thomas Bartholin how fascinated he was by the independent motions of the vena cava, which continued after the stopping of the heartbeat; they had stimulated him to carry out much research concerning the heart and respiratory organs. On April 30, 1663, in another letter to Bartholin, he described his careful investigations of the heart musculature, having boiled the heart of an ox:

> As to the substance of the heart, I think I am able to prove that there exists nothing in the heart that is not found also in a muscle, and that there is nothing missing in the heart which one finds in a muscle. Happening to have a dead rabbit at hand, I laid hold of its legs and separated its muscles. It was immediately clear in the first muscle and by the first cut that there was no fundamental difference between heart and muscle tissues.

Furthermore, he noted that the muscles of the heart were arranged in such a way that by shortening, they squeezed the chambers so that blood was propelled into the vessels. In a letter to Stensen dated August 4, 1663, Bartholin stated:

> Certainly your observations of the muscles and the heart are excellent and worthy of publication. The spirit of Hippocrates will applaud you, because you, through your outstanding observations, revive his view of the heart which we have moved away from,

> and provide clear evidence of the fact that the heart really is a muscle. Galenos and his successors will thank you because you have established the fibers of the heart are of one and the same kind.

In 1666, Stensen published *Elementorum myologiae specimen* ("Specimen of Elements of Myology"), which concerned the question of whether or not muscle increases in size during contraction. He was invited to the court of Duke Johann Friedrich of Hannover to perform several dissections and demonstrate the circulation of blood and the structure of the heart. Stensen stated that the heart is neither the seat of joy nor the source of the blood or of the *spiritus vitalis*, and regarded its automatic movement as being independent of the will.

3.7.2.2 Discovery of a Congenital Malformation of the Heart

In the short paper, "Dissection of a Monstrous Foetus in Paris," Stensen described the tetralogy, which would not be observed for another 200 years, by Fallot. Historically, some have referred to this constellation of findings as the Steno–Fallot tetralogy.

> The unusual form of the arteries arising from the heart attracted the chief attention and called for admiration. In particular, the pulmonary artery, which was much narrower than the aorta, seemed to be suggestive of something new, and hence I opened this vessel from the right ventricle to the hilus pulmonum, and then I could plainly see that the communication between the pulmonary artery and the aorta [ductus arteriosus] which usually is quite distinct in any fetus, was completely absent. When I opened the right ventricle, however, the probe that was passed forward and upward along the interventricular septum entered directly into the aorta just as readily as the probe passed from the left ventricle into the aorta. Thus, no fewer than three openings led into the right ventricle: one from the right atrium, the other two being connected with the arteries. The same aortic canal that was common to both ventricles, showed, together with the interventricular septum, a double opening. The auricles were normal. Although in this case the arteries were of uncommon structure, the resulting effect of this was in compliance with nature, as the circulation of the blood occurs in any fetus. Just as the vena cava empties into both atria [through the foramen ovale], the right ventricle empties into both arteries; just as the left ventricle receives blood from both auricles, thus the aorta receives blood from both ventricles at the same time. So, no matter whether the blood leaving the right ventricle first passes through the pulmonary artery and then is sent through its own channel [ductus arteriosus] into the aorta, or the aorta receives the blood directly as it partly straddles the right ventricle, without the blood first passing through any other channel, the movement of the blood will be the same from the right ventricle out into both arteries. As to the cause of this phenomenon, I have nothing to say … There can be no doubt that the ductus arteriosus found in infants gradually resolves itself into a ligament as the lungs expand with the establishment of respiration, and that this structure is patent only in the fetus, because all the blood coming from the right ventricle cannot pass through the pulmonary arteries. But why the blood in this case has not been able even to make its way into the pulmonary artery, but has made its way directly into the aorta, I am unable to explain. Still, no matter what the reason of this might be, I take it plainly to prove the wisdom of Nature, in as much as the effect is produced, if not in the same way, yet always somehow. Just as this fetus proves this point with regard to theat part of the blood that has to be expelled from the right ventricle in the large artery [aorta], this fetus also illustrates that the formation of the solid parts of animals does not always proceed in the same manner even though the effect obtained remains the same.

In addition to his significant contributions to the field of anatomy, Stensen is regarded as the father of geology. He is credited with developing the principles that led to the science of stratigraphy and for establishing the foundation on which Darwin's theory of natural selection would be based. Stensen's writings in this field continue to be used by geologists and paleontologists today.

While in Florence, Stensen began to question his Lutheran upbringing, and he converted to Catholicism on November 4, 1667, becoming a leading figure in the Counter-Reformation. He would later return to Florence to be ordained a priest in Florence Cathedral on Easter Eve, 1675. In Denmark, he was appointed *anatomicus regius*, as a Catholic could not hold the position of professor. He wrote, "This is the true purpose of anatomy: To lead the audience by the wonderful artwork of the human body to the dignity of the soul and by the admirable structure of both to the knowledge and love of God."

Stensen went to Rome, where he was appointed apostolic vicar of northern missions by Pope Innocent XI on August 21 and consecrated titular bishop of Titiopolis in Asia Minor on September 19, 1677. Later that year, after being appointed Apostolic Legate for Northern Germany and Scandinavia, he left Rome to minister to the minority Roman Catholic populations in northern Germany, Denmark, and Norway. As an example of his devotion to his religion, Stensen sold his bishop's ring and cross to help the needy, and he is said to have dressed like a beggar and eaten little, becoming so malnourished that one friend described him as a "living corpse."

In 1684, Stensen moved to Hamburg and was invited to Schwerin, where he spent some of the last years of his life ministering to Catholics who had survived the protracted Thirty Years' War. He wished to return to Italy, but became seriously ill and died in Germany at the age of 48. At the request of Grand Duke Cosimo of Tuscany, his body was taken by a friend to Florence and buried in the Medici tombs in the Basilica of San Lorenzo.

Stensen's piety and virtue have been evaluated with the goal of eventual canonization, and centuries after his death, Danish pilgrims appealed to Pope Pius XI to make Stensen a saint. In 1953, as part of this process, his corpse was exhumed and reburied in the Capella Stenoniana, a chapel within San Lorenzo. On October 23, 1988, Stensen was beatified (the first step to becoming a saint) by Pope John Paul II. His day (Blessed Nicolas Steno) of celebration in the Catholic Church is November 25. Stensen's name is remembered by the Steno Museum in Århus, Denmark, by craters on Mars and the Moon, by the Steno Diabetes Center, a research and teaching hospital in Gentofte, Denmark, and by the Istituto Niels Stensen, founded in 1964 in Florence and administered by the Jesuit Order.

Stensen's life was short but full of academic and spiritual discovery, and he made significant and thought-changing contributions to the fields of cardiac anatomy and pathology.

3.7.3 Pieter Camper

Pieter Camper (1722–1789) was born in Leiden, South Holland to a prosperous family in which education and the arts were highly valued. His grandfather was a physician, and his father was a minister who preached in the West Indies and a close acquaintance of the great physician Herman Boerhaave (1668–1738), a likely influence on Camper's early years and career choice. Camper was described as "A diligent, bright and easily excelling" student, who became fluent in Dutch, English, Latin, French, and German. He enrolled at the University of Leiden to pursue his doctorate of medicine and philosophy. While there, he studied medicine with Bernhard Albinus (1697–1770), one of the greatest anatomical illustrators of his time, and his eventual rival in the field. His thesis, *Dissertatio Optica de Visu* ("The Eye as an Optical Instrument"), was completed in 1746, and he received his degrees in medicine and philosophy at 24 years of age.

Camper traveled to London to study painting and enrolled in William Smellie's (1697–1763) course on midwifery. Smellie was an anatomist and artist as well as a teacher of obstetrics: the same skills Camper would later acquire. Camper visited Paris and Switzerland during the early years of his career to meet with prominent physicians and anatomists throughout Europe, including Buffon, Daubenton, Franklin, von Haller, and Blumenbach.

Camper became a professor of philosophy, medicine, and surgery at the Franke University, in the Dutch Province of Friesland. There, he continued to explore his various interests, including the graphic arts, and began to sketch zoological objects, landscapes, and people. He had a strong interest in artifacts of Greek antiquity, and expressed how esthetically pleasing he found the Greek race. Camper adamantly agreed with their philosophy that art was a valuable tool in education, and frequently incorporated his artwork into his instruction.

In 1755, Camper became a *Praelector Anatomiae* of the Amsterdam Guild of Surgeons and frequently gave public anatomy lessons in an Amsterdam theater. While in Amsterdam, he was given 46 corpses to study, from which he produced three atlases. Much of his work was presented in Dutch and Latin. From 1763 to 1773, he received a position at Groningen and organized the first surgical polyclinic.

Later in his life, Camper was involved in national politics, retiring from the practice of medicine to take the positions of Mayor of Workym and Member of the College of the Admiralty of Friesland. Later, he moved to The Hague and was appointed to the Representative of the Council of States.

Pieter Camper died of pleuritis in 1789. His interests spanned multiple aspects of medicine and anthropology, anatomy, and the arts.

3.7.3.1 The Anatomist and the Artist

Pieter Camper conveyed his anatomical knowledge through oil painting, aquatint, pastel drawings, etching, India ink, and marble busts, and he strove to create accurate proportions in his works. His understanding of anatomy was not limited to humans, and he extended his sketches to comparisons between the facial angles of apes and monkeys. His interest in this topic can be attributed to the guidance he received from the great anatomist John Hunter (1728–1793), who helped fuel his passion for comparative anatomy. Camper was also interested in the air cavities located in the bones of birds, the hearing ability of fishes, and the noises made by frogs. He was considered an expert in the field, and was frequently recruited by others to provide illustrations. In an excerpt from his writing of 1776, *Conjectures to the Peirifactions found in St. Peter's Mountain near Maefricht*, Camper recalls a colleague writing about crocodiles:

> as a friend ... I sent him a figure of the lower jaw of a crocodile, accurately done by my own hand and soon after, the skull and under jaw of a pretty large crocodile; which induced him to defer his design of writing about these antiquities of the old world until he should be better informed on the subject of cetaceous fishes.

One of the greatest examples of his artistry and anatomical mastery can be found in one of the atlases he produced, *Demonstrationum Anatomicopathologicarium Liber Primus Brachii Humani Fabricam et Morbos*, in 1760 (Figure 3.81). This 22-page atlas comprised three chapters, on the skin, the muscles and joints, and the nerves and blood vessels. It was renowned for its pioneering understanding of the chiasma tendium, a structure formed by the intersection of the flexor digitorum profundus tendons and the flexor superficialis tendons, which is considered to be the key element in the gripping action of the hand, where the two contralateral slips of the flexor digitorum superficialis pass around and compress the flexor digitorum profundus. Furthermore, the atlas illustrated Camper's comprehension that muscle is surrounded by a

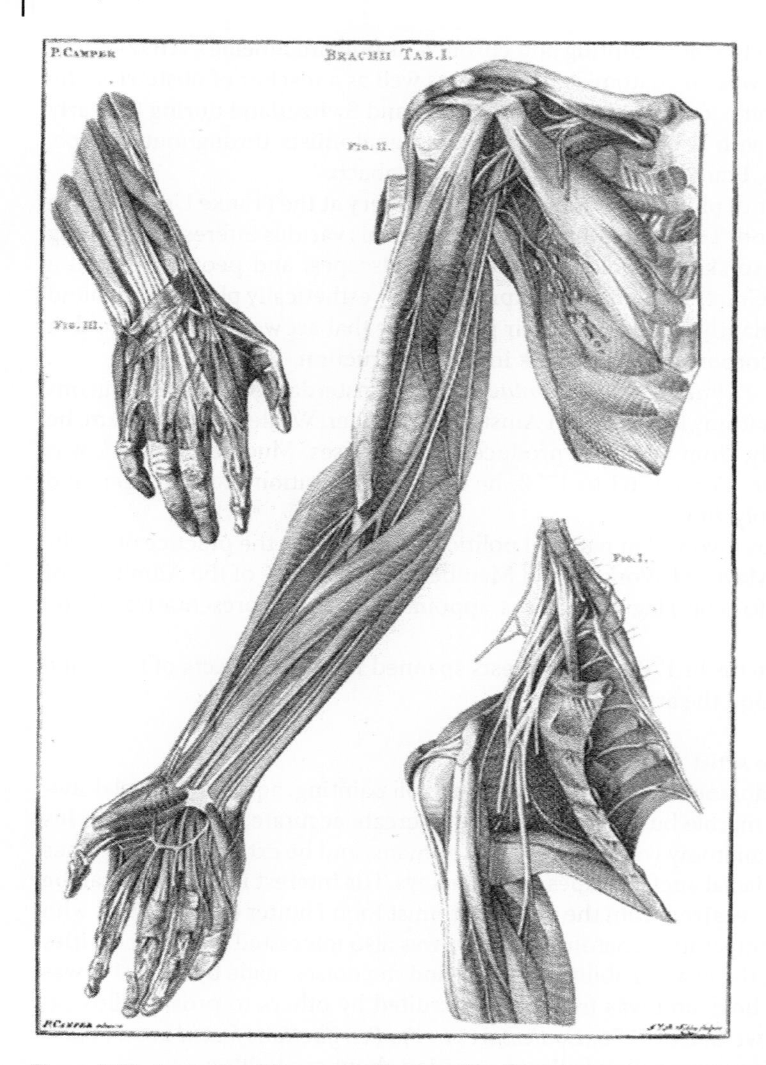

Figure 3.81 Anatomy of the arm, as illustrated by Pieter Camper in his atlas, *Demonstrationum Anatomicopathologicarium Liber Primus Brachii Humani Fabricam et Morbos*. (*Source:* Camper 1760.)

membranous sheath of fascia. He could identify and treat dislocations of the shoulder and rotator cuff with the aid of his atlas, although he believed that skin itself never completely healed if damaged.

Icones herniaum (1756–1760) was a series of 14 illustrations by Camper relating to hernias. The series was never completed, and remains untranslated from Latin. It provided an extensive study of inguinal hernias, from which Camper deduced that the hernial sac is a not a rupture of the peritoneum, as was previously accepted in medicine, but a protrusion of it.

3.7.3.2 The Anthropologist

On November 14, 1764, Camper gave a lecture "On the Origin and Color of Blacks" at the University of Groningen. He had dissected, observed, and measured the facial angle of many races (Figure 3.82). He defined the facial angle, or prognathism, to be the angle between a vertical line measured from the prominent part of the forehead to the incisor teeth and a

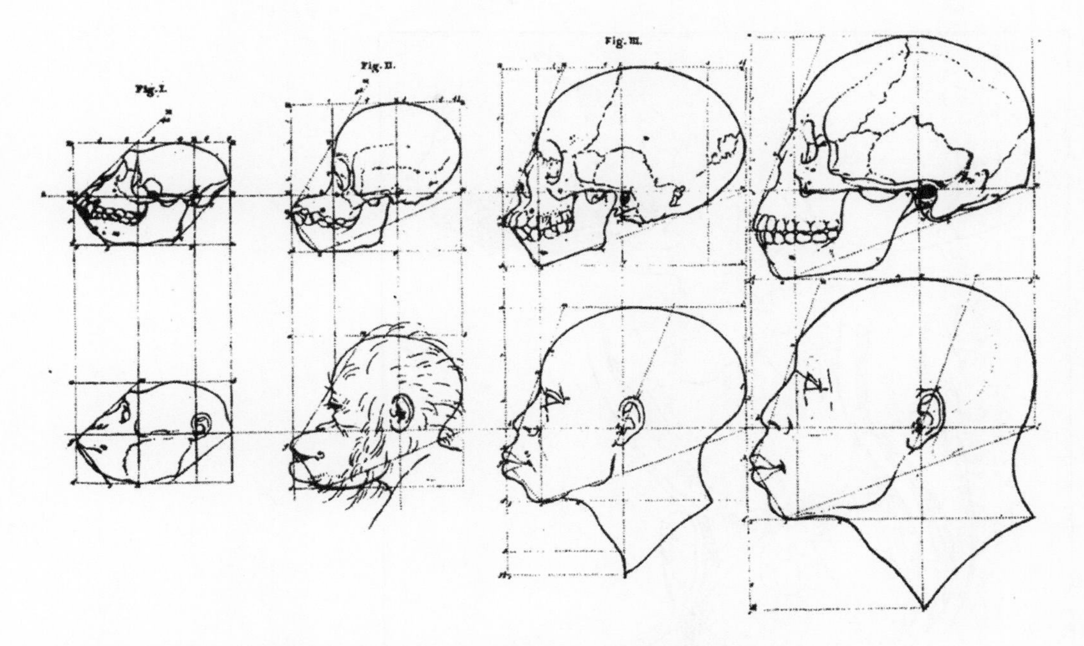

Figure 3.82 Diagram illustrating Camper's comparative study of facial angles.

horizontal line measured from external auditory meatus to external auditory meatus and traveling just beneath the bottom of the nose. Camper demonstrated the uniformity of the cranium between races, despite variable expression of the jaw. He determined that the facial angles for various races were as follows: Greek 100°, European 80°, Asian and African 70°, orangutan 58°, and tailed monkey 42°. He concluded that everything below 70° belonged to the animal kingdom and everything above 80° was human. The placement of Europeans was a concern to him, as it often fell between 70° and 80°. With his facial angle measurements, Camper could distinguish the counterfeit profiles of Romans and Greeks in engraved designs in 1767 when visiting the Count of Bentick.

Camper was accused of introducing a method of measurement that was used to demonstrate the racial inferiority of Africans. Because of this concept of "primitiveness," much of the anthropology community disapproved of him. However, Camper was opposed to slavery and did not believe in white supremacy.

As an artist, Camper thought the facial angle could be used as instruction concerning correct esthetic proportions for a particular race. He considered Greeks and other Europeans the most esthetic races. He believed that anatomical subjects should be portrayed in an organized blueprint, similar to the accepted grid technique of drawing of Dürer in the seventeenth century. This view was not shared by Albinus, his educator, and rivalry between them ensued. Camper's full cephalometric analysis was carried out using his measurements of the facial angle, the horizontal reference line, and the angle between the two. Interestingly, he discovered that as humans grow and develop, the facial angle decreases due to changing proportions.

3.7.3.3 The Physician and the Surgeon

Camper's interests outside medicine were diverse, and within the field he had a plethora of specializations and innovations, and could be considered a key founder in many areas. He was an early advocate for inoculation and vaccination. One field for which he is best known is

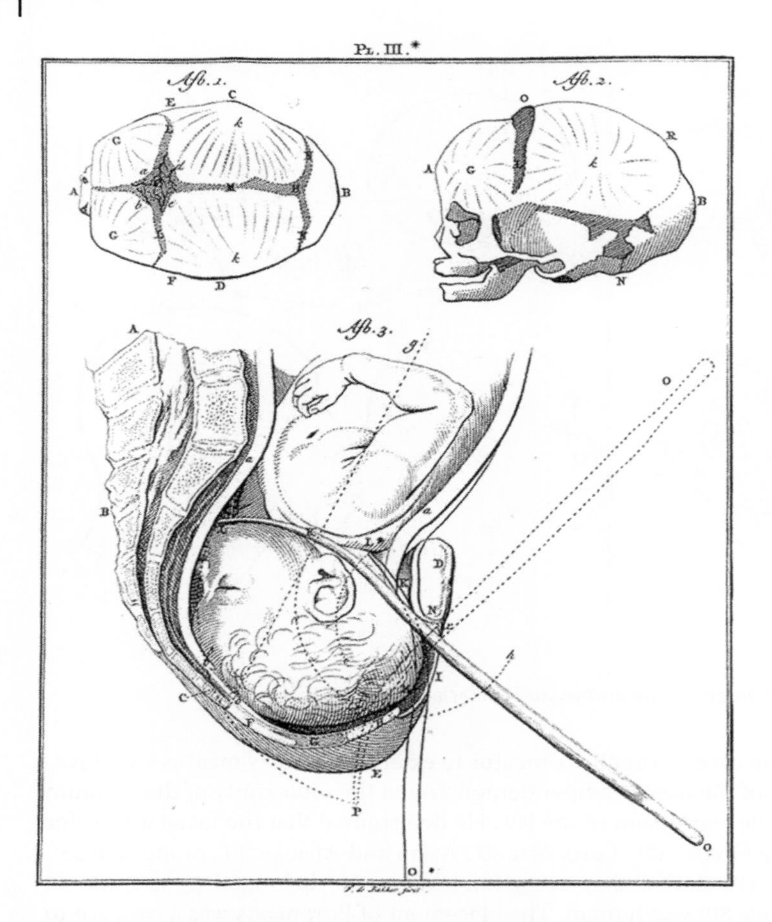

Figure 3.83 Original illustration of obstetric equipment created by Camper to assist in childbirth. (*Source:* Public domain: http://www.rug.nl/museum/geschiedenis/hoogleraren/camper.)

obstetrics, including the anatomy of the birth canal. He was a leader in this area, using forceps, early expulsion of the placenta, and cesarean sections (Figure 3.83).

Camper's extensive studies of the pelvic area include the causes and treatments of hernias. His treatment of hernias was based on the creation of trusses from the geometrical proportions of the pelvis, another application of his artistry and anatomical mastery. His trusses were useful in the treatment of inguinal hernias in children; however, in adults, surgical intervention was often required. He determined that the processus vaginalis testis was a partially open peritoneal tube that could lead to the formation of inguinal hernias. Furthermore, he studied bladder stones and was instrumental in improving surgical incisions for their removal. He described the urogenital fascia and Camper's fascia, the superficial fascia layer of the subcutaneous abdomen (Figure 3.84).

Camper carried out widespread work in orthopedics, including the anatomy of the feet, fracture treatment, amputation, and congenital dislocation of the hip; he was the first to propose symphysiotomy. In addition, he prepared detailed illustrations of the olecranon and the patella. His most notable work relating to orthopedics was determining what constituted proper footwear in his *Dissertation on the Best Form of Shoe*. Interestingly, Camper understood the anatomy of the carpal ligaments and how the tendon sheaths of the structures entering

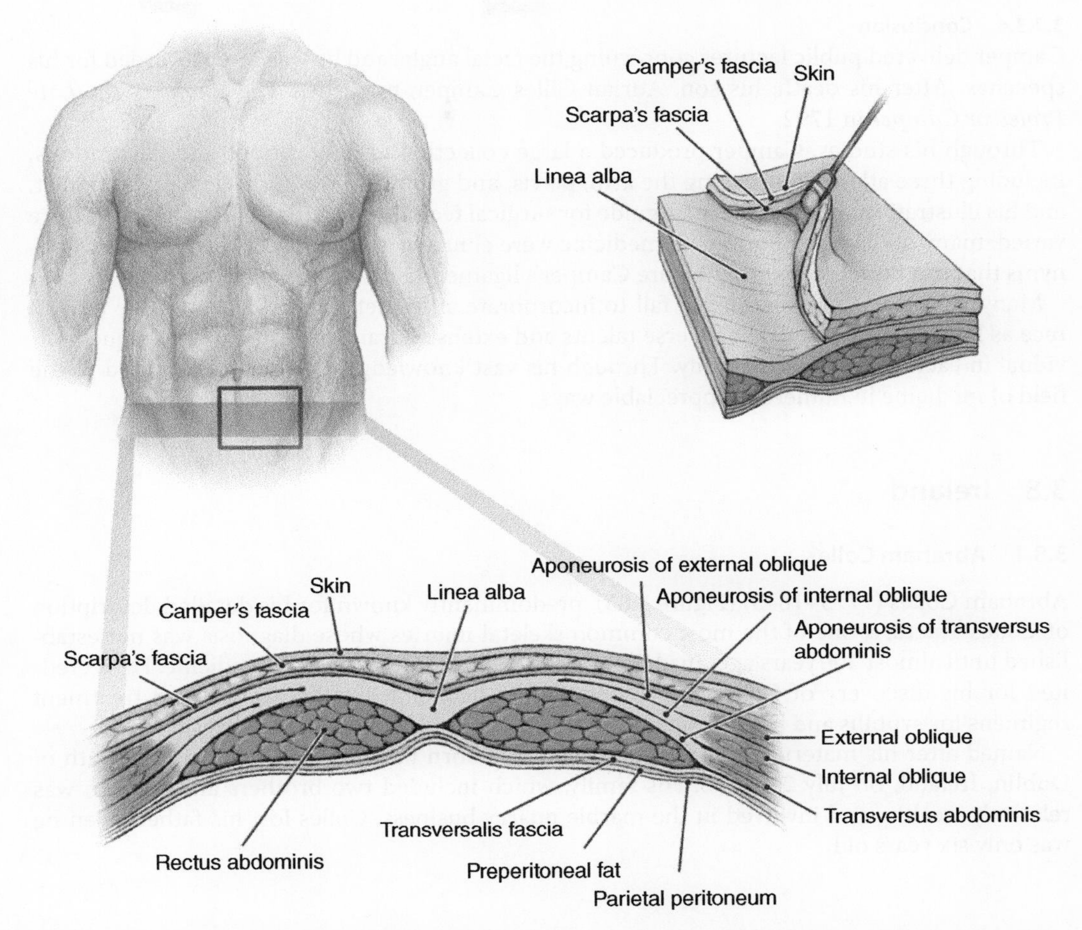

Figure 3.84 Cross-sectional image depicting the different layers of the abdominal wall, including the location of Camper's fascia.

the carpal tunnels were vulnerable to compression and pain when ganglion cysts formed in them. Among the actions he described as initiating the onset of this condition was repeated trauma caused by heavy work, similar to what is today referred to as carpal tunnel syndrome.

In neurology, Camper described nerves as fluid-filled channels that transferred signals to the brain and neuromas as white hard tumors located in nerve sheaths that were capable of causing pain. In addition, his atlas demonstrated how bloods vessels were a contiguous system and showed that severe bleeding could be stopped by vascular compression at pressure points. Furthermore, his atlas of the arm and hand demonstrated the close relationship between arteries and nerves.

As demonstrated by his thesis, Camper had an interest in the eye, and he often performed eye surgery. He was most innovative in his lacrymal duct surgeries, and created instructional texts based on other surgeons' procedures. He believed that lacrymal ducts attracted tears by capillary action, and he was successful in improving dilations of the lacrymal sac surgically, which were caused by obstruction of the nasolacrymal duct. He understood the association between light and vision, and that what is visualized is an observation of an object that is transferred by light. Anatomically, he discovered the fibrous structure of the lens.

3.7.3.4 Conclusion

Camper delivered public lectures concerning the facial angle, and he was often awarded for his speeches. After his death, his son, Adrian Gilles Camper, published *The Works of the Late Professor Camper* in 1792.

Through his studies, Camper produced a large collection of publications and illustrations, including three atlases concerning the arm, pelvis, and groin. He was a surgical revolutionist, and his illustrations often provided a guide for surgical techniques. Although his interests were varied, many of his contributions to medicine were clinically significant. The anatomical eponyms that can be accredited to him are Camper's ligament, line, chiasm, and fascia.

Many physicians and anatomists fail to incorporate an esthetic appreciation of the human race as Pieter Camper did. His diverse talents and extensive training made him a unique individual in each of his areas of study. Through his vast knowledge, Camper contributed to the field of medicine in numerous appreciable ways.

3.8 Ireland

3.8.1 Abraham Colles

Abraham Colles (1773–1843) (Figure 3.85), predominantly known for his detailed description of Colles' fracture, one of the most common skeletal injuries whose diagnosis was not established until almost 200 years ago, made other significant contributions to medicine. He is credited for his discovery of Colles' fascia and Colles' ligament, but he also offered treatment regimens for syphilis and performed the first surgery for axillary artery aneurysm.

Named after his maternal grandfather, Colles was born in Milmount, a town just south of Dublin, Ireland, on July 23, 1773. His family, which included two brothers and a sister, was relatively wealthy and involved in the marble quarry business. Colles lost his father when he was only six years old.

Figure 3.85 Abraham Colles (1773–1843) at the age of 50, from Martin Fallon's *Abraham Colles*. (*Source:* Courtesy of London Library.)

While Colles was in grammar school, a flood partially destroyed the home of the local doctor, Dr. Butler. As fate would have it, Colles came across an anatomy book from the physician's house and returned it to its rightful owner. However, Butler sensed Colles' interest in the book and allowed the young boy to keep it. This event would be the catalyst that inspired his interest in medicine and anatomy.

As a child, Colles attended school in Kilkenny, before going to the University of Dublin. He received a diploma from the Royal College of Surgeons in Dublin in 1795. He then advanced his expertise by traveling to Edinburgh, earning an MD in 1797 after defending a doctoral thesis entitled "De venesectione." Colles then walked 400 miles in eight days to study under Alexander Monro and the brothers John and Charles Bell in London. He completed his formal education there in 1797, working alongside the famous surgeon Astley Cooper. Together, Colles and Cooper studied the inguinal region through dissection, and they remained lifelong friends.

Colles returned to Dublin and was appointed to the Sick Dispensary in Meath Street, a Quaker-based charity established by the Society of Friends. During this time, he volunteered his services to the Sick and Indigent Roomkeepers' Society, putting him in close contact with the poor. In 1799, he temporarily suspended his volunteer work to become the resident surgeon at Steven's Hospital, where he earned a modest £178 4s 4.5d; in his later years, he would earn as much as £5000–6000. To develop his intellect and reputation further, Colles joined the Medico-Chirurgical Society, becoming president within two years. Furthermore, he joined the Royal College of Surgeons in Ireland in 1804 as a lecturer and professor of anatomy and surgery. He chaired both departments for the following 32 years. Although he resigned from his professorship in 1836, Colles maintained his appointment at Steven's Hospital until 1841. In 1839, he modestly declined the offer of a knighthood as a baronet. He became increasingly frail, and died of complications caused by gout on November 16, 1843.

Among his colleagues, students, and patients, Colles was considered inspirational and skilled. His teaching was clear, and it is said that his lectures were beautifully written. Furthermore, he was well regarded for his dedication to the care of his patients. In 1881, his editor, Robert McDonnell, wrote: "a finer model of every attribute and moral quality, essential to the achievement of high success in the practice of surgery and medicine, can hardly be offered for imitations." Following Colles' death, all medical schools in the Irish metropolis suspended their proceedings, and a tribute was paid to him by the College of Surgeons, the College of Physicians, and the Apothecaries' Company. Colles was laid to rest in the cemetery of Mount Jerome.

3.8.1.1 Contributions to Clinical Anatomy

For centuries prior to Abraham Colles' description of distal radius fracture, such trauma had commonly been mistaken for a dislocation of the wrist. Petit (1723) and Pouteau (1783) were probably the first to describe the fracture, but Colles' article "On the Fracture of the Carpal Extremity of the Radius" (1814) is credited as the definitive description. Now known as Colles' fracture, he described this skeletal injury as a fracture proximal to the carpal end of the radius. The posterior aspect of the extremity was characterized as deformed, with a depression of the forearm approximately 1.5 in. above the distal end of the radius. Following this depiction, Colles designed a set number of recommendations for the maintenance and treatment of the distal radius fracture. This included restoration of the limb's natural form by allowing the surgeon to "apply the fingers of one hand on the seat of the suspected fracture and make a moderate extension." Therefore, when the patient moved the hand backward and forward, the surgeon could detect the fractured ends with every movement. Moreover, Colles recommended maintaining the limb in a half-protonated, half-supinated position and using a reduction, consistent with transverse compression on the anterior surface of the limb. In addition, he suggested application of a gutter-type splint, composed of tin, to the anterior and posterior surfaces formed to the shape of the limb. This splint was probably applied to the palm and flexor surface of the hand and distal forearm, and extended around the radial side so that the upper third covered the area comprising the extensor muscles. Later, Colles modified this system by the use of wooden splints, which were narrow along the ulnar surface of the forearm.

Colles described the fascia that is continuous with the dartos layer of the scrotum and membranous layer of the superficial fascia of the anterior abdominal wall (Scarpa's fascia). Clinically, extravasation of blood through the deep penile fascia (Buck's fascia), but confined by Colles' fascia, results in a "butterfly" perineal and scrotal hematoma that can extend to the anterior abdominal wall. Furthermore, Colles described the small triangular fascia that runs from the pubic crest to the iliopectineal line and extends upwards and inwards toward the linea alba in his book, *Surgical Anatomy* (Figure 3.86).

Colles became interested in microbiology while studying venereal disease, from which Colles' law was derived. In his book, *Practical Observations on the Venereal Disease, and on the use of Mercury,* Colles stated, "a child born of a mother who is without any obvious venereal symptoms ... and shows this disease when it is a few weeks old, this child will infect the most healthy nurse, whether she suckle it or merely handle and dress it, and yet this child is never known to

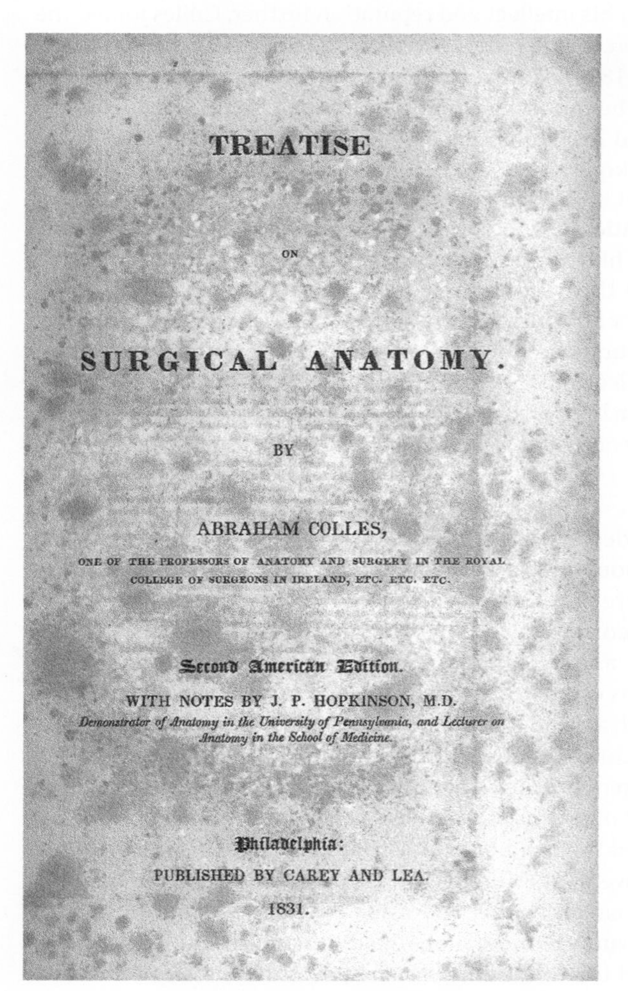

TREATISE

ON

SURGICAL ANATOMY.

BY

ABRAHAM COLLES,

ONE OF THE PROFESSORS OF ANATOMY AND SURGERY IN THE ROYAL COLLEGE OF SURGEONS IN IRELAND, ETC. ETC. ETC.

Second American Edition.

WITH NOTES BY J. P. HOPKINSON, M.D.

Demonstrator of Anatomy in the University of Pennsylvania, and Lecturer on Anatomy in the School of Medicine.

Philadelphia:

PUBLISHED BY CAREY AND LEA.

1831.

Figure 3.86 Front cover of Colles' *Treatise on Surgical Anatomy*, 2nd American edition of 1831.

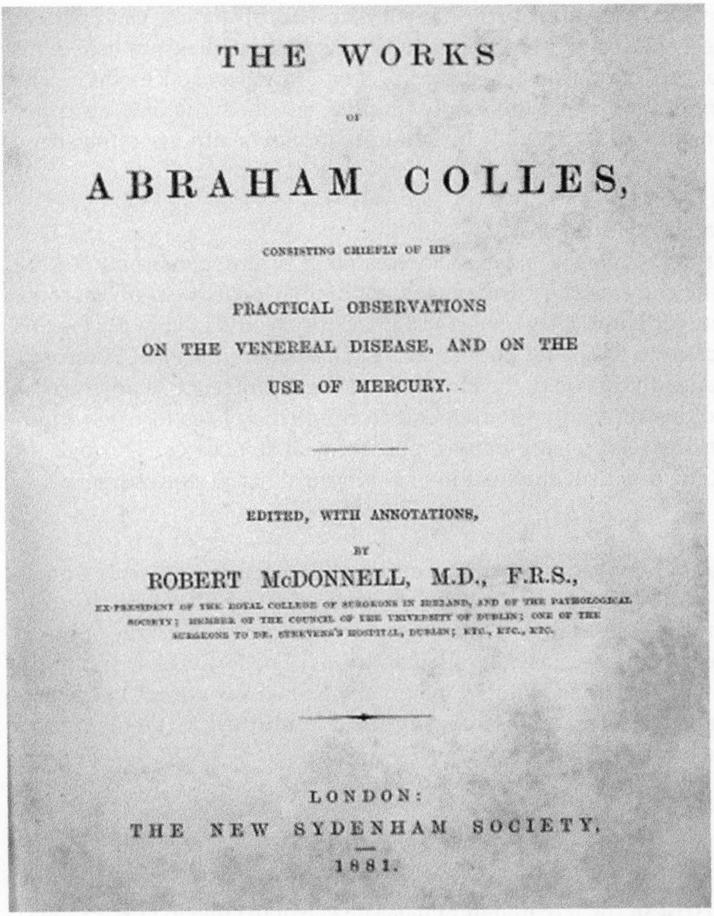

THE WORKS

OF

ABRAHAM COLLES,

CONSISTING CHIEFLY OF HIS

PRACTICAL OBSERVATIONS
ON THE VENEREAL DISEASE, AND ON THE
USE OF MERCURY.

EDITED, WITH ANNOTATIONS,

BY

ROBERT McDONNELL, M.D., F.R.S.,

EX-PRESIDENT OF THE ROYAL COLLEGE OF SURGEONS IN IRELAND, AND OF THE PATHOLOGICAL SOCIETY; MEMBER OF THE COUNCIL OF THE UNIVERSITY OF DUBLIN; ONE OF THE SURGEONS TO DR. STEEVENS'S HOSPITAL, DUBLIN; ETC., ETC., ETC.

LONDON:
THE NEW SYDENHAM SOCIETY.
1881.

Figure 3.87 Front cover of Colles' *Practical Observations on the Venereal Disease, and on the Use of Mercury*, London 1881.

infect its mother" (Figure 3.87). He made this distinction on the basis of observations regarding syphilitic infections. Of course, this law did not hold true, as Colles never considered that the mother could have already had a mild form of the infection, as was proven 70 years later by Wassermann. In the same book, Colles advocated the use of mercury to treat syphilis; he believed this to be his most important discovery. While this treatment was later shown to be ill-conceived, Colles' intentions were genuine, as syphilis was highly stigmatized at the time.

Colles has also been credited with less well known scientific findings, including providing the first account of mammary cancer spreading to adjoining lymph nodes, in his article "Three Different Situations of Scirrhous Lymphatic Glands" (1815). In addition, he was interested in the characterization and treatment of clubfoot, and published "On the Distortion Termed Varus or Clubfoot" (1818). He concluded incorrectly that the oblique position of the tarsal joint and altered form of the talus was the primary cause. Nonetheless, he treated several cases of clubfoot with a device he invented: a shoe with a resistant tin sole covered with leather and various splints to correct the equinus deformity and promote foot eversion, which was applied a few weeks after birth for a minimum of three weeks. Furthermore, he was the first physician to tie the subclavian artery in cases of axillary artery aneurysm. Unfortunately, each patient survived no longer than a week because of complications of sepsis. Colles published his findings, in which he stated that the "operation has not yet proved ultimately successful ... The history of surgery furnishes ... operation(s) now generally adopted, which, in the first few trials, failed of success." However, he displayed remarkable insight into human health by emphasizing the value of rest and open air.

3.8.1.2 Contributions to the Anatomy Curriculum

Beyond his scientific investigations, Colles was deeply interested in teaching anatomy. At the beginning of his professorship, he endured the mismanagement of the Department of Anatomy and Surgery at Trinity College under James Cleghorn (?–1826). Colles would later apply for this position himself, with no success. In 1811, he published his *Treatise on Surgical Anatomy*, which he dedicated to his students. In this book, he emphasized the importance of anatomy as applied to surgery. He was also one of the first individuals in the British Isles to teach topographical anatomy, focusing on the relationship between anatomical structures. Throughout his teaching career, he stressed the practical application of anatomical research to surgery.

3.8.1.3 Conclusion

Abraham Colles made a substantial contribution to our current knowledge concerning anatomy. He was a man who was delighted with his successes and distressed by his failures. Writing during the last hours of a patient's life, he stated: "I was forced to relinquish every ray of hope, and to prepare my mind for the speedy termination of a case, which I had watched with more fluctuations of hopes and fears than ever had agitated me on any former occasion." Therefore, while Colles may not have been a surgical giant, his integrity and commitment to medicine and anatomy have transcended time and deserve recognition.

3.9 Italy

It is well known that the great artists of the late Italian Renaissance, Michelangelo (1475–1564) and Leonardo da Vinci (1452–1518), dissected corpses – illegally for the most part – so that their work was informed by personal acquaintance with the architecture of the human body. This enabled them to create anatomically accurate representations. It is less well known that Leonardo made original contributions to the study of anatomy. In 1515, he was forbidden to continue his dissections, but by then he had set an example that others were to follow: he had shown there was more to be learned about anatomy than could be achieved by studying Galen and Avicenna, and that dissection was the way to learn it.

Opportunities to dissect human cadavers were strictly limited in sixteenth- and seventeenth-century Italy, but the Vatican allowed the practice under certain circumstances. Famously, the great anatomist Andreas Vesalius (1514–1564) wandered from his birthplace in Northern Europe in search of a place where he could dissect female as well as male bodies, and he finally found it in Padua. His work there was pivotal in breaking the monolithic grip of Galenic teaching on anatomy – and, by extension, physiology – but it is striking that the main contributions of his successors in the Padua School were to reproductive anatomy and descriptive human embryology. Three generations after Vesalius, William Harvey (1578–1657) studied medicine at Padua under Fabricius, and he went on to drive the final nails into the coffin of Galenism.

However, it would be a mistake to suppose that Vesalius and his successors in Padua were the only significant contributors to anatomy in this period. Mondino, Leonardo, Vesalius, and other pioneers directly and indirectly initiated a national tradition of anatomical investigation, which continued to flourish in the succeeding centuries, providing important foundations for the advancement of surgery and the development of pathology as a science.

3.9.1 Leonardo da Vinci

Leonardo da Vinci (1452–1518) (Figures 3.88 and 3.89) is now regarded as "the most varied genius probably who ever lived." The illegitimate son of a wealthy notary in Florence, da Vinci spent his earliest years in his mother Caterina's home in Anchiano, and at the age of five entered his father's household in Vinci, Tuscany. He was casually educated in mathematics and Latin, but showed no particular skill, although he is rumored to have given early evidence of artistic ability. When he was 14, he was apprenticed to the eminent Florentine artist Verrocchio (Andrea di Cione, 1435–1488), in whose workshop he was trained in art theory and a wide variety of technical skills. Six years later, in 1472, he attained the status of Master of the guild of artists and physicians, the Guild of St Luke. Thanks to his father's financial backing, he was then able to set up his own workshop, although he continued to collaborate with Verrocchio.

Da Vinci's activities during the following six years are somewhat obscure, but when he received his first independent commission (1478), he finally left Verrocchio's studio and no

Figure 3.88 Self-portrait of Leonardo da Vinci (1452–1518), reproduced with written permission ("Su concessione del Ministero per i Beni e le Attivita Culturali – Biblioteca Reale – Torino." *Source:* Courtesy of the Ministry of Heritage and Culture – Royal Library – Turin, Italy).

Figure 3.89 Leonardo da Vinci on a bronze medal signed by Baltazar. The reverse side of the medal reads, "Italian Sculptor, painter, musician, engineer, physicist, writer and architect."

longer lived in his father's house. In 1482, he was sent by Lorenzo de' Medici with a gift to Ludovico il Moro, Duke of Milan – a silver lyre that he had made – and he took the opportunity to persuade Ludovico of his engineering and painting skills. For the following 17 years, he was largely based in Milan, working on a number of engineering, architectural, and sculptural projects for Ludovico.

When the French invaded Milan in 1499 and overthrew the dukedom, Leonardo fled to Venice and took employment as a military engineer and architect. In 1500, he returned to Florence and set up a new workshop in the monastery of Santissima Annunziata. In 1502, he entered the service of Cesare Borgia, son of Pope Alexander VI, and traveled with him as military engineer and cartographer. He spent the years 1503–1506 in Florence, focusing on fresco work, before going back to Milan; he returned briefly to Florence to resolve difficulties arising from his father's estate.

Among the numerous published studies of da Vinci, many have focused on his contributions to medicine and natural philosophy, later dubbed "science." Plans during his life and for 50 years after his death to publish his notebooks did not come to fruition. His anatomical drawings were highly praised by William Hunter in 1784, and some were published in 1883 and 1916. All those that survive were collected and meticulously edited by Keele, who has published a number of scholarly accounts of da Vinci's contributions to anatomy. Da Vinci's name lives on eponomously for the moderator band of the heart.

Da Vinci was influenced by the writings of Mondino and Albertus Magnus, who had commented on Avicenna. His studies on the heart (1508–1513) were initially based on Mondino's, but were informed by collaboration with the anatomists at the Universities of Padua and Bologna, Marcantonio della Torre (1481–1511) and Jacopo Berengario da Carpi (1460–1530). His acquaintance and collaboration with Berengario and della Torre made him receptive to new ideas in anatomy. The death of della Torre, with whom he collaborated closely, may have prevented the publication of these studies. In any case, da Vinci was banned by papal command from conducting further dissections after 1515.

Da Vinci also made wax casts of the heart and the aorta, and on the basis of their detailed structure he constructed a glass model of the aorta and its valved opening from the ventricle to study the functions of the parts, tracking the patterns of fluid movement with dyes or suspended grass seeds. Using this model, he described the vortices formed when blood is forced between the cusps of the valve, and offered an explanation for the phenomenon based on his studies of water flow in rivers (Figure 3.90). He acknowledged that his explanation was

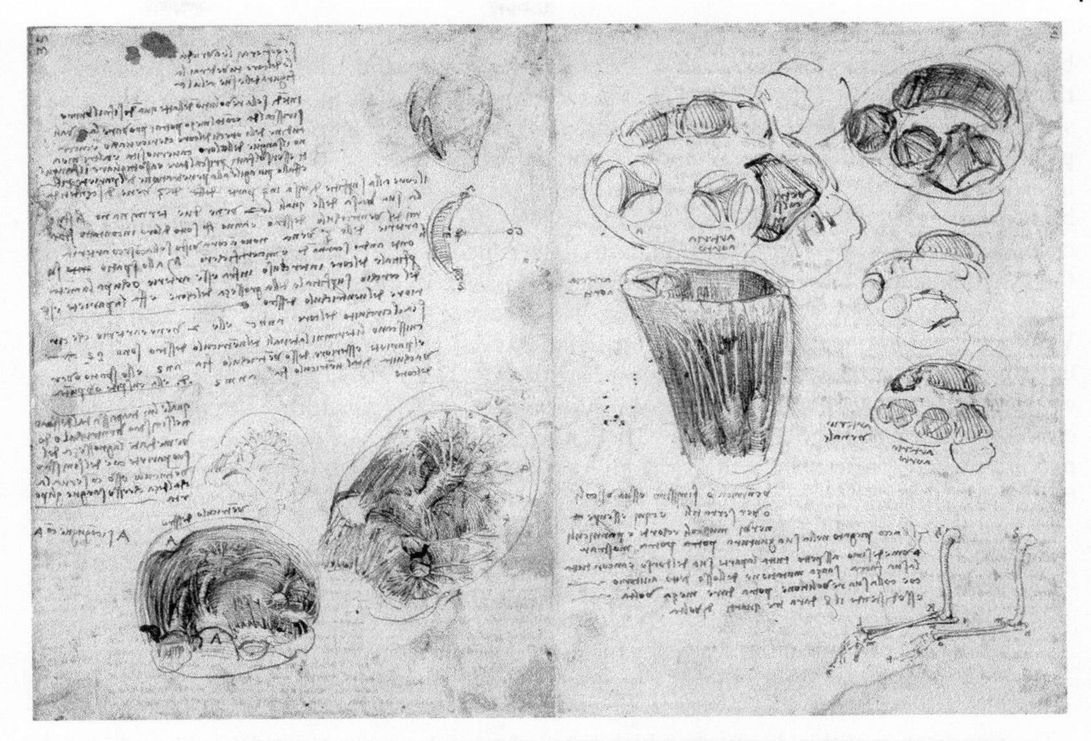

Figure 3.90 An illustration from da Vinci's collection, showing the cardiac valves, chordae tendineae, and papillary muscles. (*Source:* Supplied by Royal Collection Trust/©HM Queen Elizabeth II 2012 – reproduced with written permission.)

incomplete, but he established the importance of circular fluid motion for valve opening and closure, predicting that the inner walls of a valve would collapse if the vortices were not formed: "The interval between two beats of the pulse is half of a musical tempo. Between one and the next beat of the pulse the heart closes twice and opens once, and between one and the next opening the heart opens twice and closes once … Accordingly, in every harmonic tempo, the heart has three motions."

Da Vinci made significant contributions to knowledge of anatomy, and his work on the heart especially had implications for the subsequent development of physiology. It epitomizes the increasing willingness to question accepted beliefs that characterized European thought during and after the Renaissance.

3.9.2 Vidus Vidius (Guido Guidi)

Few details regarding the life of Vidus Vidius (Guido Guidi, c. 1509–1569) are published. He was born in Florence. His mother was the daughter of the painter Domenico del Ghirlandajo, to whom Michelangelo was apprenticed, and his father was a physician. After finishing medical school, Vidius practiced as a physician in Rome and Florence. In 1542, he traveled to Paris following a summons from Francis I and was named royal physician. He was Professor of Medicine at the University of France. In 1547, following the death of Francis I, he traveled to Pisa and became physician to Cosimo I de' Medici. At that time, he also became a priest and was given ecclesiastical benefices, including the rectorship of Pescia (Figure 3.91). He died in Pisa in 1569.

Vidius carried out important anatomical investigations at Pisa after 1548. He is most remembered for his discovery of the pterygoid (Vidian) canal, nerve, and artery. This canal is found in the floor of the sphenoid sinus and carries a composite nerve formed of preganglionic parasympathetic fibers from the facial nerve and postganglionic sympathetic fibers from the deep petrosal nerve. The deep petrosal nerve is a branch of the sympathetic plexus surrounding the internal carotid artery as it exits the carotid canal. At this point, the deep petrosal nerve combines with the greater petrosal nerve at the level of the foramen lacerum to enter the Vidian canal, seen specifically at the junction between the medial pterygoid process and the floor of the sphenoid sinus. The now-formed Vidian nerve travels more or less horizontally to exit the canal at the posterior wall of the pterygopalatine fossa. The preganglionic fibers within the Vidian nerve then synapse in the pterygopalatine (Meckel's, sphenopalatine) ganglion that rests in the pterygopalatine fossa. These fibers are disseminated to the nasopalatine glands and lacrimal gland. Damage to the Vidian nerve results in the loss of lacrimation, with potential desiccation of the cornea and a dry nose. The Vidian artery, a branch of the internal carotid or maxillary artery, accompanies the Vidian nerve while within its canal. Vidius' original description of the canal that bears his name was:

> in the neighborhood of the previously mentioned canals through which pass the carotid arteries are two other holes each of which are in the sphenoid bone and are not easy to see unless this bone is removed from the other bones. They begin like narrow canals from the rear part of the sell. They make their way forwards to the sinuses of the nostrils to provide a passage for the arteries going from the brain to the nostrils.

The significance of this discovery, which entailed careful examination of the sphenoid bone during dissection, lies in the basis it provides for the treatment of certain neuropathies. Some surgeons have obliterated the pterygopalatine ganglion or Vidian nerve in order to treat Sluder syndrome (pterygopalatine ganglion neuralgia), which is manifested as pain from the nose that radiates toward the maxilla, mastoid process, occiput, and neck with associated corneal ulceration. Good results have also been reported following the division of the Vidian nerve for chronic

Figure 3.91 Vidus Vidius c. 1547. Note the attire of a priest. (*Source:* Courtesy of Bibliotheque of Brescia.)

vasomotor rhinopathy with recurrent polyposis, senile epiphora, crocodile tears, and chronic epiphora. Transantral approaches have been performed for pterygopalatine ganglion ablation and Vidian nerve neurectomy. Sanders and Zuurmond have performed radiofrequency lesions of the pterygopalatine ganglion via an infrazygomatic approach in 66 patients with cluster headache with good results. Nine patients complained of hypesthesia of the palate, which disappeared in all cases within three months.

Vidus Vidius also made original studies on the mechanisms of articulation in the human body resulting from its vertical position in relation to the mechanisms of quadruped articulations. He gave clearer pictures and descriptions of the vertebrae than any of his predecessors.

Vidius translated a book on surgery from the Greek and added his own commentary. Multiple woodcuts in this text are attributed to François Jollat, who was working in Paris from 1502 to 1550. Other texts by Vidius were published posthumously in 1596. His anatomical treatise was printed under the title *Vidi Vidii: Florentini de anatome corporis humani libri VII* in a posthumous edition (1626), with illustrations by his nephew (Figure 3.92). These illustrations, of which there were 78, were felt to be maladroit, and were most likely not chosen by Vidius himself. Indeed, some – such as the pictures of the sympathetic nervous system – were almost certainly copied from Vesalius' *Fabrica*. Interestingly, some have stated that Vidius taught anatomy to Vesalius in Paris.

Figure 3.92 Title page from *Vidi Vidii: Florentini de anatome corporis humani libri VII* (1626).

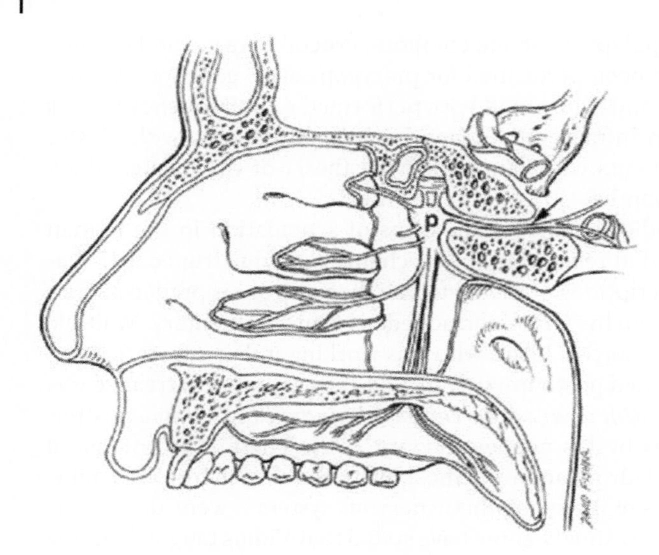

Figure 3.93 Sagittal schematic view of the Vivian nerve (arrow) and pterygopalatine ganglion (p).

Although he is remembered by many only by the canal, artery, and nerve (Figure 3.93) that bear his name, Vidus Vidius was a pioneer of scientific anatomy. It is on the works of such early anatomists and physicians that our present understanding of the intricacies of the human body, and particularly the nervous system, is based.

3.9.3 Giambattista Canano

Giambattista Canano (1515–1579) was born in Ferrara, Italy to a family of physicians and educated there at the university or studium. While his scientific studies were most likely directed by his uncle, Hippolito, he was also influenced by his master in the liberal disciplines, G.C. Giraldi. His anatomical interests were fostered by his close kinsman, Antonio Maria Canano, who had been in contact with Leonardo da Vinci in 1510. Having been inspired by da Vinci, Antonio is credited with the revival of anatomical studies in Ferrara, the spark of which was passed to Giambattista.

Of Canano's extensive anatomical works, only the first was ever published, under the title *Musculorum humani corporis picturata dissectio*. It consisted of 20 leaves with 27 illustrations. As the book remained incomplete and probably never appeared on the market, only a few copies are known to exist. One of these was held at the Royal Library in Dresden, and contained a title page that dates its printing to before 1543. In addition, the illustrations seem to indicate that the work was completed in the pre-Vesalian period of anatomy. Ludwig Choulant suggests that the appearance of Vesalius' *Fabrica* in 1543, which was widely accepted, might have discouraged Canano from completing his own works. In the face of criticism by most Galenists and Vesalian anatomists, it was Fallopius (1523–1562) who rightly credited Canano with the first description of the palmaris brevis (Figures 3.94 and 3.95). The original description, as referenced by James Douglas, is as follows:

> It arises by a membrane-like tendon from the superior and external part of the os metacarpi minimi digiti; whence ascending obliquely and adhering to the fourth bone of the carpus that lies upon the third it grows fleshy in two or three places, being separated by intervening membranes and passing under the palmaris longus. It is inserted tendinous into the ligamentum annulare and into the bone of the carpus that articulates with the

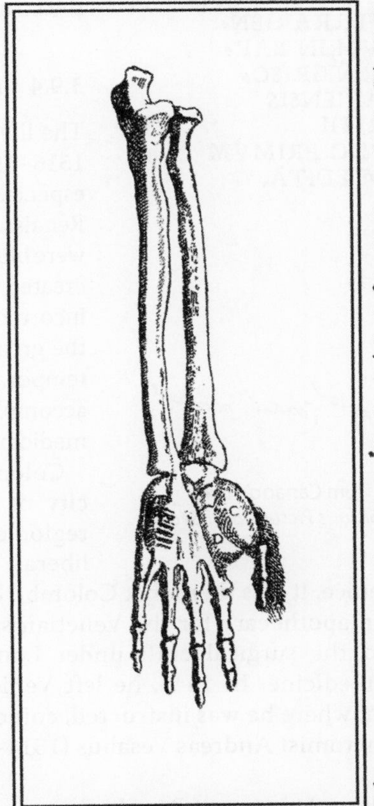

LIBER PRIMVS.
C. Musculus, oriens
à primo osse brachia-
lis ppè magnũ digitũ
cui inseritur, habet se
cundũ Galenũ abdu
cere eũ ab aliis. At vi
detur factus ad trahẽ
dum potius magnũ di
gitũ versus minimũ.
D. Musculus suo ca
pite ossi postbrachia-
lis antè mediũ digitũ
affixus, obliqs fibris in
serit primo ossi magni
digiti, habet adducere
magnũ digitũ ad indi
cẽ interius inclinãdo.
E. musculus, siue mus
culi q. siti sunt, supra
musculũ abducẽtẽ mi
nimũ digitũ ab aliis,
in parte interiore ma-
nus extremę, obliquis
fibris cuti hęrent, &
suis tẽdinibus tendini
volę iunguntur, horum musculorũ nõ meminit Galenus, qui ad
tẽdinis volę dilatationem facti videntur.

Figure 3.94 Original depiction of the palmaris brevis muscle by Canano.

thumb. The upper part of this tendon adheres to the abuductor pollicis and its lower part to the flexor fecundi internodii ejusdem. Its use if to make the palm of the hand hollow, by drawing the ball of the thumb towards the os metacarpi that sustains the little finger and so forms what they call Diogenes' cup.

Conversely, Realdo Columbo (1511–1553) failed to acknowledge Canano's contribution, and in taking the discovery for his own, proclaimed: "Mark this muscle well, for none of the ancients knew of its existence, nor did Vesalius"; however, since only four copies of Canano's work were ever published, Columbo may not have had access to it. Like so many other contributions of the times, Canano's work seems to have succumbed to the overwhelming pressure to accept and highlight "all things Vesalian." Accordingly, Roth belittles Canano in order to magnify Vesalius. Although an engraving by Bartolomeo Eustachi (1500?–1574) based upon Canano's work, depicting the palmaris brevis, was completed in 1552, it was not published until more than a century and a half later. It is perhaps this delay in publication, as well as the deification of Vesalius, that has allowed most modern references to date the first descriptions of the palmaris

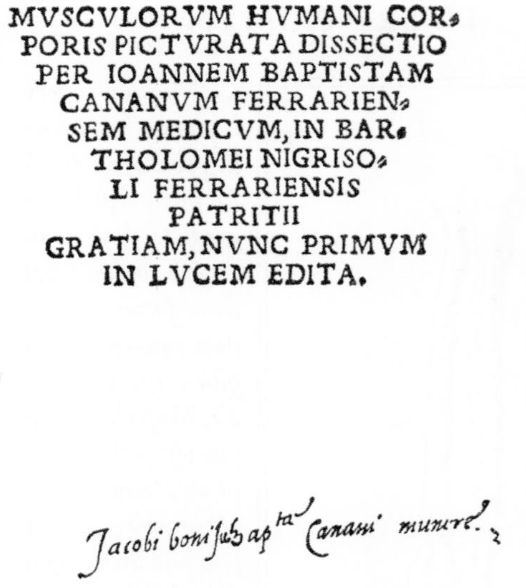

MVSCVLORVM HVMANI COR-
PORIS PICTVRATA DISSECTIO
PER IOANNEM BAPTISTAM
CANANVM FERRARIEN-
SEM MEDICVM, IN BAR-
THOLOMEI NIGRISO-
LI FERRARIENSIS
PATRITII
GRATIAM, NVNC PRIMVM
IN LVCEM EDITA.

Figure 3.95 Title page from Canano's text, *Musculorum Humani Corporis Picturata Dissectio.*

brevis incorrectly to the eighteenth century. It is striking that Columbo's apparent rediscovery of the structure is similarly overlooked.

3.9.4 Matteo Realdo Colombo

The life and work of Matteo Realdo Colombo (c. 1516–1559) (Figure 3.96) reflect the cultural and especially religious upheavals that followed the Renaissance. Born in an era when distinctions were blurred and orthodoxies destroyed, Colombo created a lasting reputation by both challenging incorrect medical convention and influencing the great artwork of the time. Although his contemporaries are often held in greater esteem, his accomplishments helped set the stage for modern medicine.

Colombo was born in the northern Italian city of Cremona but moved throughout the region during his education. He completed his liberal arts education in Milan before his family moved to Venice. It was there that Colombo began his surgical training, in 1533. His father worked as an apothecary for the Venetian surgeon Giovanni Antonio Lonigo, and Colombo practiced the surgical craft under Lonigo's tutelage for seven years before deciding to study medicine. In 1540, he left Venice to study anatomy at the acclaimed University of Padua, where he was instructed, coincidentally, by another apothecary's son: the great Belgian anatomist Andreas Vesalius (1514–1564), who held the University's Chair of Anatomy.

Colombo proved to be one of Vesalius' most talented dissectors. In fact, Vesalius appointed the former surgeon his prosector, and in 1541 his assistant. In 1543, Vesalius traveled from Padua to Basel, Switzerland to assist in the publication of his monumental text, *De Humani Corporis Fabrica* ("On the Fabric of the Human Body"), and Colombo was temporarily awarded the departmental chair. In 1544, Vesalius informed the university that he would be leaving academia permanently, and Colombo's position became permanent. Despite their relatively short period of direct interaction, Colombo's relationship with Vesalius would go on to shape his career.

Vesalius is perhaps most famous for his iconoclasm – particularly his attacks on Galenic medicine – and Colombo would prove to be no different. Like Vesalius and other post-Renaissance scientists, Colombo did not blindly accept anatomical convention. Instead, he personally investigated matters before subscribing to a school of thought. Such direct observations allowed him to identify errors in prevailing wisdom. For example, he noted that the position of the right kidney was lower than that of the left (previously, the opposite was believed to be true, as demonstrated in the *Fabrica*'s erroneous illustrations), he observed the correct function of the cardiac valves, he showed that the lens lies at the anterior aspect of the eye (previously, it was thought to occupy the central portion, as also incorrectly shown in the *Fabrica*), and he disproved the existence of calvarial sutures configured in the shape of the Greek letter X (still another ancient error propagated in the *Fabrica*).

Figure 3.96 Matteo Realdo Colombo (c. 1516–1559), from his *De Re Anatomica* (1559).

Colombo's greatest correction was his accurate description of the pulmonary circulation, contradicting Galen's traditionally accepted septal-pore theory. Galen had proposed that blood in the left ventricle bypasses the lungs and crosses into the right ventricle via pores in the intraventricular septum. In this theory, both the lungs and the pulmonary veins serve as conduits for air, and the mixing of air and blood first occurs within the heart itself. While he was not the only scientist to express disbelief in Galen's concept of the pulmonary circulation, Colombo's work had the widest impact, for various reasons (e.g., the proposal of a viable and mostly correct alternative theory, the broad availability of his treatise, his reputation, religio-political considerations, etc.). For such achievements, it would seem that Colombo should be forever mentioned alongside the storied names of medicine; however, he was impolitic enough to reveal Vesalius' errors, not just Galen's, and Vesalius proved to be much more capable of delivering criticism than receiving it.

When Vesalius returned to Padua following the successful publication of the *Fabrica,* he became aware of disparaging remarks made by Colombo regarding aspects of his work. This placed an immense strain on their relationship; where there was once friendship and admiration, there was now animosity. Vesalius struck public blows against the young Italian in the 1546 "Letter on the China Root," in which he stated that Colombo "learned something of anatomy by assisting me in my work, although he was incompletely educated." This is in contrast to the assessment he made in the first edition of the *Fabrica,* where he referred to "my friend

Colombo, skilled professor at Padua, most studious of anatomy." All mentions of Colombo were removed in the second edition. It is notable that Vesalius belittled his own instructor, the French anatomist Sylvius: he not only criticized Sylvius' Galenist leanings, but also remarked that his teacher's motives for entering the medical field were material rather than noble. The relationship between Sylvius and Vesalius deteriorated in a manner similar to that between Vesalius and his own student.

The latter feud – in conjunction with Vesalius' prominence and reputation – led to a bias against Colombo, effectively tarnishing his reputation in some circles. This bias still exists in contemporary discussions of sixteenth-century medicine. Some historical commentaries have incorrectly labeled Colombo a Galenist, while others have chosen to ignore his discovery of the pulmonary circulation, attributing this feat to the Englishman William Harvey (1578–1657).

However, all was not lost for Colombo. The anatomist attracted the attention of the Duke of Tuscany, Cosimo de' Medici. De' Medici offered Colombo a teaching position at the University of Pisa, which the anatomist accepted in 1547. Despite the university's lesser reputation, accepting the duke's offer aligned Colombo with one of the most powerful families in Europe. This proved to be an advantageous move for him, since he was offered a teaching position at the Papal Medical School in Rome just one year later. While this position required Colombo to move from his native northern Italy, the intellectual opportunities presented by Rome proved irresistible; Colombo moved to this southern epicenter of Italian culture in 1548.

Living and working in Rome provided Colombo with interesting collaborative opportunities, most notably with the artist Michelangelo. Colombo first met Michelangelo owing to the great artist's recurrent bouts of urolithiasis. The anatomist provided treatment for this painful condition, and a close friendship developed between the two. They shared a common interest in anatomy, since Michelangelo used the anatomical knowledge he gleaned from attending and performing dissections to enhance his artwork. Colombo furthered this interest throughout their friendship, even providing the artist with a cadaver for dissection. He eventually approached Michelangelo to work jointly on an illustrated anatomical text; unfortunately, nothing came of this collaboration prior to Colombo's death in 1559. Colombo's *De Re Anatomica* ("On Anatomy") was published posthumously the same year, with only one illustration: the title page's beautiful dissection-scene woodcut. Although lacking the visual component typically desired in an anatomical work, *De Re Anatomica* went on to become an influential text in its time, perhaps even more so to ordinary physicians than the more heralded and richly illustrated *Fabrica* of Vesalius.

Colombo certainly achieved greatness during his time, yet his story remains tragic. Capable, bright, and brave, he questioned and illuminated where others blindly accepted. While not as well known as his contemporaries, he sacrificed his fame for a greater cause. If Colombo and others had never challenged Vesalius, then Vesalius' teaching likely would have petrified into a new medical canon, supplanting Galenic teachings with equally static Vesalian ones. The fact that Vesalius' own student and friend found truth and the progress of science more compelling than currying favor validated Vesalius' own principles and emphasized the Renaissance's triumph of reality over orthodoxy.

3.9.5 Costanzo Varolio (Constantius Varolius)

In 1593, Pope Paul III instituted the Chair of Anatomy (associated with that of Surgery) in Rome. Many great physicians and anatomists hailed from this chair, including Arcangelo

Piccolomini (1525–1586), Bartolomeo Eustachio (1520–1574), and Domenico Mistichelli (1675–1715), as well as Costanzo Varolio (1543–1575) (Figure 3.97). Although Varolio made many discoveries, little is known of his life.

Varolio was born in Bologna and died in Rome. A professor of anatomy (1569–1572) and papal physician, he was the first to examine the brain from its base up, in contrast with previous top-down dissections. This allowed the cranial nerves to be observed more clearly. Varolio named the pons the pons cerebelli, and even today this structure is known to many as the "bridge of Varolius" or pons Varolli.

Varolio first studied philosophy, and then medicine. He studied anatomy under Julio Caesar Aranzio (1530–1589) at the University of Bologna, and received his degrees in 1567. Interestingly, his teacher Aranzio was a pupil of Vesalius at Padua, and at the young age of 27 was elected Professor of Medicine and Surgery at the University of Bologna. He later became physician to Pope Gregory XIII, a role that Varolio may have also performed, although some have debated this.

In 1569, the Senate of Bologna created an extraordinary Chair in Surgery for Varolio, with responsibility for teaching anatomy as well. Varolio went to Rome in 1572 and may have taught at the Sapienze. It was in Rome that he honed his trade as a surgeon, and he is referenced on a plaque as "having great skill in removing stones."

Varolio's text *De Nervis Opticis Nonnullisque Aliis Praeter Communem Opinioneu in Humano Capite Observatis*, consisting of a letter to Mercurialis dated April 1, 1572, Mercurialis' answer, and Varolio's reply, was published in 1573. Choulant stated that it was published without the approval of Varolio. The work contained three woodcuts pertaining to the brain and created by Varolio himself (Figure 3.98). Although elementary in essence, these woodcuts were considered distinctive and instructive for their day. The human basilar pons was first described in this text. Additionally, Varolio's publication made comments

Figure 3.97 Drawing of Costanzo Varolio (1543–1575). (*Source*: From Biblioteca Comunale di Bologna.)

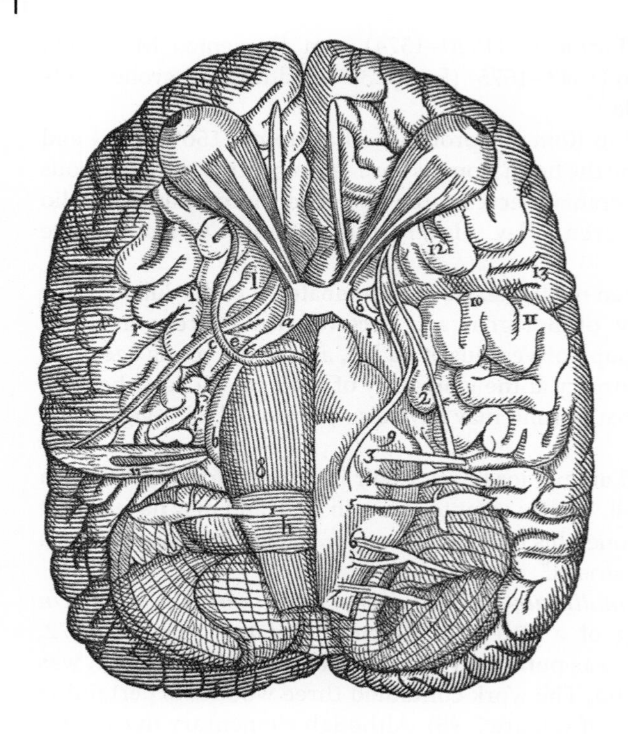

Figure 3.98 Woodcut illustration from Varolio's *De Nervis Opticis Nonnullisque Aliis Praeter Commune Opinioneu in Humano Capite Observatis*, published in 1573.

regarding the hippocampus, cerebral peduncles, and the so-called valve of Varolius (ileocecal valve). Interestingly, Sappey (1810–1896) reported plagiarism by Bauhin (1560–1624), stating that "Bauhin's valve" was actually the valve of Varolius discovered in 1573. A second work by Varolio, a teleological physiology of man, was published posthumously in 1591: *Anatomiae Sive de Resolutione Corporis Humani ad Caesarem Mediovillanum libri IV* (Figure 3.99).

Although he died at just 32, Varolio contributed much to our current understanding of the brain. Neuroanatomy was very much in its infancy during the late sixteenth century. Vavolio's approach to the dissection of the brain, his specific discoveries (notably the pons), and his refined descriptions of a number of brain structures laid an important foundation for subsequent contributions.

3.9.6 Giulio Cesare Casseri

Giulio Cesare Casseri (1552–1616) (Figure 3.100) was born in Piacenza, Italy. Like many early authors, he is known by several names, including Julius Casserius, Piacentino, Ivlii Casserii Placentini, and Julius Casserius Plancentinus, which means "Giulio Casserio of Piacenza." As a young man, he moved to Padua while in the service of an Italian student. Shortly thereafter, he was hired as a servant in the household of Gerolamo Fabrici d'Acquapendente, also known as Fabricius, who served as Public Lecturer of Anatomy and Surgery for the Universitá Artista. Casserius was so inspired by Fabricius that he began to study anatomy in his free time, including teaching himself Greek and Latin in order to review classical texts.

Although the exact dates are unknown, it is clear that Casserius enrolled in Padua's School of Medicine and obtained a degree in medicine and philosophy around 1580. Following his graduation,

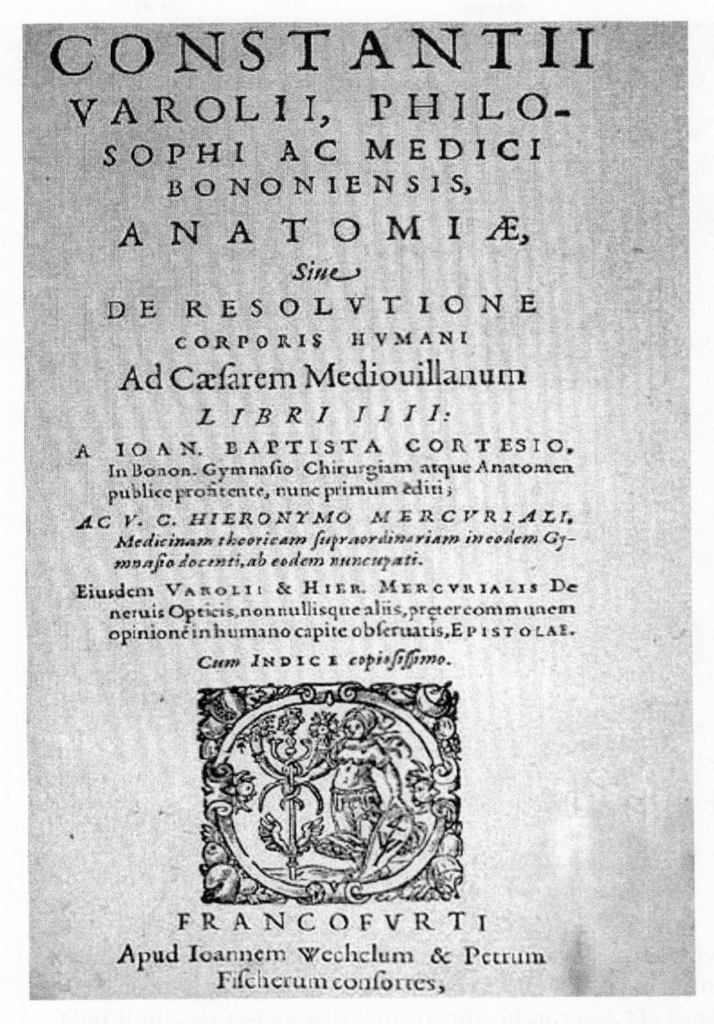

CONSTANTII
VAROLII, PHILO-
SOPHI AC MEDICI
BONONIENSIS,
ANATOMIÆ,
Siue
DE RESOLVTIONE
CORPORIS HVMANI
Ad Cæſarem Mediouillanum
LIBRI IIII:
A IOAN. BAPTISTA CORTESIO,
In Bonon. Gymnaſio Chirurgiam atque Anatomen
publice profitente, nunc primum editi;
AC V. C. HIERONYMO MERCVRIALI,
Medicinam theoricam ſupraordinariam in eodem Gy-
mnaſio docenti, ab eodem nuncupati.
Eiusdem VAROLII & HIER. MERCVRIALIS De
neruis Opticis, nonnullisque aliis, preter communem
opinionē in humano capite obſeruatis, EPISTOLAE.
Cum INDICE *copioſiſſimo.*

FRANCOFVRTI
Apud Ioannem Wechelum & Petrum
Fiſcherum conſortes,

Figure 3.99 Title page of Varolio's *Anatomiae Sive de Resolutione Corporis Humani ad Caesarem Mediovillanum Libri IV* (1591).

he remained in Padua, and began giving private lectures to medical students, while also working as Fabricius' anatomical dissector and starting what became a successful surgical practice.

Casserius' prestige rose quickly, and in 1584 he took Fabricius' place as an examiner on the board for surgical licensing; at the time, surgery was considered a lesser branch of medicine. However, Fabricius grew to disapprove of Casserius' teaching methods, and tensions soon grew between them. In 1595, Casserius acted as a substitute lecturer while Fabricius took a leave of absence for illness. Sterzi describes how upon his return, Fabricius recognized and resented the enthusiasm with which his students thanked Casserius. In response, he resumed his position on the board for surgical licensing, which had been Casserius' for the previous 11 years.

From 1597 to 1598, Fabricius was forced to shorten his public course on anatomy due to a shortage of cadavers. At the same time, Casserius hosted a five-week private course in his home, during which he dissected a monkey, several dogs, and nine human cadavers. His students were so grateful that they presented him with an expensive silver chandelier. Fabricius,

Figure 3.100 Portrait of Giulio Cesare Casseri (1552–1616), from the frontispiece to *De Vocis Auditus que Organis*. (*Source:* Picture in public domain.)

no doubt inflamed by his former pupil's success, responded by appealing to the academic authorities to enforce a statute prohibiting private anatomy lectures. This appeal may have been successful, as there is no record of Casserius hosting further private lectures until 1604.

The feud between Casserius and Fabricius festered as they both aired grievances in their works. They published anatomical works on the sense organs almost simultaneously: Casserius' *De Vocis Auditusque Organis Historia Anatomica* (1600–1601) and Fabricius' *De Visione Vocis Auditus*. Neither acknowledged the works or contributions of the other. Wright believes that the absence of tribute was due to the timing of experimentation and publication. However, Wright's work is concerned with the history of the medicine of the nose and throat, and not with the deteriorating relationship between Casserius and his former mentor. For his work, Casserius had the German painter Josias Mauer stay in his home for the purpose of "painting anatomic illustrations." Both anatomists claimed to be preparing texts complete with colored plates, a resource that was not yet available to medical students. Casserius, in particular, felt that a comprehensive treatise on anatomy would be his life's work. According to Sterzi, Casserius boasted in a letter of 1613 that 150 figures engraved in copper were ready for publication.

Fabricius experienced recurrent bouts of illness, and in 1604 the Rector of Padua asked Casserius to take over the anatomy course, most likely in response to the requests of his medical students. Casserius accepted the position but refused to teach in Fabricius' public theater, and instead continued to work from his home. From 1605 to 1608, he gave successful private lectures, while Fabricius' courses were reduced to a minimum. This continued until 1609, when the University of Padua separated the teaching of anatomy and surgery into two disciplines; anatomy was given back to Fabricius and surgery to Casserius. The two taught their respective departments until 1613 when, after 50 years of lecturing, Fabricius retired from the University of Padua and recommended Giulio Cesare Sala as his replacement. Despite achieving a lectureship at last, Casserius still refused to teach in the public theater and continued to host his dissections at home.

In January 1616, Casserius finally taught dissection in a public theater at the request of the academic authorities. The three-week course was so successful that his students, in another display of gratitude, published a laudatory booklet thanking him for his teaching.

Shortly after the conclusion of the course, and at the apex of his fame, Casserius' career was cut short: he contracted a fever and died on March 8, 1616. Fabricius held the Chair of Anatomy until his death in 1619, when the departments of Anatomy and Surgery were reunited under the leadership of Adriaan van den Spieghel (1578–1625).

In his will, Spieghel (Spigelius) would ask the German physician Daniel Rindfleisch to publish his posthumous work *De Humani Corporis Fabrica* ("On the Fabric of the Human Body"), which had no illustrations. Rindfleisch approached Casserius' heirs and asked permission to include his anatomical plates in Spieghel's work. Eventually, 77 plates from this collection were added to the text (Figures 3.101–3.106).

Casserius' ambition to create a comprehensive anatomy text was not realized during his lifetime, but *Iulii Casseri Placentini Tabulae Anatomicae* was published posthumously in 1627, with anatomical illustrations drawn by Odoardo Fialetti and etched on copper plates. In this volume, Casserius gave several novel anatomical descriptions and often attempted to explain human anatomy by reference to lower animals. He was the first, for example, to accurately describe the muscles of the abdomen and the back. Casserius also discovered the lumbrical muscles of the hand, the transverse head of the adductor hallucis, and the inguinal fossae, and was the first to illustrate the circular folds of the small intestine, the vascular supply of the appendix, the adrenal glands, the musculature of the urinary bladder, the navicular fossa of the urethra, the superior and inferior laryngeal nerves, and the prostate gland as a single organ. Fallopius (1523–1562) was the first to describe the adrenal glands, and Casserius confirmed his finding with illustrations. These illustrations were so well received that they continued to be reprinted in anatomy textbooks throughout the seventeenth and eighteenth centuries, and are even appreciated today as baroque-inspired art. The English anatomist and surgeon John Browne (1642–1702) would use copies of the anatomical drawings from the *Iulii Casseri Placentini Tabulae Anatomicae* for his own anatomy text, and would go on to be labeled a plagiarist. Choulant would say of the artwork used by Casserius:

> Casserius' plates mark a new epoch in the history of anatomic representation, owing to the correctness of the anatomic drawing, their tasteful arrangement, and the beauty of their technical execution. And this all the more, since they cover the whole field of anatomy and have become the models for anatomic illustrations in copper, just as the Vesalian representations had been for anatomic woodcuts.

Julius Casserius was a popular and inspirational lecturer and a gifted anatomist. He is remembered eponymously by Casserio's muscle and nerve, the coracobrachialis muscle and musculocutaneous nerve, respectively. The nerve is also often referred to as the "perforans Casseri." Unfortunately, his legacy, like his life, is dominated to an extent by his feud with his former mentor, Fabricius. However, this should not overshadow the substantial contributions that he made to anatomy. During his life, Casserius saw the publication of two major works, *De Vocis Auditusque Organis Historia Anatomica* (1600–1601) and *Pentaestheion* (1609). *Iulii*

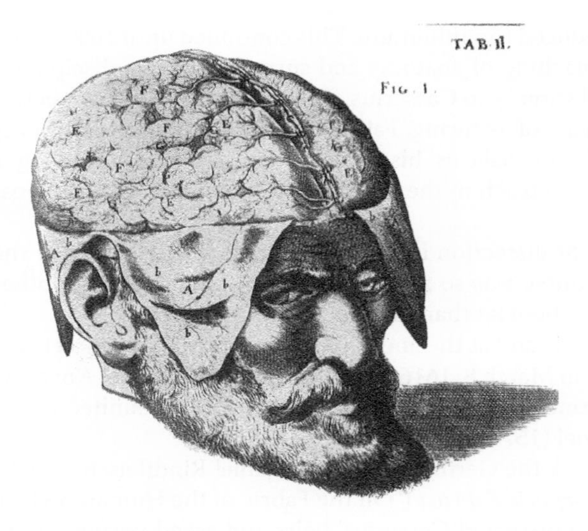

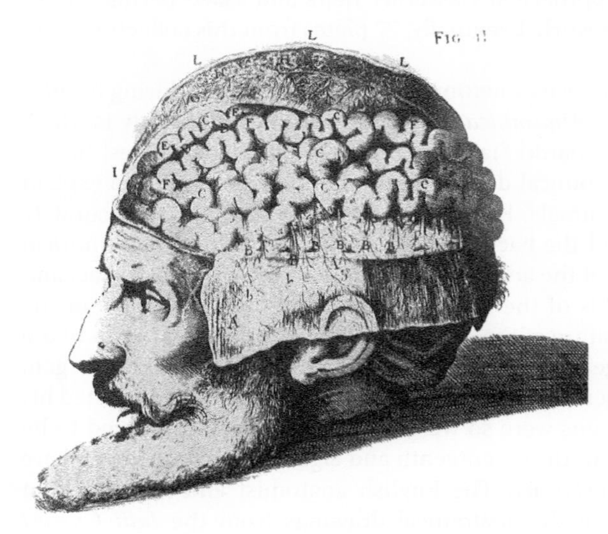

Figure 3.101 Illustration from Casserius' *Tabulae Anatomicae* (1627), depicting the brain and meninges. (*Source:* 1971, Editions Medicina Rara, New York, Facsimile of the 1627 Venice edition.)

Casseri Placentini Tabulae Anatomicae (1627) contained numerous discoveries and innovative anatomical descriptions, cementing his position as one of the leading anatomists of his day.

3.9.7 Giovanni Maria Lancisi

Giovanni Maria Lancisi (1654–1720) (Figure 3.107) was born in Rome, but his mother died during parturition and he was raised by her sister in Orvieto, Italy. At the age of 12, he returned to Rome to study theology at the Jesuit Roman College. His academic interests apparently evolved, as he switched from theology to physical sciences before settling on studying medicine at Sapienza University. During his medical studies, Lancisi developed an interest in patients and post-mortem examination while working at hospitals throughout Rome. This experience culminated in his earning a doctorate in medicine degree at 18 years of age, and the position of assistant physician at Santo Spirito Hospital at 22.

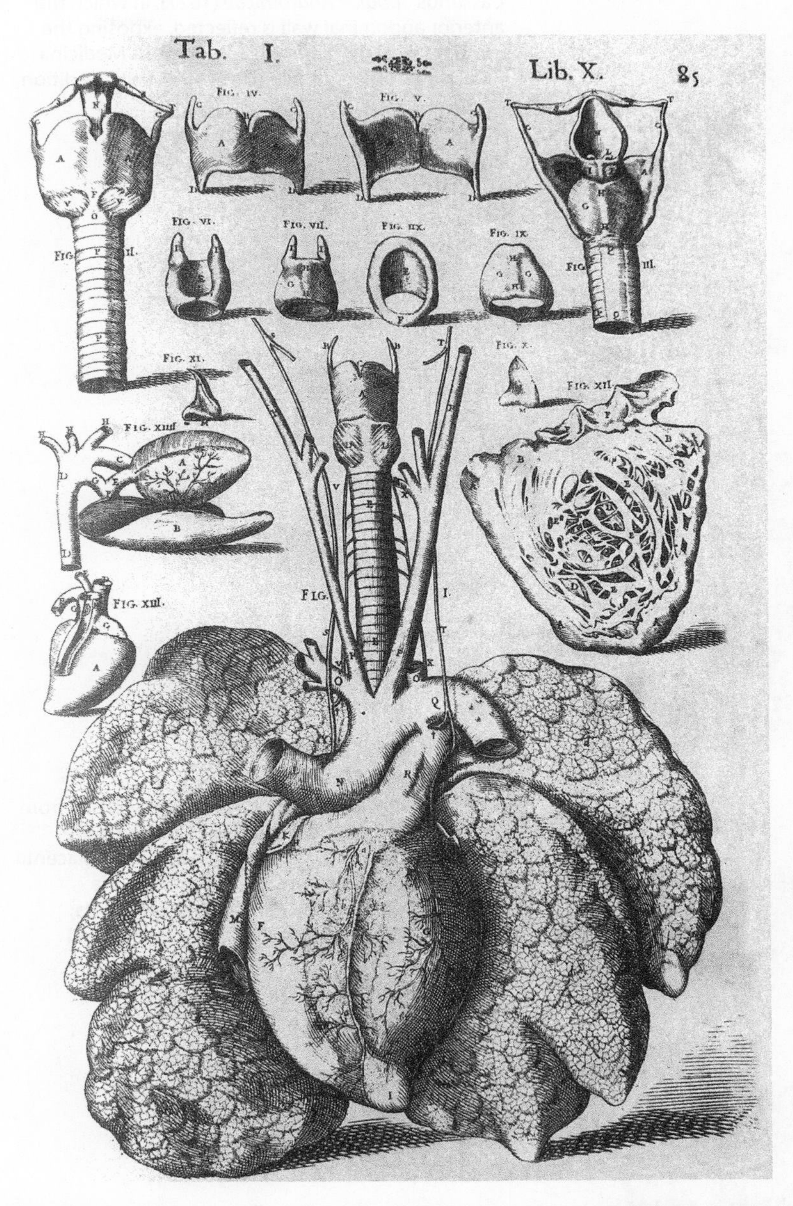

Figure 3.102 Illustration from Casserius' *Tabulae Anatomicae* (1627), depicting the larynx, the heart, and the lungs. (*Source:* 1971, Editions Medicina Rara, New York, Facsimile of the 1627 Venice edition.)

Lancisi developed his knowledge and skills as a clinician during his tenure at Santo Spirito Hospital. In 1676, his contemporary Assalto noted, "[Lancisi is] attending at the bedsides of patients, carefully not[ing] their signs and symptoms, [and using] his skill to explore the causes of their diseases." After only two years at Santo Spirito, he sought further training in medicine at Picentine College, at the same time studying chemistry, botany, geometry, and mathematics. In 1684, at the age of 31, he moved back to Sapienza University, where he was appointed Chair of Anatomy, a position he held for 13 years.

Figure 3.103 Illustration of a pregnant woman from Casserius' *Tabulae Anatomicae* (1627), in which the anterior abdominal wall is reflected, exposing the peritoneal cavity. (*Source:* 1971, Editions Medicina Rara, New York, Facsimile of the 1627 Venice edition.)

Figure 3.104 Illustration of a pregnant woman from Casserius' *De Formato Foetu* (1627), in which the anterior abdominal wall is reflected and the placenta with the surrounding vessels and the fetus is exposed. (*Source:* 1971, Editions Medicina Rara, New York, Facsimile of the 1627 Venice edition.)

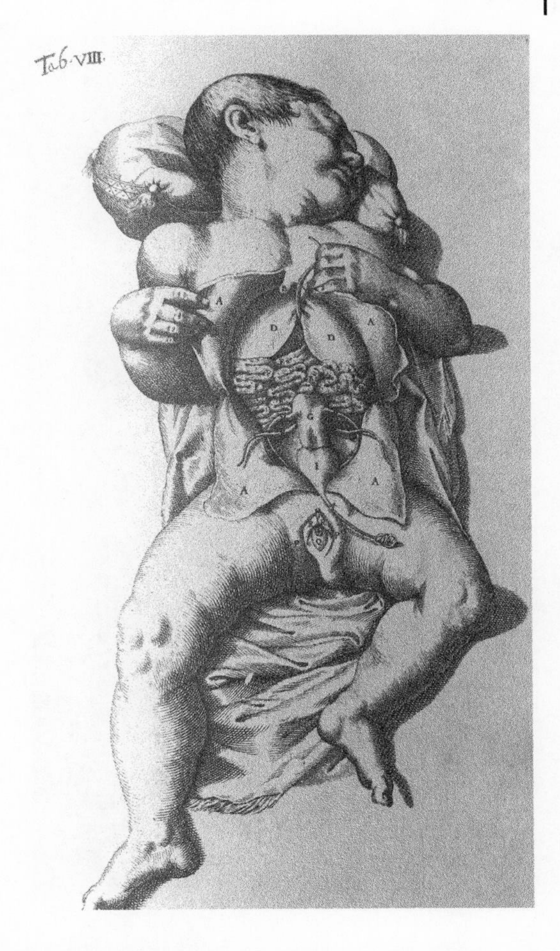

Figure 3.105 Illustration of a fetus from Casserius' *De Formato Foetu* (1627), in which the contents of the abdominal cavity are exposed. (*Source:* 1971, Editions Medicina Rara, New York, Facsimile of the 1627 Venice edition.)

Lancisi served three different popes. While studying at Picentine College and as Chair of Anatomy at Sapienza, he caught the attention of Pope Innocent XI. When the pope's personal physician died in 1688, Lancisi was appointed to the position, at the same time becoming his *intimus cubicularius* (secret waiter). In addition to being the pontifical doctor, Lancisi was also appointed a representative of Cardinal Altieri, leading the pontifical committee for conferring degrees in the medical college of Sapienza. When Innocent XI died in 1689, Lancisi performed the autopsy. He then left the Vatican until the health of Innocent XI's successor, Innocent XII, deteriorated, and he was summoned back to care for him until his death in 1700. The subsequent election of Clement XI kept Lancisi at the Vatican for the rest of his life, as he was appointed archiater of the Papal States.

With his position at the Vatican came new medical and scientific opportunities. During the early eighteenth century, when an epidemic broke out in Rome, causing sudden deaths among the population, Clement XI issued Lancisi authority to perform autopsies on selected patients. A year later, in 1707, he published the first autoptic manuscript, *De Subitaneis Mortibus* ("On Sudden Death") (Figure 3.108), providing a potential link between myocardial disease and sudden death. Lancisi formulated an exhaustive list of different types of death:

> Indeed this absolutely complete cessation of animal movements and this departure of the soul from the body, even though it happens at all times more swiftly than thought

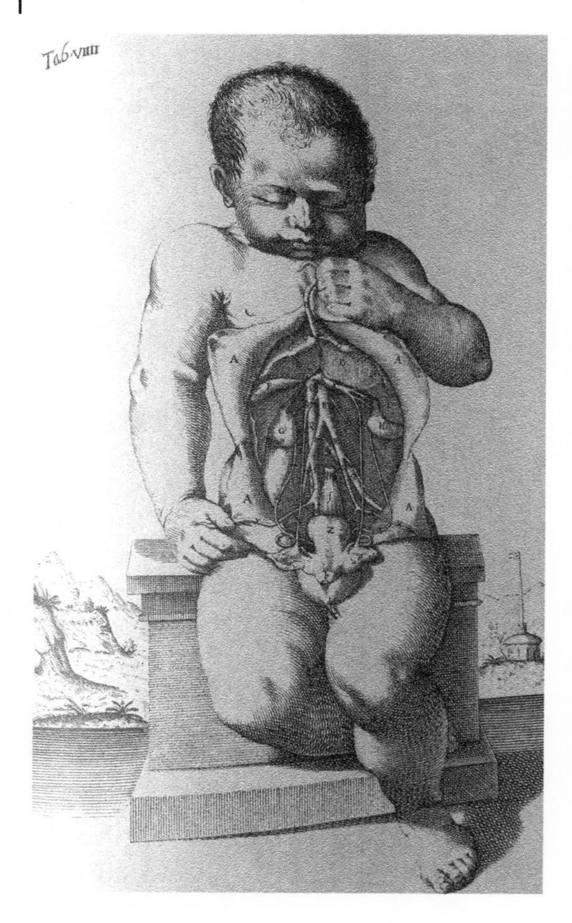

Figure 3.106 Illustration of a fetus from Casserius' *De Formato Foetu* (1627), in which the abdominal aorta and the inferior vena cava are exposed. (*Source:* 1971, Editions Medicina Rara, New York, Facsimile of the 1627 Venice edition.)

itself, is nevertheless divided for the sake of common parlance and for greater clarity of teaching, into natural, untimely and violent death, and those again individually into slow and sudden death, into those that are foreseen and forefelt and finally into such as unforeseen, imperceptible and unexpected.

Lancisi's interest in autopsy and post-mortem reporting developed early in his medical career, when he was still a student, and led to his fascination with sudden death. He noted that such death could occur when any of the three major fluids (air, blood, and nervous "fluid") or three major solids (the respiratory, cardiovascular, and nervous systems) was disrupted. Additionally, he believed sudden death could result from abnormalities of the heart or great vessels. Lancisi thought that any serious disorder impairing either the systolic or the diastolic function of the heart could lead to death. His pathological reports point toward these dysfunctions as underpinning the association between myocardial disease and sudden death in Rome in 1706.

Lancisi's interest in the anatomy and function of the heart extended beyond *De Subitaneis Mortibus*. As Chair of Anatomy at Sapienza University, he collaborated with Marcello Malpighi (1628–1694) in heart embryology studies. During this time, Lancisi was also associated with Tozzi and Galliani, and corresponded with such eminent medical men as Bellini, Boerhaave, Morgagni, and Heister. Lancisi was also the first to describe the chest pain commonly known as angina pectoris, although it was 50 years before that term was coined. Lancisi referred to Hippocrates' Coan prognostications concerning the association of chest pain with sudden death in old men, and quoted Hollerius, who recorded a case of sudden death in a 70-year-old man suffering from chest pain and dyspnea. Describing his own experience, he noted that "pains in the interior of the chest accompanied by difficulty in breathing, especially on going uphill, 'cause' instantaneous death." It is clear from the original Latin text, however, that it was the dyspnea and not the pain that was related to going uphill. Lancisi also described in detail the case of a 50-year-old man from Genoa who developed difficulty in breathing and a painful tension of the intercostal muscles toward the sternum after violent exercise while hunting. These symptoms subsided with rest, but, subsequently, he experienced constriction of the thorax whenever he went upstairs or uphill. However, comparatively little space is given to discussion of chest pain in Lancisi's writings, and it is ranked third in his list of the symptoms associated with sudden death, which included dyspnea on effort, orthopnea, severe coughing, irregularities and undue slowness of the pulse, palpitations, and paroxysmal disorders of consciousness.

Figure 3.107 Portrait of Giovanni Maria Lancisi (1654–1720).

Lancisi's work on aneurysms is also well documented in the posthumous publication of *De motu cordis et aneurysmatibus* ("On the Motion of the Heart and on Aneurysms"), eight years after his death. In this manuscript, he referred to dilatation of the heart as an aneurysm, distinguished between true and false aneurysms, and delineated the relationship between syphilis and aneurysms. More specifically, he described specimens with wart vegetations and thickening of the valve cusps, the occurrence of so-called "aneurysma gallicum" (a chronic dilatation of the heart as a sequel of "lues," i.e., syphilis), the relationship between valvular stenosis and cardiac alterations, and the differentiation between cardiac hypertrophy and dilatation. Alongside his classification of aneurysms, Lancisi described the underlying etiological factors as heredity, mechanical obstruction, calcified/incompetent valves, large vessel calcification, palpitation, nervous disorders, and asthma.

Furthermore, Lancisi contributed to the knowledge of heart failure by describing a patient with aortic valve disease and mitral regurgitation. He also made the very important observation that dilatation of the right side of the heart was regularly associated with prominent pulsation of the jugular veins. He attributed this to regurgitation through the tricuspid orifice, and laid great emphasis on its importance as a physical sign. Lancisi also mentioned that a cause of cardiac enlargement was calcified coronary arteries, because of the obstructive nature of the disease. He recognized the mechanical barrier produced by narrowing of the valvular orifices, having noted that mercury appeared in the chambers of the heart after being injected directly into the coronary arteries. He speculated that the mercury passed into the ventricular chambers via venous channels.

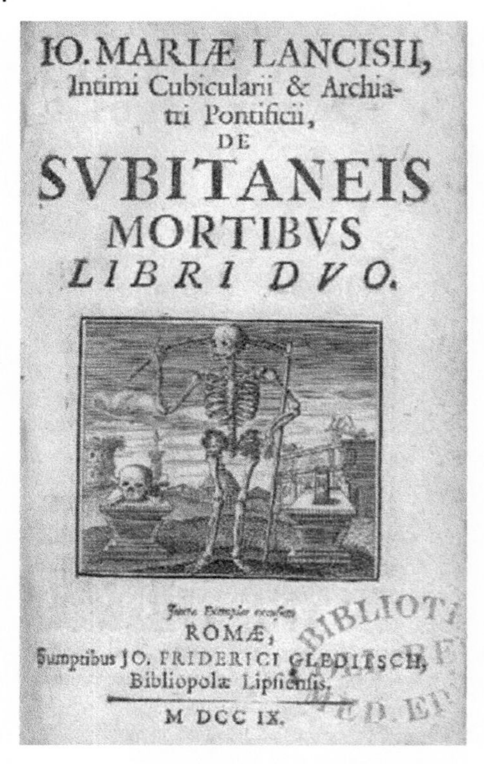

IO. MARIÆ LANCISII,
Intimi Cubicularii & Archiatri Pontificii,
DE
SVBITANEIS
MORTIBVS
LIBRI DVO.

ROMÆ,
Sumptibus JO. FRIDERICI GLEDITSCH,
Bibliopolæ Lipsiensis.
M DCC IX.

Figure 3.108 Original cover page of Lansici's *De Subitaneis Mortibus*.

In addition to cardiology, Lancisi had many medically related interests, including nephrology and urine formation. In his 1696 book *Tractatus de Urinis* ("Treatise on Urine"), Lancisi was the first to write on the physiological mechanisms of urine formation. He was interested in urine being the product of "processed" blood, its relationship to ingested liquid, and the anatomical and functional description of the kidneys and bladder. He believed urine was a watery medium consisting of phlegma, volatile salts, fixed salts, an earthy substance, and fetid oil.

Arguably, Lancisi's most famous anatomical contribution was his discovery of striae longitudinales mediales corporis callosi (medial longitudinal striae of the corpus callosum, or nerves of Lancisi). These nerves are myelinated fibers that form ridges in the indusium griseum on the superior aspect of each half of the corpus callosum. They connect the septal area to the hippocampus in the temporal lobe, potentially making them a component of higher cognitive functions of the limbic system and involving them in the acquisition of new data for memory. A study by Arroyo-Guijarro et al. analyzed Lancisi's nerves in bulls, depicting them as a possible focus for controlling aggressive behavior. Regardless of their function, the medial longitudinal striae are still an enigma to neurophysiologists, 300 years after Lancisi described their anatomy in meticulous detail.

Lancisi made other contributions to the field of neurology. In 1710, he wrote *Dissertatio Physiognomica* ("Dissertation on Physiognomics"), discussing mimicry of the frontal muscle and fibers of the meninges. This was followed in 1712 by *Dissertatio de Sede Cogitantis Animae*, in which he wrote about cerebral localizations and specific cortical functions. A major theme of this latter book was Lancisi's quest for the structure related to superior psychic functions. He believed the organ in question was the corpus callosum, disagreeing with Descartes, who believed it was the pineal gland.

Beyond the fields of anatomy and medicine, Lancisi made a major contribution to veterinary medicine and the epidemiological study of disease. During the early eighteenth century, rinderpest (cattle plague) was rampaging through Europe. In response to this plague, Lancisi published *De bovilla peste* in 1715, illustrating its characteristics and detailing measures to control its spread. Notably, Lancisi suggested "stamping out" as a control measure: killing and burying affected animals, prohibiting animal movement, and implementing hygienic and political measures during the time of crisis. He collected all the available facts and information, and used them to illustrate the etiology of epidemic and of endemic diseases in general. Consequently, he impressed upon the papal government the necessity of adopting sanitary measures. The principle of stamping out, first introduced by Lancisi, is still used today when dealing with animal-related epidemics.

Lancisi's greatest contribution to public health may have been his epidemiological work on malaria. In 1717, he published *De Noxiis Paludum Effluviis* ("Of the Noxious Effluvium of

Marshes"), in which he supported the miasmatic etiology of the disease. He considered two intuitive theories: (i) the noxious effluvia of marshes inoculated into host organisms lead to fevers in humans who come into contact with these organisms; and (ii) mosquitoes and other biting insects induce the fever in humans by introducing the effluvia into their bodies. As papal physician, Lancisi was so influential that nearly all physicians accepted his view that "bad air" around marshes was responsible for the periodic fevers (Figure 3.109).

Lancisi was interested in all aspects of learning, developing a fascination for language and literature; he was fluent in Greek and Italian. His diverse academic interests are illustrated by the pleasure he took in applying his scientific and medical knowledge to a historical mystery. In 1717, he exchanged a series of letters with his friend Giovanni Battista Morgagni (1682–1771), discussing the death of Cleopatra, Queen of Egypt. Historically, it is accepted that Cleopatra and her two handmaidens were killed by the venom of two asps, which was Morgagni's point of view when he wrote his initial letter to Lancisi. In Lancisi's reply, he noted that more than two snakes would have been necessary to kill all three women, and that Cleopatra probably drank poison mixed with a narcotic to induce death. In a final rebuttal to Morgagni, he argued that since no literary sources contain a first-hand testimony of Cleopatra's death, conjecture must be used.

Like many lifelong learners, Lancisi was a lover of books, and he authored many during his professional career. When his medical library, the Bibliotheca Lancisiana, opened in the Santo Spirito Hospital on May 14, 1714, two important documents were likely among the vast collection he donated: reprints of the *Collection of Anatomical Plates* of Bartolomeo Eustachius

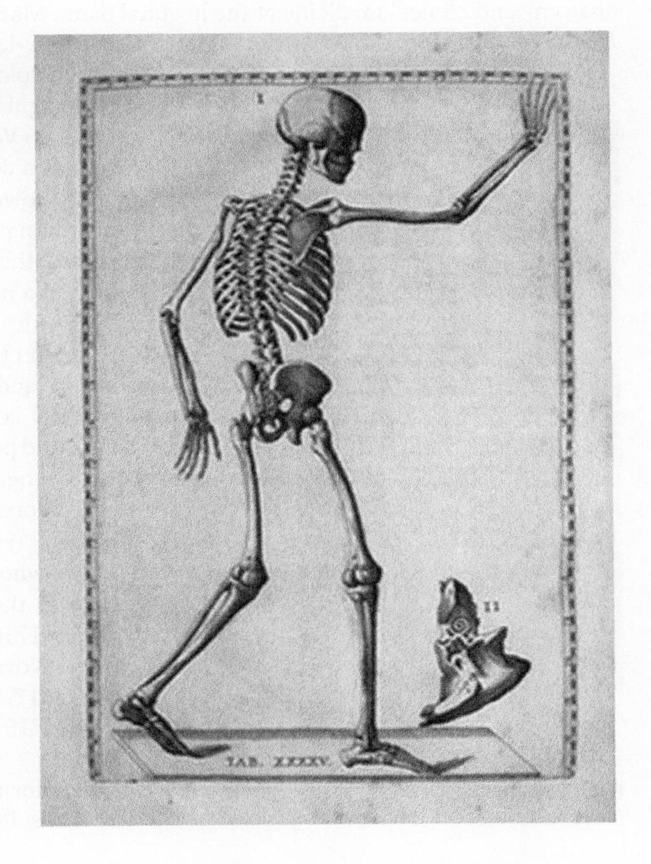

Figure 3.109 One of 47 copperplates by Eustachius hidden in the Vatican Library until the early eighteenth century, when they were presented by Pope Clement XI to his physician, Lancisi. Lancisi published them in 1714, together with his own notes.

(1520–1574) and of the *Metallotheca* of Angela Mercati (1541–1593). Both documents had been written by previous papal physicians. Lancisi discovered 47 copper plates of the *Bartholomaei Eustachii: Opuscula Anatomy. Venet. Vincent. Luchinus Excudebat, 1564.* He edited and published 39 of these in 1714 under the title, *Bartholomaei Eustachii, Quas e Tenebris Tandem Vindicatas... Praefalione, Notisque Illustravit ac Publici Juris Fecit Jo. Maria Lanxisius.* Lancisi wanted to offer young physicians and surgeons a medical education with a broad selection of books, creating "A place where professors and physicians can gather." The library still exists today, with 23 000 volumes and 375 manuscripts.

Giovanni Maria Lancisi, author and physician, died in Rome on January 21, 1720 at 65 years of age. He remains arguably the most influential Italian physician of the seventeenth and eighteenth centuries. His interest in all aspects of medicine enabled him to make significant contributions to cardiology, nephrology, neurology, malarial research, and epidemiological disease control.

3.9.8 Giovanni Battista Morgagni

Giovanni Battista Morgagni (1682–1771) (Figure 3.110) was born in Forli, 40 miles southeast of Bologna, near the Adriatic Sea; his parents were Fabrice and Marie Fornielli. This period marked the onset of the Spanish domination of Italy. From a very early age, Morgagni was interested in science and literature, as well as philosophy. After completing his early studies at Forli, he began studying medicine at the University of Bologna at the age of 16, receiving his doctorate in philosophy and medicine in 1701. He went on to work at the three hospitals in Bologna, studying anatomy and clinical medicine at the hospital Santa Maria della Morte as prosector to Antonio Maria Valsalva (1666–1723), who was one of the better-known students of Malpighi, the "Father of Histology." Valsalva's influence encouraged Morgagni to study pathology.

Figure 3.110 Giovanni Battista Morgagni (1682–1771).

When Valsalva left Bologna, Morgagni replaced him as a demonstrator in anatomy. After a short time, however, he gave up that post, and at the beginning of 1707 he moved to Venice, where he stayed through May 1709. In Venice, he conducted a number of dissections of human cadavers with the anatomist Giovanni Domenico Santorini (1681–1737), who was at that time dissector and lector in anatomy at the Venetian medical college. In 1709, Morgagni returned to Forli and practiced as a physician until September 1711, when he was asked to be the Second Chair of Theoretical Medicine at the University of Padua, replacing Antonio Vallisnieri (1661–1730), who had been promoted to First Chair following the death of Domenico Guglielmini (1655–1710). Morgagni gave his inaugural lecture, "Nova Institutionum Medicarum Idea," on March 17, 1712.

In 1715, he was appointed First Chair of Anatomy (a very prestigious position at that time: former chairs had included Andreas Vesalius, Realdo Colombo, Girolamo Fabricius

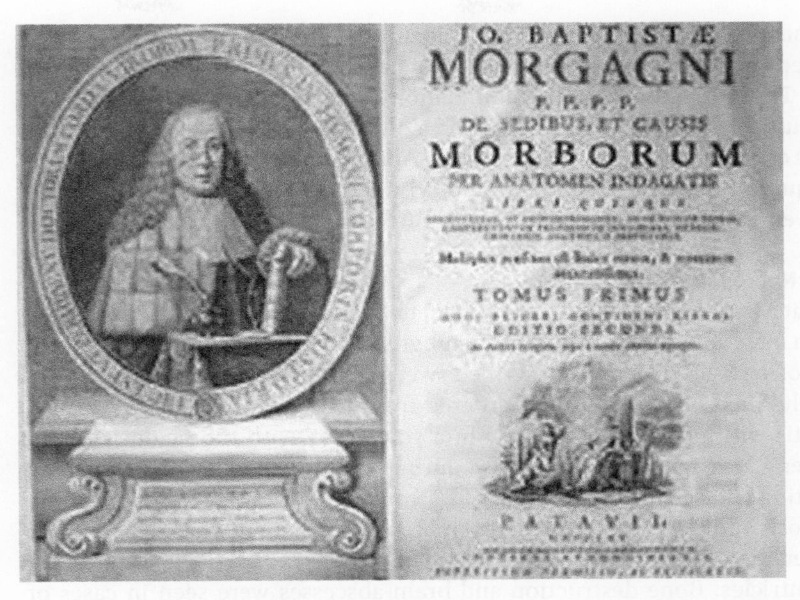
IO. BAPTISTÆ
MORGAGNI
P. P. P. P.
DE SEDIBUS, ET CAUSIS
MORBORUM
PER ANATOMEN INDAGATIS
TOMUS PRIMUS
EDITIO SECUNDA
PATAVII

Figure 3.111 Title page of Morgagni's *De Sedibus et Causis Morborum per Anatomen Indagatis* (1761).

of Aquapendente, and Gabriello Fallopio). He began his teaching of this subject with an inaugural address on January 21, 1716, and from 1726 he was the only professor of the discipline at Padua. He held this position until his death in 1771. Morgagni was well respected by his students and colleagues, for both his academic work and his personality. He was affectionately called "His Anatomic Majesty" by his students. His reputation attracted new students to Padua from all over Europe.

Morgagni was admitted to the Academia degli Inquieti in 1699, and became its head in 1704. He reformed the academy on the model of the Paris Académie Royale des Sciences and accepted an invitation to hold meetings in the mansion belonging to Luigi Ferdinandino Marsili, paving the way for its incorporation into the Istitutio delle Scienze founded by Marsili in 1714. It was to the Inquieti, in 1705, that Morgagni presented his series of lectures, "Adversaria Anatomica Prima." He published his first major work, the *Adversaria Anatomica* ("Notes on Anatomy"), based on his lecture series, which was the foundation for his reputation as an anatomist. Over the course of his career, Morgagni became a member of numerous prestigious academic societies, including the Academia Naturae Curiosorum, the Royal Society, and the Academy of St. Petersburg.

After his *Adversaria Anatomica*, Morgagni published no major works until 1761, at the age of 79, when he produced what is undisputedly his most important work: *De Sedibus et Causis Morborum per Anatomen Indagatis* ("Seats and Causes of Disease Investigated by Means of Anatomy") (Figure 3.111).

3.9.8.1 De Sedibus et Causis Morborum per Anatomen Indagatis

In 1679, Theophilus Bonetus had published the *Sepulchretum Sive Anatomica Practica*, considered the first book to describe signs and symptoms and correlate them with findings at dissection. Almost 3000 cases were described in this treatise, including clinical histories and reports of dissections. A larger second edition was published in 1700. In 1740, the idea for Morgagni's *De Sedibus* was born. A young friend of Morgagni, Lelius, asked him to record his findings and comment on them for Bonetus' book. Morgagni complied with Lelius' request, and over the next 20 years wrote 70 letters comprising documentations of his dissections and the corresponding patient histories. The letters also contained comments and critiques on Bonetus' *Sepulchretum*. These 70 letters were the foundation for *De Sedibus*, and documented the results of over 700 autopsies. Each volume addressed a different organ system: (i) Diseases of the Head; (ii) Diseases of the Thorax; (iii) Diseases of the Abdomen; (iv) Diseases of a General Nature and Disease requiring Surgical Treatment; and (v) a Supplement. The book became very successful, with seven editions and translations into three languages.

3.9.8.2 Contributions to Neuropathology

Many of the cadavers that Morgagni dissected were known to have neurological diseases. Some of his reports focused on infectious diseases such as meningitis, which he theorized often arose in the sinuses. In a case of syphilis, Morgagni described swelling of the dura, which adhered to the brain in a patient who died in 1717 and had suffered from fever and headaches. He also reported basal meningitis with optic nerve atrophy in a young boy with tuberculosis of the lungs, who had died after a long course of his disease during which he had become confused and suffered from vomiting and seizures. Morgagni saw a connection between findings in the lung due to the tuberculosis and the meningitis. In several cases, he assumed that an abscess of the brain might have been the cause of the meningitis, and he also described the possible entry of infection into the ventricles. Bone destruction and brain abscesses were seen in cases of chronic otitis media, as well as basal meningitis and thrombosis of the transverse sinus. Morgagni saw another possible cause of meningitis in head traumata, even minor ones. He observed several patients with encephalitis who died within a few days. He was the first to describe cerebral gumma, and he proved Valsalva's contention that cerebral lesions result in paralysis on the opposite side of the body.

3.9.8.3 Contributions to Neurotrauma

Morgagni's work and findings on neurotrauma contributed significantly to the field. He noted bleeding from the nose and the ear in skull base fractures. He described several cases of depressed skull fractures in which the sutures were frequently diastatic with injury to the underlying dura mater. He recognized subdural and epidural hematomas as being potentially fatal. He found that epidural hematomas occurred especially after injury of the superficial temporal artery and that damage to the meningeal vessels could also cause a hematoma. He found head trauma of the occiput to be exceptionally dangerous, as the medulla oblongata and underlying venous sinuses were often injured. Another common result of head trauma, according to Morgagni, was epileptic seizures. An injury of the heart could also harm the brain, owing to air in the brain vessels. Following head traumata, Morgagni often found only the tabula externa injured, but sometimes several foci of contusion could also be seen (coup and contre-coup). He described contralateral palsies and aphasia in left-sided head traumas. He found an epidural hematoma with intact skull and only superficial soft-tissue injury caused death within four hours after a left temporal head trauma in a 50-year-old alcoholic with syphilis. He distinguished emphysema in cerebral vessels due to injury from gas formation due to putrefaction – a distinction he considered important in forensic questions.

3.9.8.4 Stroke

Stroke played a major role in Morgagni's neuropathological work. According to him, stroke might occur after the excessive imbibing of wine or after accidents, and even during sleep. When dissecting the cerebral vessels of the elderly, he often found rigid deposits that he believed decreased blood flow to the brain. He distinguished between water stroke (intraventricular) and blood stroke (intracerebral). He reported the case of a 55-year-old cardinal who developed dizziness and drowsiness, followed 20 days later by repeated vomiting and headache with a left-sided palsy, then seizures and tachycardia followed by unconsciousness shortly before death. Dissection of the brain showed an intracerebral hemorrhage (blood stroke) with bleeding into the right ventricle (water stroke). Morgagni discussed a possible connection of the stroke to an accident on the basis of the case of a 60-year-old man who fell and suffered a contusion of the frontal bone, as well as epistaxis; shortly thereafter, a palsy of the left arm developed, and four days later also a speech disorder. Dissection of the brain showed an intracerebral hematoma with accompanying intraventricular hemorrhage and extensive destruction of the corpus striatum.

3.9.8.5 Cerebral Aneurysms

Morgagni found aneurysms in the cerebral circulation to be especially dangerous, and in the event of rupture to lead to sudden death. He found that ruptured aneurysms of the vertebral artery might cause hemorrhage in the cerebellum. He also found that hemorrhages in the subarachnoid space appeared mainly in the middle-aged and were fatal if the bleeding extended to the ventricle; that subdural hematoma might be present if the arachnoid mater was torn; and that pain in the neck might be caused by the accumulation of blood around the brainstem. In subdural hematomas, Morgagni discovered that the underlying brain was compressed; the patients were hemiplegic on the contralateral side and unconscious. He mentioned that these patients might be saved by trepanation. On the basis of several cases, he inferred that a vascular dissection in the cerebellum quickly leads to respiratory dysfunction and anal incontinence, with subsequent death. The death of a 40-year-old female alcoholic in 1740 led Morgagni to describe thrombosis of the superior sagittal sinus.

3.9.8.6 Brain Tumors

Morgagni also made important observations regarding brain tumors. He described the case of a patient who was amaurotic. He found a tumor the size of a fist between the cerebellum and the cerebrum, with optic nerve damage. Another brain tumor between the cerebrum and cerebellum impaired the acoustic nerve. According to his observations, an enlarged pituitary gland frequently led to hydrocephalus. He noted that diseases of the optic chiasm led to scotoma of both eyes.

3.9.8.7 Epilepsy

Epilepsy was considered a dangerous brain disease whose complications often caused death. Convulsions in a 50-year-old man led to death after seven days, and Morgagni's dissection showed hyperemia of the brain, a malformation in the frontal gyri, and a softening of the medulla oblongata and of the crura cerebri. Morgagni often identified hydrocephalus in patients with epilepsy.

3.9.8.8 Spine

Morgagni observed atrophy of the musculature of the neck and a laterally directed motion restriction due to the formation of block vertebrae in the cervical spine with a shortening of the dens and occipitalization of the atlas in an older man. He also described injuries of the spine that caused tetraplegia with respiratory insufficiency, loss of sensitivity below the lesion, and palsies of the bladder and the rectum. He stated that malformations in the nervous system were often associated with anomalies in the spinal column (spina bifida) and cystic formations of the spinal cord (Syringomyelia). He believed that the underlying pathology of myelomeningocele was hydrocephalus that caused excess spinal fluid to bulge out through a bony spina bifida; it is now known that myelomeningocele is a developmental defect, but 90% of cases are associated with hydrocephalus. He also reported that spinal hemorrhage could cause paraplegia.

3.9.8.9 Neurosurgery

Treating patients with head trauma by performing a trepanation, Morgagni saw an appearing hyperemia of the brain and its meninges while the patient was holding breath or exerting pressure. He noted that one had to watch out for brain vessels when trepanning a patient and must be careful not to harm them.

3.9.8.10 Conclusion

Virchow considered that Morgagni introduced modern pathology: "with him begins modern medicine." Morgagni's contribution to the understanding of disease may well rank with the contributions of Andreas Vesalius (1514–1564) in anatomy and William Harvey (1578–1657) in physiology. Certainly, his contributions to neuropathology were wide-ranging and highly influential, covering observations about strokes, cerebral aneurysms, epilepsy, brain tumors, and birth defects such as spina bifida, and he was a pioneer of neurosurgery. He died in Padua on December 6, 1771, at the age of 89.

3.9.9 Ruggero Ferdinando Antonio Guiseppe Vincenzo Oddi

Ruggero Oddi (1864–1913), baptized Ruggero Ferdinando Antonio Guiseppe Vincenzo, was born to an archivist and secretary for the Perguria Hospital, Filipo Oddi, and the Florentine Zelinda Pampaglini; he was the youngest of five children. Little is known of his life before the discovery that made him famous, except that the Oddi family lived a middle-class existence, and that his mother died in 1880, three years before Ruggero entered the University of Perugia.

It was at Perugia that Oddi undertook his research, earnestly sectioning the duodenal walls of various species – horse, sheep, cow, pig, hen, and human. His fervor earned him a place in the group of researchers at the Physiological Institute of the University headed by his mentor, Arturo Marcacci (1855–1915). In 1887, the 23-year-old medical student described his findings on the sphincter in an article written in French, published in the *Archives Italiennes de Biologie* (Figure 3.112), and simultaneously published in the Italian journal *Facolta medica-chirurgica*. Since the university could not award a medical degree owing to a papal decree set in 1825, Oddi continued his medical education at the University of Bologna. Under the tutelage of Pietro Albertoni (1849–1933), Arturo Marcacci's mentor, Oddi furthered his work on the physiological properties of the sphincter, publishing his findings in *Archivio Per Le Scienze Mediche* in 1888 (Figure 3.113). This article focused on his cannulation of the bile duct and his ability to measure the pressure generated by the sphincter through the introduction of mercury into a manometric tube connected to the cannula – a process almost identical to the technique used today for biliary tree radiomanometry. Through this process, Oddi determined a sphincter tone of 675 mm water, 475 mm higher than the normal pressure of bile secretion.

Oddi's discovery of the sphincter in humans was not accidental, but rather a confirmation of his previous work in dogs. Although many researchers, such as Vesalius in 1538, Glisson in 1654, Vater in 1721, and Cage in 1879, had identified the sphincter before him, Oddi was the first to measure its resistance.

In 1889, Oddi moved again to Florence to complete his education and receive his degree in medicine and surgery at the Reale Instituto Superiore di Perfezionamento. There, he defended his thesis on "Common Bile Duct Sphincter Tone" and two oral theses on "Ataxia" and "Antipiretics." (His original written thesis was lost during the Arno flood of 1966.) Upon completion, he was offered the position of Assistant to the Director, Prof. Luigi Luciani, at the Institute of Physiology, teaching a free course on physiological chemistry.

During this time, Oddi studied nervous contributions to the regulation of bile flow by transecting the vagus nerve in dogs, publishing two articles on the subject (Figure 3.114). He also

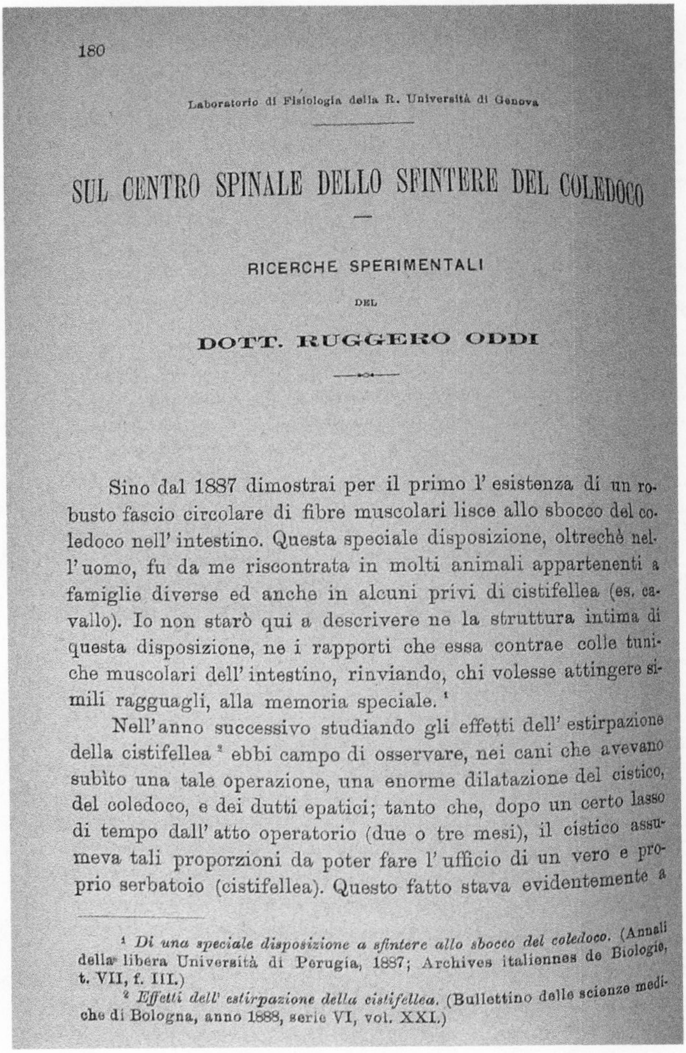

180

Laboratorio di Fisiologia della R. Università di Genova

SUL CENTRO SPINALE DELLO SFINTERE DEL COLEDOCO

RICERCHE SPERIMENTALI

DEL

DOTT. RUGGERO ODDI

Sino dal 1887 dimostrai per il primo l' esistenza di un robusto fascio circolare di fibre muscolari lisce allo sbocco del coledoco nell' intestino. Questa speciale disposizione, oltrechè nell' uomo, fu da me riscontrata in molti animali appartenenti a famiglie diverse ed anche in alcuni privi di cistifellea (es. cavallo). Io non starò qui a descrivere ne la struttura intima di questa disposizione, ne i rapporti che essa contrae colle tuniche muscolari dell' intestino, rinviando, chi volesse attingere simili ragguagli, alla memoria speciale.[1]

Nell' anno successivo studiando gli effetti dell' estirpazione della cistifellea[2] ebbi campo di osservare, nei cani che avevano subito una tale operazione, una enorme dilatazione del cistico, del coledoco, e dei dutti epatici; tanto che, dopo un certo lasso di tempo dall' atto operatorio (due o tre mesi), il cistico assumeva tali proporzioni da poter fare l' ufficio di un vero e proprio serbatoio (cistifellea). Questo fatto stava evidentemente a

[1] *Di una speciale disposizione a sfintere allo sbocco del coledoco.* (Annali della libera Università di Perugia, 1887; Archives italiennes de Biologie, t. VII, f. III.)

[2] *Effetti dell' estirpazione della cistifellea.* (Bullettino delle scienze mediche di Bologna, anno 1888, serie VI, vol. XXI.)

Figure 3.112 First page of Ruggero Oddi's "Sul centro spinale dello sfintere del coledoco. Lo Sperimentale, sec. biologica, Florence, in 1894," where he describes how he isolated a spinal center to control the tone of the common bile duct sphincter.

developed theories on the importance of the sphincter in the development of biliary diseases and published another series of articles regarding this entity. During the years 1887–1894, he published a total of 13 articles on his research, and in 1897 he presented a review entitled "Physiological Pathology of the Biliary Tract."

In 1893, Oddi was awarded a government scholarship to study abroad in the Laboratory of Experimental Pharmacology of Strasbourg University, France. His advisor was the German pharmacologist Oswald Schmiedeberg (1838–1921), who had isolated chondroitin sulfate from cartilage. During his sabbatical, Oddi isolated chondroitin sulfate from amyloid – which was later to be the topic of his inaugural lecture at the Genoa Royal Academy in 1894.

D'UNE DISPOSITION A SPHINCTER SPÉCIALE
DE L'OUVERTURE DU CANAL CHOLÉDOQUE (1)

par RUGGERO ODDI.

Laboratoire de Physiologie de l'Université de Pérouse.
Prof. A. MARCACCI, directeur.

Glisson est le seul des anatomistes qui ait certainement entrevu ou supposé la disposition spéciale dont je vais m'occuper ici. Il s'exprime, en effet, comme suit dans son « Anatomia hepatis » (2): « Denique regressus omnis in ductum communem praepeditur a fibris anularibus, quae non modo orificium ipsum, sed et totum obliquum tractum obsident... ». Si l'on tient compte cependant qu'au temps de Glisson l'existance des fibro-cellules n'était pas même soupçonnée, et que la présence de ces fibres annulaires n'a été confirmée, en suite, par l'observation directe de faits convainquants, il est facile de se persuader qu'elles ne seraient, en tout cas, qu'une hypothèse, heureuse il est vrai, de ce célèbre auteur.

On peut donc affirmer que, jusqu'à présent, aucun auteur n'a démontré l'existence d'une disposition musculaire spéciale à l'ouverture du cholédoque, qui puisse agir comme un sphincter proprement dit.

J'ai fait des recherches à ce sujet sous la direction et suivant les conseils du prof. Marcacci, et je puis démontrer clairement l'existence d'une disposition musculaire, que, pour être bref, j'appellerai *sphincter du cholédoque.*

Méthode employée pour mettre en évidence le sphincter du cholédoque; sa position, sa forme et ses rapports.

Pour me faire une idée de l'ensemble de cette disposition musculaire, je me suis servi de la même méthode qu'employa le prof. Marcacci pour isoler le muscle aréolaire du mamelon (3).

(1) *Annali della Università libera di Perugia.*
(2) FRANCISCI GLISONII, *Anatomia Hepatis*, 1681.
(3) A. MARCACCI, *Le muscle Aréolo-Mamelonnaire.* — Archives ital. de Biologie, t. IV, fasc. III.

Figure 3.113 First page of Oddi's paper "D'une disposition a sphincter speciale de l'ouverture du canal choledoque," where he describes his findings on the probable functions of the sphincter.

Oddi's investigations were widespread, and he published extensively on such topics as pregnancy, obesity, respiratory exchange upon exertion, and the pathways of afferent nerves in the spinal cord. He had earned substantial recognition in the scientific community, so it was expected when the Chair in Physiology in Genoa became vacant that it would be offered to him. Consequently, at the age of 29, he was appointed Acting Director of the Physiological Institute of Genoa. He held that position for seven years, before scandal and catastrophe removed him from his post and from Italy.

On October 15, 1891, Oddi married 23-year-old Teresa Bresciani Bartoli, who, four months previously, had borne him a daughter, Adelaide. In March of 1893, his son Enrico was born in Perugia. For reasons that are unclear, there are no records of his wife or children after 1900. It is unknown whether it was trouble surrounding his marriage that led to the financial ruin of the institute and his quarrel with the Academic Senate of the Faculty of Medicine, or whether it was his addiction to narcotics or his turning to Indian mysticism (both facilitated by his friendship with Stefano Capranica, Head of Physiological Chemistry Teaching). Whatever the original cause of the trouble, he left his post and fled to Brussels, Belgium in 1901.

There was still the possibility of redemption. During this time, he finished a book entitled *Foods and their Function in the Economy of the Individual and Social Organism,* which was

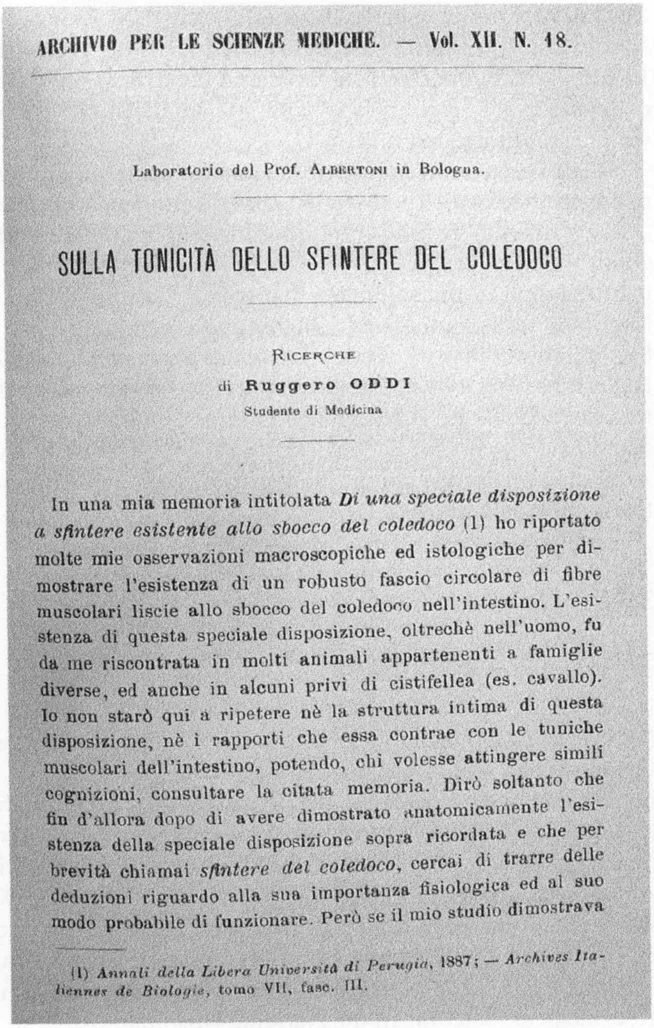

Figure 3.114 First page of Oddi's "Sulla tonicitàdello sfintere del coledoco," where he illustrates the sphincter's physiological properties.

ARCHIVIO PER LE SCIENZE MEDICHE. — Vol. XII. N. 18.

Laboratorio del Prof. ALBERTONI in Bologna.

SULLA TONICITÀ DELLO SFINTERE DEL COLEDOCO

RICERCHE

di **Ruggero ODDI**

Studente di Medicina

In una mia memoria intitolata *Di una speciale disposizione a sfintere esistente allo sbocco del coledoco* (1) ho riportato molte mie osservazioni macroscopiche ed istologiche per dimostrare l'esistenza di un robusto fascio circolare di fibre muscolari liscie allo sbocco del coledoco nell'intestino. L'esistenza di questa speciale disposizione, oltrechè nell'uomo, fu da me riscontrata in molti animali appartenenti a famiglie diverse, ed anche in alcuni privi di cistifellea (es. cavallo). Io non starò qui a ripetere nè la struttura intima di questa disposizione, nè i rapporti che essa contrae con le tuniche muscolari dell'intestino, potendo, chi volesse attingere simili cognizioni, consultare la citata memoria. Dirò soltanto che fin d'allora dopo di avere dimostrato anatomicamente l'esistenza della speciale disposizione sopra ricordata e che per brevità chiamai *sfintere del coledoco*, cercai di trarre delle deduzioni riguardo alla sua importanza fisiologica ed al suo modo probabile di funzionare. Però se il mio studio dimostrava

(1) *Annali della Libera Università di Perugia*, 1887; — *Archives Italiennes de Biologie*, tomo VII, fasc. III.

published in Turin in 1902 and was well received. His escape to Belgium was motivated by the intention of traveling to the State of Congo on a three-year contract as a second-class doctor, yet during his stay in Brussels, Oddi was once again struck with adversity. He fell victim to a psychic depression, and as a result of his treatment by one Dr. Mersh, became obsessed with a homeopathic preparation called vitaline, composed of pure glycerin, sodium borate, ammonium chloride, and alcohol.

Upon his departure from Antwerp to the port of Boma on December 12, 1901, Oddi suffered from a series of illnesses of unknown origin and made unrestrained use of narcotics. He remained at Boma for five months, until he was declared unsuitable for colonial service and returned to Belgium. From 1902 until his death in 1913, Oddi became a vagabond, first traveling to Spain with a patient he had been treating with vitaline, then to the small town of Torgiano near Perugia, and finally returning to Perugia to practice as a homeopathic physician whose main form of treatment was the prescription of vitaline. Despite his problems with drug addiction, Oddi continued to write. In 1906, he even wrote a pamphlet on the miraculous properties of vitaline entitled *New Therapy for Infectious Diseases and Malignant Terminal Diseases: Method for their Prevention and Treatment: Gatchkowskj's Vitaline.* Oddi, once an esteemed and illustrious physiologist, was now reduced to a medical charlatan, and charges of voluntary manslaughter and abusive use of medical products were brought against him.

He left Perugia for Tunisia in 1911 in the hopes of joining the Foreign Legion, and died there of unspecified causes on March 22, 1913 at the age of 48. The site of his burial place is unknown.

Ruggero Oddi lived a tumultuous life, full of grief and tragedy, which culminated in a lonely death while exiled in a foreign land. He gained fame at the tender age of 23 years with his identification of the sphincter that would later be named in his honor, but he is now all but forgotten in his motherland. Ongoing research on the sphincter of Oddi serves to reclaim the honor in his life's work. The sphincter, as described by Oddi and his predecessors (i.e., Glisson, Haller), has since been found to control secretions from the liver, pancreas, and gall bladder into the duodenum. Furthermore, Oddi's manometry techniques are still used to this day in the diagnostic process for patients with sphincter dysfunction and dysfunction of the gall bladder. A brilliant anatomist and physiologist, he was finally honored with a monument in his native town of Perugia in 1984.

3.10 Norway

3.10.1 Johan Georg Ræder

Johan Georg Ræder (1889–1956) (Figure 3.115) was the son of Dr. Anton Henrik Ræder (1855–1941), a philosopher, historian, and educationalist who studied in Germany, Italy, and Greece. Johan Georg was born in Kristiania (now Oslo). In 1912, while still a medical student, he became a candidate in the Department of Ophthalmology at the National Hospital (Riskhospitalets) for three months. He took his medical exam in 1915 and graduated with the highest grade (*Laudabilis præ ceteris*). Shortly after graduation, he traveled to Angola, Western Africa, and served as a physician for the South Atlantic Whaling Company. After returning home, he completed his candidate services in the Department of Ophthalmology (1916), Ullevaal Hospital's Department of Medicine (1917), and the National Hospital's departments of Neurology (1917), Surgery (1918), and Ophthalmology (1919). He also practiced at the Neurological Policlinic in the National Hospital.

In Copenhagen, on December 23, 1918, he married Marie Cathrine (1894–1930), the daughter of Rittmeister Jens Jacob Jensen (1863–1919) and the widow of a first lieutenant of the cavalry, Ulf Lars Sparre Thams (1889–1917). From January to August 1919, he studied in Copenhagen with a grant from the Rosencrone-Hjelmstierne legacy, and from July 1920 to September 1921 he studied in Paris, Freiburg im Breisgau and Halle in Germany, and Graz in Austria with a grant from the Hencrichsen legacy.

From 1920 to 1925, Ræder held a scholarship in ophthalmology and physiological optics at the University of Oslo. From 1921 to 1924, he worked as an assistant physician in the Department of Ophthalmology at the National Hospital. He was assistant to Hjalmar Schiotz (1850–1927), a famous physician in Norway who invented the tonometer, which measures ocular pressure, and was nominated for the Nobel Prize for his attempts to advance the management of glaucoma. In 1924, he was approved as a specialist in ophthalmology, and in April of the same year, he was promoted to Dr. med. (thesis: "Examinations on the Position and Thickness of the Lens inside the Human Eye in Physiological and Pathological Conditions, Measured by a New Method"). In the fall of 1924, he began his practice as an ophthalmologist in Oslo, and in 1926 he became a consultant ophthalmologist at the Ullevaal Hospital.

Figure 3.115 Portrait of Johan Georg Ræder (1855–1941).

In 1930, Ræder treated Edvard Munch (1863–1944), the famous Norwegian painter, who had suffered from an intraocular hemorrhage. Later, he wrote that "Mr. Edvard Munch suffered six years ago from a very serious hemorrhage in his right eye. There now has occurred a similar condition in his left eye, so that he is now threatened with complete blindness in both eyes." In 1931, Ræder was hired as Chief Physician at the newly established Department of Ophthalmology at Ullevaal Hospital.

A big-game hunter and lover of nature, Ræder traveled several times to Africa and wrote two books in Norwegian about his experiences (*Paradisiske Afrika* and *Ville dyr og sorte mennesker. Med børse og kamera i Vest-Afrika*). Some of his big-game trophies are exhibited in the Zoological Museum of Oslo. Ræder suffered from Dupuytren's contracture and arthritis, which forced him to resign from his positions in 1942. In later years, his painful disabilities could only be combated with morphine. He died in 1956, at the age of 71 years.

3.10.1.1 Publications and Legacy

Ræder's papers were written primarily in German and Norwegian, and focused on various ophthalmological topics. In 1918, he developed optical equipment for measuring the depth of the anterior chamber (distance between the anterior surface of the cornea and the edge of the pupil). He noted a decrease in depth of 0.6 mm from the second to the sixth decades of life, and attributed this finding to age-related thickening of the lens. In 1920, he introduced the concept of pupillary block in patients with closed-angle glaucoma, and explained why iridectomy was not effective in chronic, open-angle glaucoma. With the Norwegian physician Francis Gottfred Harbitz (1867–1950), Ræder reported secondary glaucoma in association with carotid artery disease, currently known as Takayasu's arteritis. For this reason, some also refer to carotid arteritis as Ræder–Harbitz syndrome. Ræder's landmark paper on paratrigeminal neuralgia, which is now known as Ræder's syndrome, was published in *Brain* in 1924 (see later). A list of his international publications (per journal) is as follows:

1) *Norsk Magazin for Lægevidenskaben*: "A Case of Intracranial Sympathetic Paralysis" (1918); "A New Instrument for Measuring the Depth of Anterior Chamber (Thalamometer)" (1918); "Facial and Optic Atrophy (Pre-senile Cataract and Glaucoma), caused by Symmetrical Carotid Disease" (with Francis Harbitz, 1926).

2) *Norsk tidsskrift for Militærmedicin*: "A Measure for the Depth of Anterior Chamber in the Emmetropic Eye" (1919).
3) *Ophth Foren i Kristiania*: "A Couple of Experiments with Miotics" (1919).
4) *Klinische Monatsblätter für Augenheilkunde*: "The Optical Shortcomings of Eyeglasses" (1921); "A Correction to M. Rohr's 'Eyeglasses as Optical Instrument'" (1922); "Some Cases of Inversion of Intraocular Pressure Variation in Retinal Detachment With Secondary Glaucoma" (1925).
5) *Albrecht von Graefes Archiv fur Ophthalmologie*: "Examinations on the Position and Thickness of the Lens inside the Human Eye in Physiological and Pathological Conditions, Measured by a New Method. I. Position and Thickness of the Lens in Case of Emmetropia, Hypermetropia, and Myopia" (1922) and "II. Position of the Lens in Glaucoma Patients" (1923).
6) *Tidsskrift for Den norske lægeforening*: "On Diplopia" (1924); "Preliminary Notification on a New Glaucoma Operation" (1927 and 1928)
7) *Brain*: "Paratrigeminal Paralysis of Oculopupillary Sympathetic" (1924).

3.10.1.2 Paratrigeminal Sympathetic Paralysis

In 1924, Ræder described five patients with intracranial oculopupillary sympathetic paralysis. The first case, in a Norwegian, had been reported previously by him in 1918. This was an 18-year-old male with left ophthalmic pain, radiating neck pain, epiphora, variable extraocular muscular paralysis, ptosis, miosis, and ocular sympathetic paralysis without vasomotor changes. On post-mortem examination, Ræder found that the patient had a middle fossa tumor (endothelioma) encasing and infiltrating the Gasserian ganglion and sympathetic plexus, which connect the internal carotid artery plexus with the ganglion and the trigeminal and oculomotor nerves (Figure 3.116). He indicated that "the coincidence of oculopupillary paralysis of the sympathetic and trigeminal symptoms makes the designation of 'paratrigeminal', which was suggested to me by Professor Monrad-Krohn, appear suitable for this type of sympathetic lesion." The Norwegian neurologist Georg Herman Monrad-Krohn (1884–1964) was Professor and Chair of the Neurological University Clinic of Oslo and the author of *The Clinical Examination of the Nervous System*, a textbook widely used abroad as a reference. Professor Monrad-Krohn is also known for his report of the so-called "foreign accent syndrome," which is a rare acquired speech disorder characterized by the loss of normal phonemic and phonetic contrasts of native dialect.

The second of Ræder's five cases had trigeminal neuralgia, hypoesthesia in the distribution of the ophthalmic nerve, miosis, ptosis, and paresis of cranial nerves IV and VI. The third and fifth cases had histories of skull base fractures with optic nerve and sensory or motor trigeminal injuries in addition to incomplete Horner's syndrome; however, neuralgia was not present.

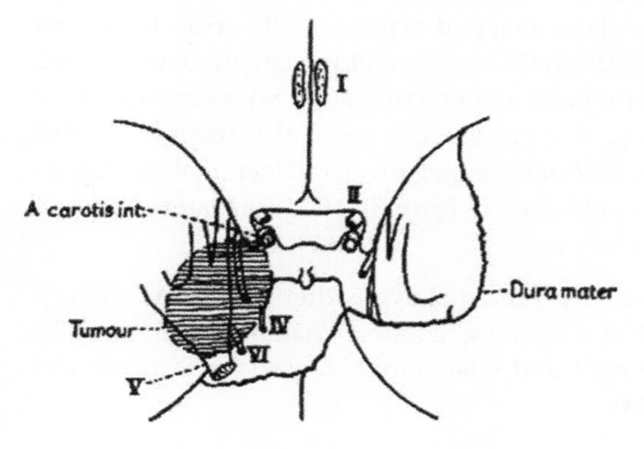

Figure 3.116 Illustration of the middle cranial fossa tumor involving the Gasserian ganglion and other cranial nerves. (*Source:* From Ræder 1924).

The fourth case, which according to Ræder resembled Herpes zoster ophthalmicus, presented only with sympathetic ptosis, miosis, and trigeminal neuralgia, but no proof of a middle fossa lesion. Solomon asserts that this patient may have been misdiagnosed, as such manifestations may also occur with other diseases (Figure 3.117).

The true nosological indication of Ræder's syndrome has been debated in the modern imaging era. The majority of patients with manifestations of this syndrome may not have a definite middle fossa lesion. Boniuk and Schlezinger differentiated the typical Ræder's syndrome with parasellar cranial nerve involvement (group I) from a relatively benign and self-limited form of neuralgia and oculosympathetic paralysis (group II). Group I Ræder's syndrome with an organic cause is rare, and most reports have been benign and idiopathic (group II). Solomon and Lusting demonstrated that patients from the latter group may have lesions of the carotid artery such as dissection. Furthermore, the manifestations of cluster headache (characterized by periodic painful Horner's syndrome) may resemble those of idiopathic Ræder's syndrome. Notably, Goadsby found clear evidence of neuralgia in only two of Ræder's patients, and suggested that the term "paratrigeminal oculosympathetic syndrome," rather than "paratrigeminal neuralgia," should be used.

The importance of Ræder's original report lies in the fact that he attempted to localize the sympathetic lesion paratrigeminally (or in the middle cranial fossa) on the basis of parasellar cranial nerve (III–VI) involvement. He stated:

> It is seen that the sympathetic symptoms were accompanied by disturbances pointing to a lesion of one or more cranial nerves. The nerve most constantly involved was the trigeminal; this syndrome has been consequently termed paratrigeminal.

In fact, Ræder's syndrome may appropriately reflect a lesion of the middle fossa with affection of oculopupillary sympathetic fibers. The diagnosis of this syndrome solely on clinical grounds

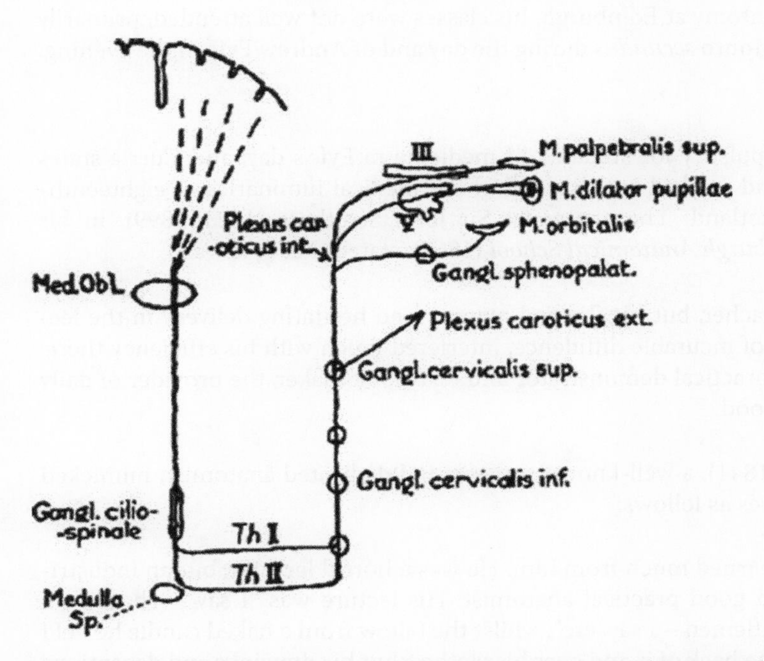

Figure 3.117 Diagram of the central nervous system, cervical, and intracranial sympathetic pathway. (*Source:* From Ræder 1924).

may be misleading, as painful Horner's syndrome also occurs with such diseases as cluster headache and carotid artery dissection or aneurysm.

3.11 Scotland

3.11.1 Andrew Fyfe the Elder

Andrew Fyfe the Elder (1752/54–1824) was born in Corstorphine, the second son of John Fyfe and Agnes Alexander, who were married in 1750. Information regarding Fyfe's early years is scarce. However, he i known to have been intrigued by the human body from childhood. To further his interest, he began a surgical apprenticeship with one Mr. Anderson, and numerous documents exist in the University of Edinburgh's archive that attest to his participation in medical classes during the academic years of 1775, 1776, and 1779–1781. However, for unknown reasons, he did not finish his medical studies and never earned a degree. On October 19, 1787, he married Agnes Williamson. They had at least nine children. Three died during infancy, and four entered the medical profession.

In 1777, at the age of 25, Fyfe was appointed by Professor Monro *secundus* (1733–1817) conjointly with John Innes as "Dissector" at the University of Edinburgh. Innes died shortly afterwards, and Fyfe succeeded him as principal janitor and macer, a post usually held by a student that offered free living accommodation. With his focus on human anatomy, Fyfe caught the eye of staff within the universtiy, particularly Professor Monro *secundus*, who was well regarded throughout Europe. When Monro noticed Fyfe's talent and interest in anatomy, he appointed him as his personal assistant. Although this position did not allow him to pursue professorial advancements in anatomy, he remained in charge of various dissections and demonstrations in the medical school under the guidance of Monro *secundus* and his son, Monro *tertius* (1773–1859), for more than 45 years. Fyfe was dedicated to the Monros. When John Barclay (1758–1826) attempted to teach anatomy at Edinburgh, his classes were not well attended, primarily owing to the opposition of Monro *secundus* during the day and of Andrew Fyfe in the evening.

3.11.1.1 Teaching

Studying anatomy was compulsory for students of medicine in Fyfe's day, and Currie states that he must have taught and worked with several of the medical luminaries of eighteenth- and nineteenth–century Scotland. The anatomist Sir John Struthers (1823–1899), in his *Historical Sketch of the Edinburgh Anatomical School* (1867), stated that Fyfe was

> a most painstaking teacher, but his flurried manner and hesitating delivery in the lecture-room, the result of incurable diffidence, interfered much with his efficiency there. He was the plodding practical demonstrator and text-book maker, the provider of daily common anatomical food.

Sir Astley Cooper (1768–1841), a well-known surgeon and dedicated anatomist, mimicked Fyfe and described his lectures as follows:

> Fyfe I attended, and learned much from him. He was a horrid lecturer, but an industrious, worthy man, and good practical anatomist. His lecture was "I say – eh, eh, eh, gentlemen; eh, eh, gentlemen – I say, etc"; whilst the tallow from a naked candle he held in his hand ran over the back of it and over his clothes; but his drawings and deceptions were well made and very useful.

Sir Astley Cooper's nephew, Bransby B. Cooper (1792–1853), also a student at Edinburgh, recalled:

> Mr. Fyfe was a tall thin man, and one of the most ungainly lecturers I ever knew. He had been assistant to Dr. Monro, and by hard study, and dissecting for the doctor's lectures, became an excellent anatomist. Sir Astley used to mimic very admirably the awkward style of delivery and primitive habits which distinguished Mr. Fyfe in the lecture room, even when he was in Edinburgh, and invariably excited much laughter.

Sir Robert Christison (1797–1882), who attended the University of Edinburgh from 1815 to 1816, commented on Fyfe:

> one of the last in Edinburgh to wear the pigtail ... every afternoon going over what every student had done with his dissected part ... duty over, we all gathered round him at the fireside, where he entertained us with anecdotes of the departed medical worthies who had adorned the University in his day.

In 1904, L.M.A. Liggett, a former student of Fyfe's, stated:

> Fyfe gave private lectures and demonstrations useful to the tyro in anatomy. He was a sharp looking man, excellent with the scalpel.

3.11.1.2 Fyfe's Contributions to Anatomical Illustration

Fyfe was a brilliant artist, and was awarded the Annual Prize Medal by the Board of Trustees in 1775 for his drawings. Rock referred to him as a "fine draughtsman." Fyfe's intention throughout his life was to produce a set of illustrations that would be convenient to handle and affordable for his anatomy students. He published several books and articles, one of his most influential of which was *Views of the Bones, Muscles, Viscera and Organs of the Senses*, published in Edinburgh in 1800. Thanks to his publications, Fyfe became extremely well known throughout Europe. The majority of these publications had several editions owing to demand from students, who required such books to accompany their anatomical dissection classes.

A Compendium of the Anatomy of the Human Body was another well-known publication of Fyfe's. It was published several times between 1800 and 1823. This book was illustrated with 160 tables and approximately 700 figures. Interestingly, Fyfe's wife and daughters were given the task of coloring the plates used in his texts.

Fyfe did not create all of the engravings and plates that he published. It is likely that he collected them, and reprinted or republished them in his *Compendium* (Figures 3.118–3.120). Rock said of this publication that Fyfe had a reckless style, as formatting took precedence over presentation, with several plates having arms and legs trimmed off.

He was the editor of *Views of the Bones, Muscles, Viscera and Organs of the Senses*, which consisted of 23 folio tables with short explanations. However, he gave no credit to others who contributed to this title. In fact, many of the pictures appear to have been the design of Richard Cooper Senior (1701–1764), who had illustrated anatomical works for Monro *primus* (1697–1767). Interestingly, Cooper would go on to found the first Scottish academy of artists, the Edinburgh School of St. Luke. Rock stated that some of the illustrations in Fyfe's *Views of the Bones* may have been drawn by Cooper's pupils, (Sir) Robert Strange (1721–1792) and Thomas Donaldson (1755–1800). Fyfe himself probably trained under the painter Alexander Runciman (1736–1785), who was also a pupil of Cooper. Fyfe was a proponent of the Ruyschian (quicksilver) method of injecting anatomical specimens, and devoted an entire section of the 1802 edition of his *Compendium* to this technique. It allowed greater detail to be observed during dissections

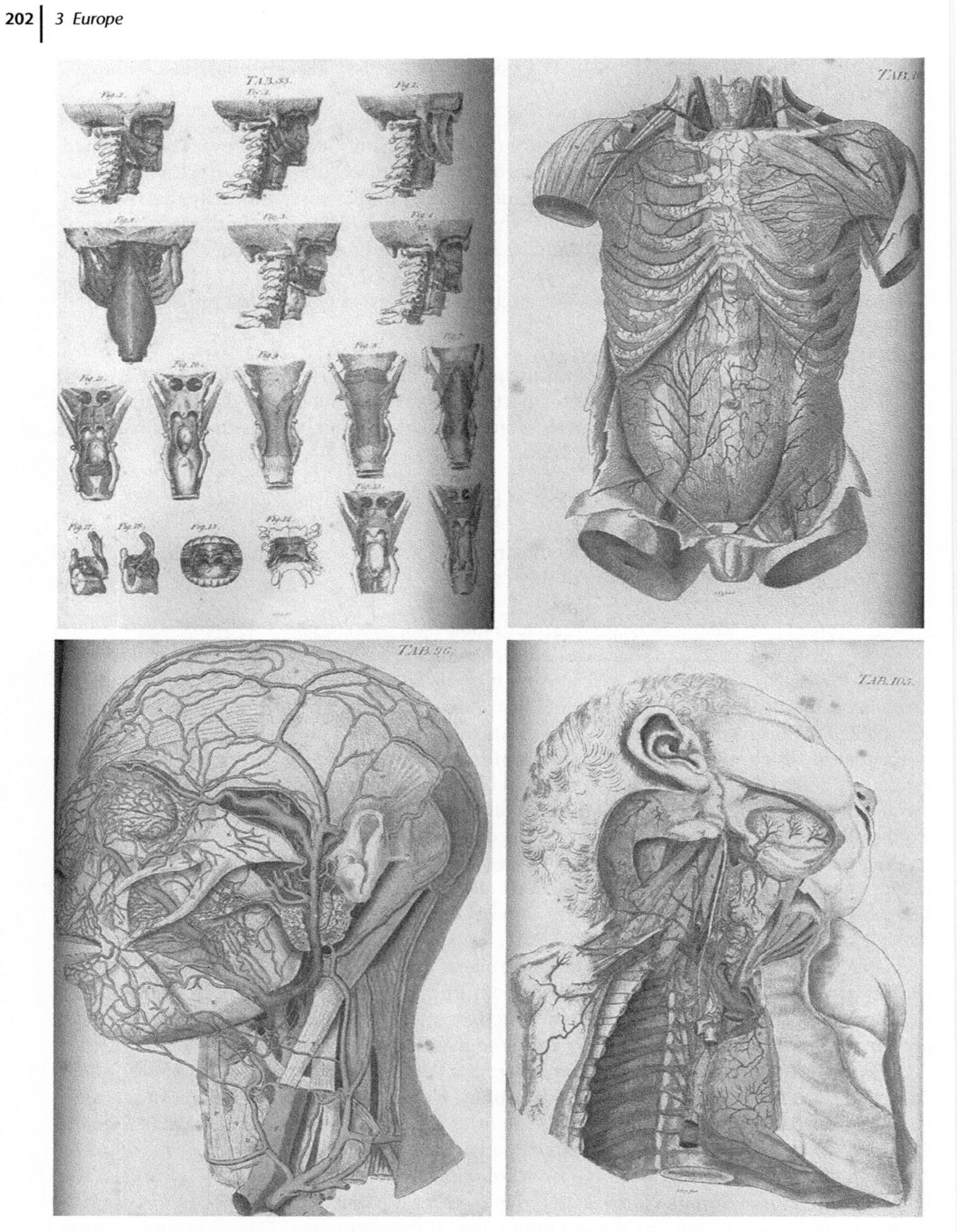

Figure 3.118 Artwork drawn by Andrew Fyfe and used in his 1800 *Compendium of the Anatomy of the Human Body*.

Figure 3.119 "Child's Skull," from Fyfe's *Compendium of the Anatomy of the Human Body*.

Figure 3.120 Cover page of the 1802 American edition of Fyfe's *Compendium of the Anatomy of the Human Body*, demonstrating his use of the Ruyschian method.

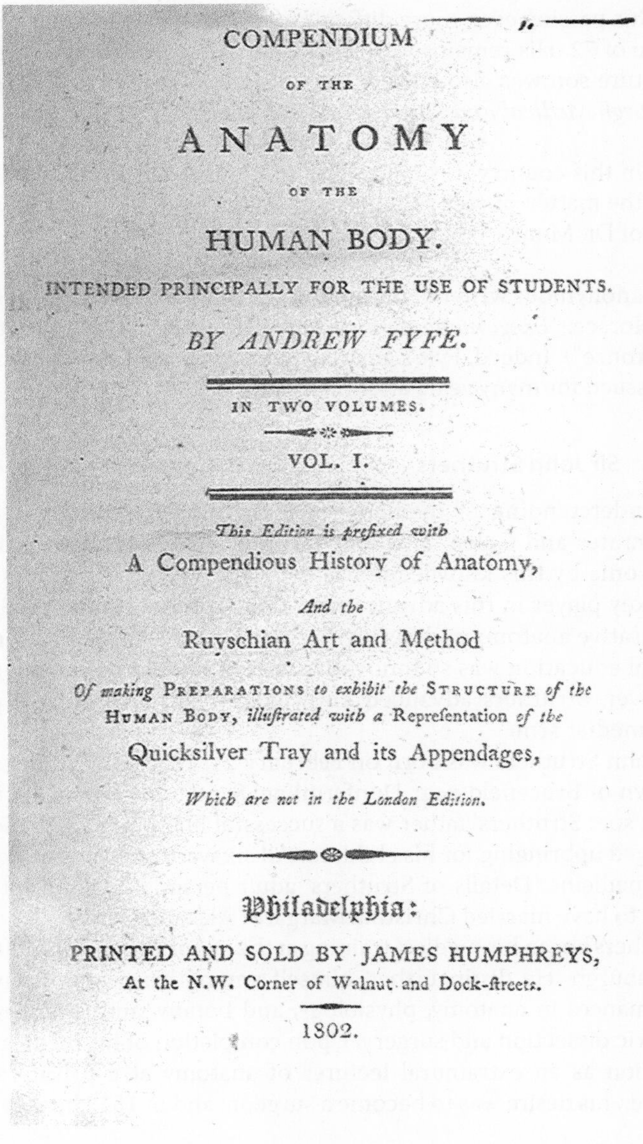

COMPENDIUM

OF THE

ANATOMY

OF THE

HUMAN BODY.

INTENDED PRINCIPALLY FOR THE USE OF STUDENTS.

BY ANDREW FYFE.

IN TWO VOLUMES.

VOL. I.

This Edition is prefixed with

A Compendious History of Anatomy,

And the

Ruyschian Art and Method

Of making PREPARATIONS *to exhibit the* STRUCTURE *of the* HUMAN BODY, *illustrated with a* Repreſentation *of the*

Quickſilver Tray and its Appendages,

Which are not in the London Edition.

Philadelphia:

PRINTED AND SOLD BY JAMES HUMPHREYS,

At the N.W. Corner of Walnut and Dock-ſtreets.

1802.

and thus in subsequent anatomical illustrations. Fyfe was one of many anatomists who contributed to anatomical illustration during the eighteenth and nineteenth centuries.

3.11.1.3 Later Life

Fyfe spent most of his life increasing knowledge concerning the human body and helping to make anatomy one of the most important medical science courses. Later in life, he acted as curator of anatomical figures given to the university by Monro *secundus* in 1800. He became a Fellow of the Royal College of Surgeons of Edinburgh on October 23, 1818, a few weeks before the entry of his eldest son, Andrew (1792–1861), who graduated as an MD from Edinburgh in 1814 and later became president of the Royal College in 1842–43. Andrew the younger provided private chemistry and pharmacy lectures at Edinburgh for several years. In 1844, he became Professor of Chemistry at the University of Aberdeen, which position he retained until his death. Fyfe the Elder's grandson was also named Andrew, and became a London physician.

Andrew Fyfe the Elder was fortunate to have enjoyed a constant state of good health, which made his passing even more difficult for his friends and family. He died on March 31, 1824 at the age of 72. His remains were buried at Calton New Burying-place, the cemetery where his premature son was also buried. Currie states that the consensus of the day, as quoted in the *Edinburgh Medical and Surgical Journal*, was that

> in this country it is impossible to number any other works professedly systematic than the matter-of-fact volumes of Fyfe, the work of John and Charles Bell, and the Outlines of Dr. Monro.

The anonymous writer of his obituary proclaimed that Fyfe's last words may have been taken from Horace: "*Exegi monumentum aere perennius*" ("I have erected a monument more lasting than bronze"). Indeed, Fyfe's anatomy texts were published in multiple editions and continued to be issued for many years after his death.

3.11.2 Sir John Struthers

Our understanding of medicine and anatomy has progressed greatly since the time of Hippocrates and Galen. Not only has our knowledge grown, but the way in which physicians come by this knowledge has evolved. Sir John Struthers (1823–1899) (Figure 3.121) was a key player in this advancement. He is often remembered for his research concerning comparative anatomy and his involvement in the "Tay Whale" incident. His work relating to medical education was so innovative that he should be considered as a pioneer in this field. Moreover, Struthers advanced knowledge regarding sites of potential nerve entrapment in the medial arm.

Sir John Struthers was born on February 21, 1823 to Alexander Struthers and Mary Reid in the town of Brucefield near Dunfermline, Scotland. He was the third of six children, and the second son. Struthers' father was a successful linen merchant and mill owner who provided a privileged upbringing for his children. All six were educated at home, and each son went on to study medicine. Details of Struthers' adult personal life are not well documented, but he is known to have married Christina Margaret Alexander in 1857.

Struthers began his medical training at the age of 18, when he left Brucefield for the University of Edinburgh. He distinguished himself among his peers, receiving multiple awards for his performances in anatomy, physiology, and botany, and receiving accolades for his ability in cadaveric dissection and surgery. Upon completion of his medical degree in 1845, he accepted a position as an extramural lecturer of anatomy at a private anatomy school in the area. However, his desire was to become a surgeon, and in 1847 he received a fellowship at the Royal

Figure 3.121 Sir John Struthers (1823–1899).

College of Surgeons in Edinburgh. Following this training, he continued as an anatomy lecturer, but also worked as an assistant surgeon at the Royal Infirmary. In conjunction with several other anatomy lecturers, he formed the new School of Anatomy at Surgeon's Hall in 1849. He was later promoted to full surgeon, and remained at the infirmary until his appointment as the Regius Chair of Anatomy at Aberdeen in 1863. After receiving this coveted position, he gave up his work as a surgeon. He believed such a position to be worthy of his full attention, and disapproved of colleagues who attempted to balance a career in anatomy with another in surgery or medicine.

Struthers' distinguished career continued following his retirement from the Chair of Anatomy at Aberdeen in 1889; he remained active in local and national medical affairs for several years. He received an honorary Doctor of Laws (LLD) degree from the University of Glasgow in 1885, served on the General Medical Council (GMC) from 1883 to 1891, served as president of the Royal College of Surgeons from 1895 to 1897, and was knighted by Queen Victoria in 1898, one year prior to his death. While serving on the GMC, he was appointed Chairman of the Education Committee, where he was the driving force behind several changes to the education system.

As an anatomical researcher, Struthers' early research interest was the innervation of the extraocular muscles. At that time, bloodletting was considered a treatment for congestive heart failure, and his work helped to end this practice. He demonstrated its ineffectiveness by describing how the removal of blood from the body wall could not relieve congestion in the abdominal viscera. In the early 1860s, he became one of the first prominent anatomists in Scotland to embrace Darwin's controversial theory of natural selection and adaptation. Following this, he shifted his research focus to describing vestigial structures. He published works describing the carpal and tarsal bones in horses and the development of a rudimentary hindlimb in the Great Fin Whale. Darwin and Struthers communicated regarding their research. In fact, Darwin compared a rudimentary ligamentous structure first described by Struthers in describing a supracondylar foramen present in, for example, marsupials. This structure is known today as the ligament of Struthers (Figure 3.122), and is one of two anatomical structures named after him.

Struthers first described this ligament in 1854, as a ligamentous or muscular structure (often an extension of the humeral head of the pronator teres muscle) passing between the medial epicondyle and supracondylar processes of the humerus. It is present in approximately 1% of individuals, and can result in median nerve and/or brachial artery entrapment. Tiedemann actually first described such a supracondylar process in 1818, followed by Knox in 1841. Darwin wrote of Struthers' interest in such an anomaly:

> In some of the lower Quadrumana, in the Lemuridae and Carnivora, as well as in many marsupials, there is a passage near the lower end of the humerus, called the supra-condyloid foramen, through which the great nerve (median) of the fore limb and often the great artery (brachial) pass. Now in the humerus of man, there is generally a trace of this passage, which is sometimes fairly well developed, being formed by a depending hook-like process of bone, completed by a band of ligament. Dr. Struthers, who has closely attended to the subject, has now shewn that this peculiarity is sometimes inherited, as it has occurred in a father, and in no less than four out of his seven children. When present, the great nerve (median) invariably passes through it; and this clearly indicates that it is the homologue and rudiment of the supra-condyloid foramen of the lower animals. Prof. Turner estimates, as he informs me, that it occurs in about one per cent of recent skeletons. But if the occasional development of this structure in man is, as seems probable, due to reversion, it is a return to a very ancient state of things, because in the higher Quadrumana it is absent.

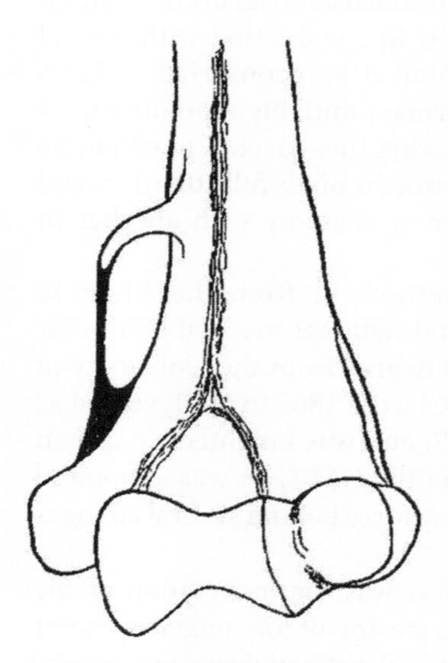

Figure 3.122 Original drawing by Struthers of a ligament (the ligament of Struthers) that can connect an anomalous supracondylar spur of the humerus to the medial epicondyle. Such an anomaly creates a canal, which the median nerve and/or brachial artery may course through, becoming entrapped.

The second eponym for which Struthers is remembered is the arcade of Struthers, which is present in approximately 70% of individuals on the medial aspect of the distal third of the humerus. The arcade, a thickening of the anteromedial aspect of the medial intermuscular septum, forms a myofibrous sheath around the ulnar nerve after it passes from the anterior to the posterior compartment of the arm, and may be responsible for ulnar nerve entrapment (Figure 3.123). Interestingly, Struthers never mentioned this structure in his writings, and this region was only attributed to him later.

Struthers used whales, which were often washed ashore near Aberdeen, to satisfy his interest in comparative anatomy. They were brought to the city, where Struthers would perform a dissection and give lectures to medical students concerning vestigial structures in mammals. Through this work, he published several papers concerning the anatomy of whales, particularly the humpback.

Struthers' most famous dissection was that of the "Tay Whale" in December 1883 (Figure 3.124). This humpback whale was so-named because it was killed by whalers in the Tay Estuary and later washed ashore. Struthers was outbid for the carcass by an oil merchant, "Greasy" John Wood, from nearby Dundee. However, Wood allowed Struthers to perform a dissection, for which he sold tickets so that the public could observe the spectacle. Struthers commented that the circus-like affair was a

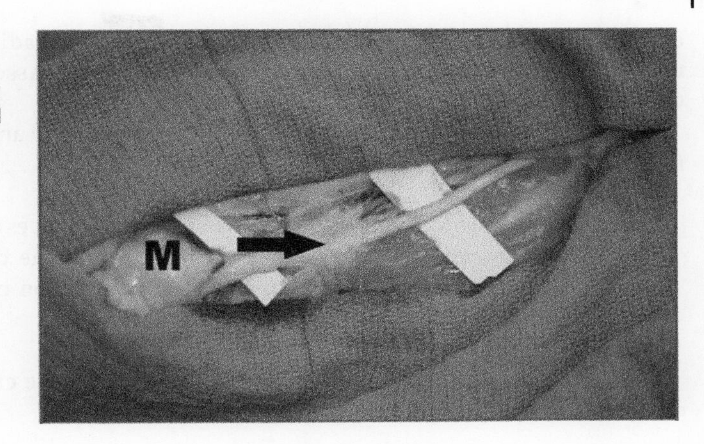

Figure 3.123 Cadaveric example of the ulnar nerve traveling from the anterior to the posterior compartment of the arm through a thickened medial intermuscular septum known as the arcade of Struthers (arrow). M, right medial epicondyle.

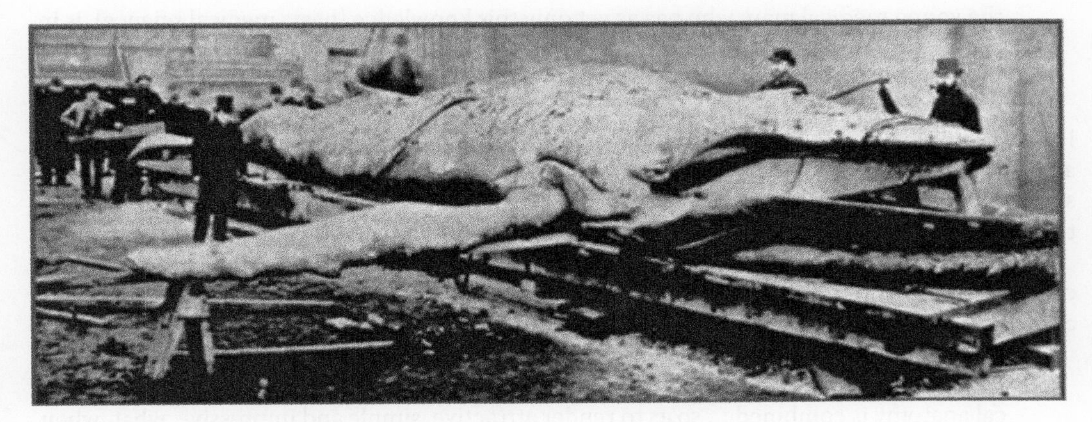

Figure 3.124 Photograph of the Tay Whale incident.

hindrance to his scientific objectives. Despite being considered one of the worst poets in Scotland's history, William McGonagall ensured the event would not be forgotten by describing the details of the whale's demise in his poem, "The Famous Tay Whale."

On his arrival at Aberdeen, Struthers was dismayed at the university's facilities. He felt them so inadequate that he commissioned an entire new department, at a cost of £4000. The buildings Struthers constructed are still used for anatomical instruction today. The new department included a museum of anatomical specimens, hailed as one of the finest in the country; Struthers prepared many of the museum's specimens personally. As a Darwinian, Struthers designed several of these specimens to illustrate support for Darwin's controversial theory. He published a novel method for the preparation of specimens during winter months that involved apparatus designed to maintain stable summer temperatures for effective maceration of tissues. This method allowed him to prepare new specimens for the museum throughout the year.

During his time at Aberdeen, Struthers formulated ideas regarding medical education, which he was able to put into action as Chairman of the GMC Education Committee. With these ideas, he spearheaded a substantial reformation of medical education in the United Kingdom. Perhaps the most important of these changes concerned an extension of the medical curriculum from four to five years, with the fifth year being devoted strictly to clinical training. Struthers' ideas relating to medical education are still reflected in the educational guidelines set forth by two of the world's leading medical education authorities: the GMC, of which he was a member, and the Association of American Medical Colleges (AAMC).

As a researcher, Struthers believed the basic sciences held an irreplaceable position in a good medical education:

> Unless you are well informed in the foundation sciences and principles, you may practice your profession, but you will never understand the disease and its treatment; your practice will be routine, the unintelligent application of the dogmas and directions of your textbook or teacher.

Struthers' propensity for the basic sciences led him to be critical of premature exposure to clinical medicine. He wrote:

> the young medical man, who has as yet only this knowledge [basic medical science], is in every way more hopefully situated than the other who has neglected it in the endeavor to grasp prematurely at knowledge of the living phenomena and treatment of disease.

He was averse to placing students in clinical settings to teach them about patient interaction and the physical treatment of patients early in their training. He believed this type of teaching should be carried out in the newly developed fifth year of education, after successful completion of a rigorous basic science curriculum.

While opposing premature exposure to clinical medicine, he did not discount the benefits of clinically oriented teaching in the basic sciences. Struthers described the need to incorporate clinical correlations as supplements to hard science lectures:

> The lecture is rendered doubly interesting and useful when ... both scientific and surgical anatomy is combined ... so as to render attractive, simple and impressive, what, when otherwise treated, have been, and with sure truth might be called the dry details of anatomy.

To achieve this end, he oversaw a dramatic restructuring of anatomical lectures; he removed details he deemed irrelevant to clinical practice from his lectures, as he felt they did not improve the performance of clinicians and were prone to cause frustration and disdain for anatomy among students.

Struthers was respected by many at Aberdeen, and in his honor an anatomy lecture was given every third year from 1911 until 1974. The first of these was given by one of Struthers' top students, Arthur Keith (1866–1955), who in 1907 first described the sinoatrial node. In 1893, Keith received the Struthers Prize for his demonstration of various ligaments in humans and apes. A monetary prize is still awarded annually to the student authoring the best dissertation. Furthermore, to commemorate his accomplishments, the Struthers Medal and Award was instituted in 1893. For physicians trained outside the University of Aberdeen, the only memory of Sir John Struthers is to be found in their knowledgeable and effective practice of medicine, and it is to him and his training model that they owe much of their ability and success.

3.11.3 Charles Bell

Sir Charles Bell's (1774–1842) (Figure 3.125) intellect and drive were manifested at an early age. Born in Edinburgh to Rev. William Bell, he demonstrated a keen interest in knowledge

Figure 3.125 Painting of Sir Charles Bell (1774–1842), by John Stevens.

acquisition, and with the help of his brother, an anatomist and surgeon, he gained a vast understanding of medicine. At the age of five, Bell mourned the death of his father, leaving him and his two brothers alone with their mother. She was sensitive and spiritual, and instilled in her boys understanding and compassion for others. Ultimately, she nurtured Bell's passion for art. Bell graduated from high school with distinction and continued to excel at the University of Edinburgh, where he used his gift for art to depict accurate anatomical figures for his medical professors and wrote his first publication, *A System of Dissection Explaining the Anatomy of the Human Body*. In this work, Bell used his artistic talents to display his vast knowledge regarding human anatomy. Immediately, the university commended him for his exceptional attention to detail in his artistic depictions, and colleagues demanded his demonstrations be published. His innovative designs made him one of the most influential authors of his time. In view of his reputation, Bell was elected to the College of Surgeons, and was later offered work at Edinburgh's only hospital, the Royal Infirmary. In 1802 and 1803, the third and fourth volumes of his *Anatomy of the Human Body* were published.

In an attempt to avoid competition with his older brother and jealousy from the staff at the university, Bell moved to London in 1808, where his geniality and passion enabled him to excel beyond the city's most noble physicians. On his journey, he carried the manuscript of his *Essays on the Anatomy of Expression in Painting*, in which he outlined the anatomical and psychological nature of facial expressions for artists. Bell reveled in outstanding accolades for his work. However, his most prominent accomplishment was his work as a surgeon at the Middlesex Hospital, which began in 1814. During the Battle of Waterloo in 1815, Bell cared for and operated on injured soldiers, portraying the brutality and gore, including neurological injuries, in his anatomical artwork. At the battle, Bell operated on the wounded until "his clothes were stiff with blood and his arms powerless with the exertion of using the knife."

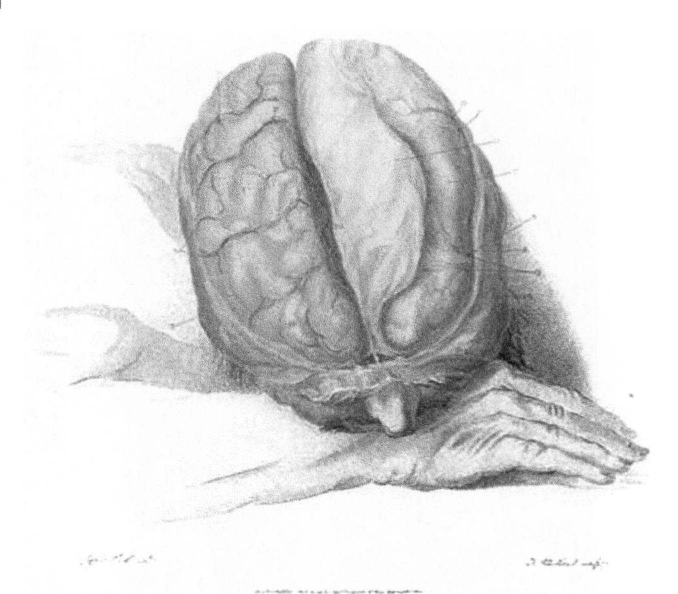

Figure 3.126 Painting from Bell's *The Anatomy of the Brain, Explained in a Series of Engravings.*

He received his highest accolade for his 1811 book, *An Idea of a New Anatomy of the Brain* (Figure 3.126), which has been described as the "Magna Carta of neurology." In 1826, Bell was granted the Royal Society's first ever medal, and in 1831 he was knighted by King William IV (1765–1837).

3.11.3.1 Contributions to Neuroanatomy

Bell advanced neurology, surgery, and anatomy; his precise drawings vividly depicted various anatomical structures to explain the preferred surgical methods of the time better. Bell was one of the first physicians to study details of neuroanatomy and apply this knowledge to the clinical and surgical setting. Therefore, he was an early pioneer of neurology. As a surgeon and neurologist, he should be regarded as a founder of neurosurgery, as exemplified by his depictions of and operations on the nervous system. Information regarding the brain had been readily available prior to his work, but multiple structures and functions had yet to be discovered. For example, the various areas of the spinal cord controlling sensory and motor modalities were not fully understood. In his *A New Anatomy of the Brain,* Bell highlighted the differences between the cerebrum and cerebellum, and stated that bundles of distinct nerves span the body and conjoin in a single filament. In regard to the spinal cord, he claimed that the anterior portion controls motor function while the posterior helps to elicit a sensory response.

Although they serve distinct functions, Bell showed that two sets of nerves from a single side of the spinal column assemble in the same bundle. In *A New Anatomy of the Brain,* he supported his notion with a groundbreaking observation during vivisection:

> On laying bare the roots of the spinal nerves, I found that I could cut across the posterior fasciculus of nerves, which took its origin from the posterior portion of the spinal marrow without convulsing the muscles of the back; but that on touching the point of the knife, the muscles of the back were immediately convulsed.

From his observations, Bell inferred that the function of the posterior roots of the spinal nerve differed from that of the anterior roots (Figure 3.127). He demonstrated that disturbance of the ventral roots of the spinal nerves induced cramps, while agitation of the dorsal roots elicited no symptoms. In *A New Anatomy of the Brain,* he recognized that for a nerve to relay multiple functions, its roots must connect to corresponding yet different areas of the brain or spinal cord. He perceived the nerves of sensation as "entering" the brain and the nerves of motion as "passing out" from it (Figure 3.128). He claimed that each portion of the brain serves a specific function, and reiterated that not all portions drive sensibility, with various nerves conveying distinct yet specific functions. In this work, he also stated that the anterior root elicited contraction of the muscles, and through further experimentation he determined that the posterior root served motor control function.

In 1821, Bell expounded upon the long thoracic nerve, which innervates the serratus anterior muscle and is now named after him (external respiratory nerve of Bell). He proved that the abducens and hypoglossal nerves serve as motor nerves to the eye and tongue, respectively. While working in London, Bell operated on patients with facial paralysis, and he is credited with distinguishing peripheral from central facial paralysis. His findings regarding the innervation

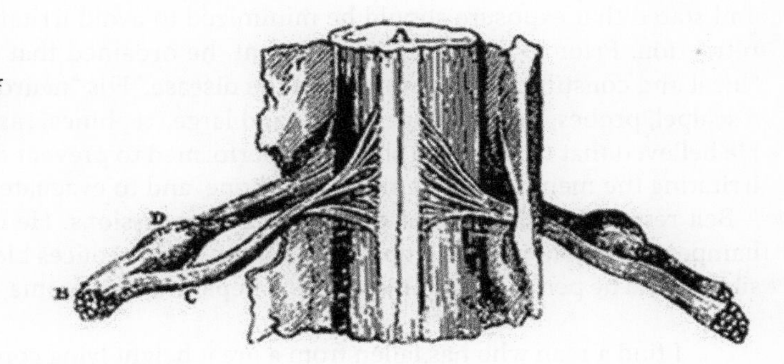

Figure 3.127 Illustration of the anterior and posterior roots of the spinal cord from Bell's *An Exposition of the Natural System of the Nerves of the Human Body*.

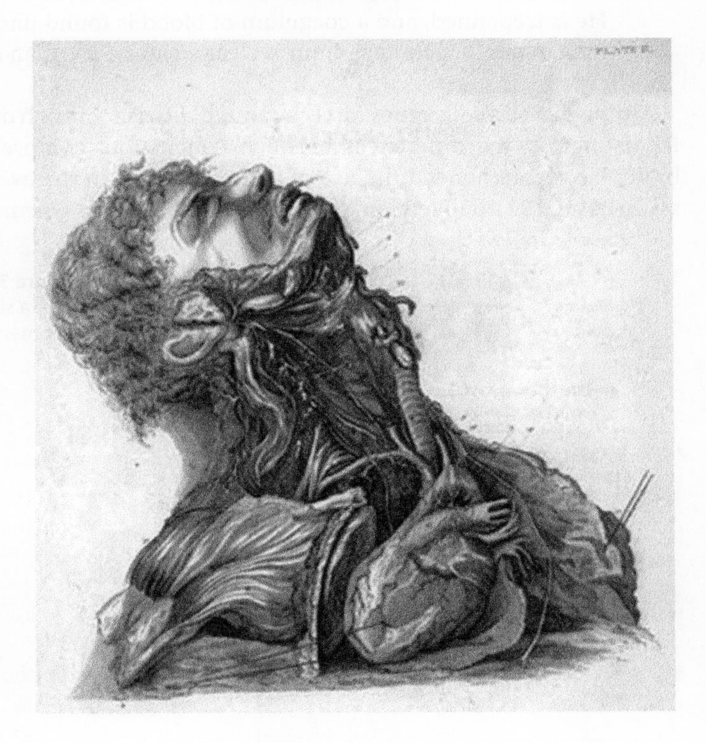

Figure 3.128 Illustration from Bell's work on the peripheral nerves, demonstrating the details of the nerves of the neck and thorax.

and paralysis of the seventh cranial nerve provided insight into the condition now known as Bell's palsy. He noted that the eyeball corresponding to the side of paralysis rotated upward as the patient attempted to close their eyes. He witnessed an identical rotation of the eyeball in normal individuals, calling it the palpebraloculogyric reflex, which is now known as Bell's phenomenon. Furthermore, Bell produced clinical descriptions of trigeminal nerve injury. He experimented with the portio minor and found that it served a motor function, while the portio major served a sensory one. His work, together with that of François Magendie regarding the motor and senory modalities of the ventral and dorsal roots, became known as the Bell–Magendie law. Bell went on to discover that sensory nerves originate in the gray matter of the cerebrum, spinal cord, and medulla oblongata, while the motor nerves originate in white matter. He helped to explain the fundamentals of the sympathetic nervous system and found that visceral nerves aid motion and sensation, and secretion. He discovered that sympathetic nerves anastomose with motor and sensory nerves, and that respiration is an involuntary action.

With his keen understanding of medicine, Bell proposed proper surgical protocols for efficient treatment of his patients. He highlighted the dangers of bone exposure during surgery and stated that exposure should be minimized to avoid irritation, inflammation, and further retraction. Prior to operating on a patient, he ordained that the surgeon must evaluate the "local and constitutional symptoms of the disease." His "neurosurgical" instruments included a scalpel, probes, quill, sponges, small and large trephines, rasp, forceps, saws, and elevators. He believed that trepanation should be performed to prevent depressed bone fragments from irritating the meninges, to remove dead bone, and to evacuate epidural blood (Figure 3.129).

Bell researched the clinical evaluation of concussions. He claimed that a concussion may hamper sensation, as the compression on the brain reduces blood supply and diminishes sensibility, and he performed craniotomies for epidural hematoma, as exemplified by the following:

> I find a man who has fallen from a great height lying comatose, with a very feeble pulse. He is trepanned, and a coagulum of blood is found under the skull (above the dura), an inch in depth. The coagulum is cleaned away; the man considerably revives.

Although Bell's colleagues rarely analyzed disease through dissection, Bell utilized dissection to distinguish the differences between concussion, compression, and inflammation of the brain. He recommended the use of adhesive straps in the event that the scalp is cut from the skull. In his 1821 publication, *Illustrations of the Great Operations of Surgery: Trepan, Hernia,*

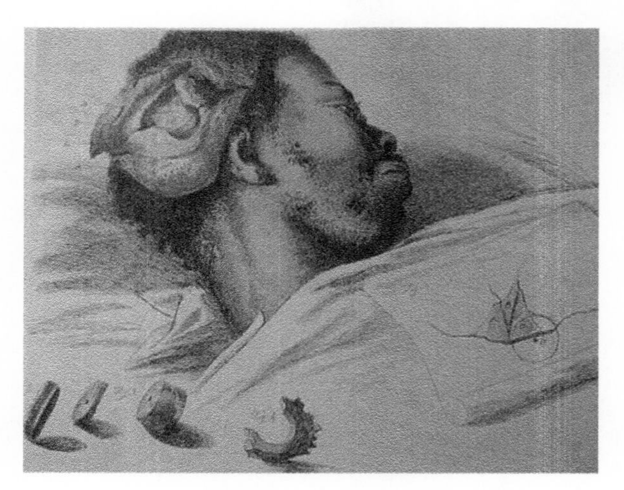

Figure 3.129 Painting by Bell of a patient with a skull fracture following removal of fragments via trepanation. (*Source:* Bell 1821.)

Amputation, Aneurysm, and Lithotomy, he depicted and discussed methods for craniotomy. He characterized diseased calvaria as being separated from the pericranium in the event that the pericranium is yellow or white in color. As a surgeon, he utilized the term "puffy tumor of the scalp" to describe a direct blow to the scalp, causing danger to the patient. He posited that blood located between the bone and the dura mater is a direct result of blood seeping through small vessels in the event of a perturbed dura mater (Figure 3.130). Although he did not operate, he described a patient with a ruptured middle cerebral artery. In addition, he described three types of tumor of the skull: the first was a fungus-like excrescence from the dura mater; the second was a state of protrusion caused by effusion of blood; and the third was fungus cerebri, which he advocated excising. However, he advised that the pericranium should not be left exposed for a prolonged time, to prevent the bone from being "deprived of its nourishing vessels."

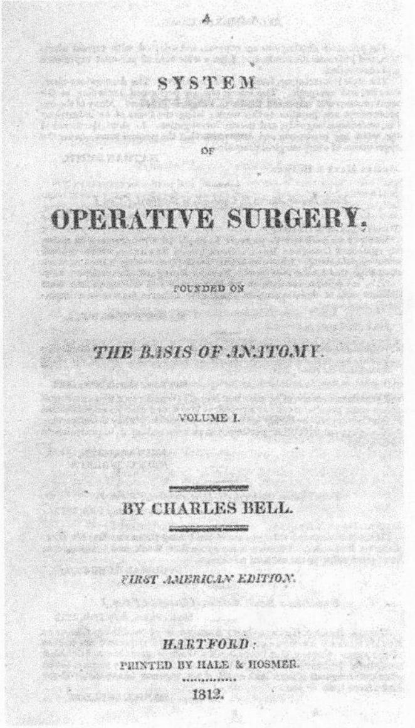

A SYSTEM OF OPERATIVE SURGERY, FOUNDED ON THE BASIS OF ANATOMY. VOLUME I. BY CHARLES BELL. FIRST AMERICAN EDITION. HARTFORD: PRINTED BY HALE & HOSMER. 1812.

Figure 3.130 Cover page of Bell's *A System of Operative Surgery Founded on the Basis of Anatomy* (1812).

3.11.3.2 Conclusion

Six years before his death, Bell left London for Edinburgh, stating that "London was a good place to live in but not to die in." After achieving vast success as an artist, scientist, and physician, Bell passed away on April 28, 1842 at the age of 68. His early advances in anatomy, physiology, and surgery ultimately instilled in him an interest in neurology and the nervous system, sparking his vast successes and his remarkable reputation as an innovative surgeon. It is on such contributions to neuroanatomy and surgery that our current discipline of neurosurgery is based.

3.12 Spain

3.12.1 Antonio de Gimbernat y Arbós

Don Manuel Louise Antonio de Gimbernat y Arbós (1734–1816) (Figure 3.131) was born in Cambrils, Tarragona, one of four Spanish provinces of Cataluña (Catalonia). His parents funded his six-year education at the University of Cervera, where he received a Bachelor of Arts in Latin and philosophy. In 1756, he matriculated into the College of Surgery founded by King Fernando VI, and because the number of Spanish surgeons in the military was limited, Pedro Virgili, the chief naval surgeon, is believed to have recruited him. Located in the wealthy port city of Cádiz, where the Spanish Navy headquarters were located, the College of Surgery used modern techniques that revitalized surgery.

Demonstrating great skill in his dissections, Gimbernat was appointed substitute Professor of Anatomy by the faculty in Cádiz before his graduation, filling the position left vacant by Virgili's son-in-law, Lorenzo Roland. Roland and Virgili left the institution in Cádiz to found a second college of surgery in Barcelona.

Gimbernat graduated from the College of Surgery in 1762. Following his graduation, Virgili nominated him for a professorship at his new school in Barcelona, but the administration in Madrid opposed this appointment owing to Gimbernat's young age. As a result, Gimbernat

Figure 3.131 Portrait of Antonio de Gimbernat y Arbós (1734–1816). (*Source:* From the Museum of Modern Art, Barcelona.)

was named an honorary professor in 1763, and appointed Chair of Anatomy at the college in Barcelona in 1764.

Described as a bright young mind, Gimbernat was confident and audacious, and he occasionally had to develop new instruments and tools to match his creative surgical procedures and methods, as demonstrated by his improvised renal lithotome, which became popular in 1773. It was during 1772 that his creativity and devotion to surgery led him to develop a new surgical technique for the treatment of femoral hernias.

By 1774, Gimbernat was a recognized surgeon with a famed reputation, and King Charles III of Spain sent him abroad to study surgery and health care delivery methods. He began his long trip in Paris, studying at l'Hôpital de la Charité and l'Hôtel Dieu (the oldest hospital in Paris). He continued his studies at several institutions in London, where he attended lectures and training sessions conducted by the renowned surgeon and professor of anatomy, John Hunter (1728–1793). Gimbernat's legendary contribution to hernia surgery took place on April 25, 1777, following Hunter's clinical tutorial concerning current hernia repair techniques (Figure 3.132). Gimbernat approached Hunter and requested permission to display a new technique, which involved incising the lacunar ligament to reduce strangulation of a femoral hernia. Hunter, amazed by Gimbernat's creativity, approved the procedure, acknowledging that he would use it and share it with his students and colleagues. Furthermore, Hunter named the lacunar ligament "Gimbernat's ligament" in his honor. Gimbernat commented, "I am encouraged to make my method public by the decided approbation it received from Dr. Hunter to whom I explained it on a preparation in 1777." Gimbernat continued his journey, and studied in Edinburgh and Leiden. He returned to Spain in 1778.

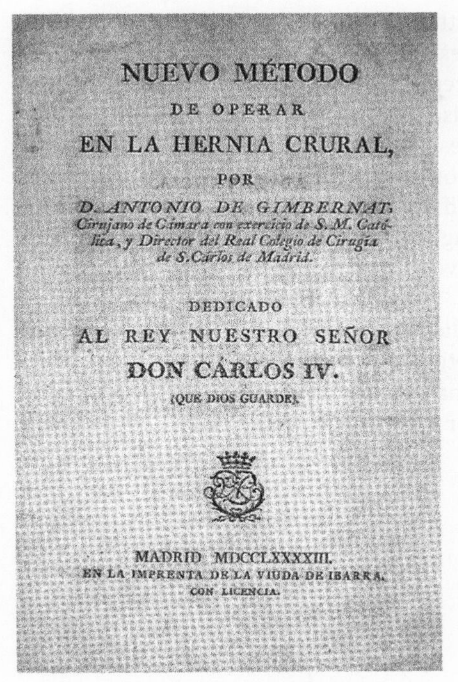
NUEVO MÉTODO
DE OPERAR
EN LA HERNIA CRURAL,
POR
D. ANTONIO DE GIMBERNAT,
Cirujano de Cámara con exercicio de S. M. Católica, y Director del Real Colegio de Cirugía de S. Cárlos de Madrid.
DEDICADO
AL REY NUESTRO SEÑOR
DON CÁRLOS IV.
(QUE DIOS GUARDE).
MADRID MDCCLXXXXIII.
EN LA IMPRENTA DE LA VIUDA DE IBARRA.
CON LICENCIA.

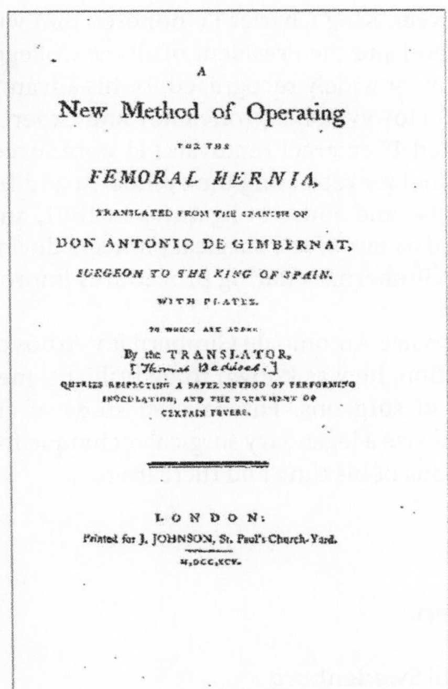
A
New Method of Operating
FOR THE
FEMORAL HERNIA,
TRANSLATED FROM THE SPANISH OF
DON ANTONIO DE GIMBERNAT,
SURGEON TO THE KING OF SPAIN.
WITH PLATES.
TO WHICH ARE ADDED
By the TRANSLATOR,
[Thomas Beddoes.]
QUERIES RESPECTING A SAFER METHOD OF PERFORMING INOCULATION, AND THE TREATMENT OF CERTAIN FEVERS.
LONDON:
Printed for J. JOHNSON, St. Paul's Church-Yard.
M,DCC,XCV.

Figure 3.132 Left: Title page of *Nuevo Método de operar en la Hernia Crural*. Right: Title page of *A New Method of Operating for the Femoral Hernia, Translated from the Spanish of Don Antonio de Gimbernat.*

In 1793, Gimbernat published *Nuevo Método de operar en la Hernia Crural* ("A New Method of Operating for the Femoral Hernia"), dedicated to King Charles IV, which was translated into English in 1795 by Thomas Beddoes. In this publication, Gimbernat provided a detailed anatomical description of the femoral canal and Gimbernat's ligament. He claimed to have discovered this ligament in 1768, and later to have devised a surgical procedure for treating strangulated femoral hernias. He proposed reducing the hernia before resorting to an invasive surgical procedure. He explained how surgeons could push and compress the herniated tissue using their fingers, and that for this reason they must maintain close-cut fingernails to avoid further injuring the patient. As much as he advocated non-invasive treatment, Gimbernat stated that in certain cases non-invasive reduction could be painful and potentially fatal. In such cases, he recommended the judicious use of his surgical repair technique.

It is important to note that in his book, Gimbernat described the lymph node that in later years would become known as Cloquet's or Rosenmüller's gland. Jules Germain Cloquet (1790–1883) was a professor of anatomy and surgery in Paris who realized that this node could be mistaken for a femoral hernia. Johann Christian Rosenmüller (1771–1820) was a distinguished embryologist, anatomist, and surgeon, and professor of anatomy and surgery in Leipzig from 1802 until 1820.

A new college of surgery was being established in Madrid during the late 1780s, which would be named the College of San Carlos. Before knowing who would become the director of this institution, Gimbernat moved his family to the capital. He was appointed Director of San Carlos on January 27, 1787, and the college was inaugurated on October 1, 1787. In 1789, Gimbernat received the title of Royal Surgeon and was relieved from surgical instruction.

Later that same year, King Charles IV honored him with noble status. In 1801, he became the First Royal Surgeon and the President of all the Colleges of Surgery in Spain.

Gimbernat is most widely recognized for his advancement of the treatment of strangulated femoral hernias. However, his proficiency and expertise in surgery extended to ophthalmology. He performed 47 cataract removals (41 were successful), and his creative abilities allowed him to design a tool for retracting the eyelids. In addition, he published a valuable dissertation concerning the use and abuse of ligatures in 1801, and a paper concerning corneal ulcers in 1802. Recognized as one of the surgical pioneers during "the age of dissection" (the late eighteenth century), Gimbernat's daring procedures improved the safety and efficiency of hernia-related surgery.

Don Manuel Louise Antonio de Gimbernat y Arbós died on November 17, 1816. A man with great determination, he was known to be brilliant, meticulous, and creative and to exemplify the importance of surgeons. His devoted study of anatomy, particularly inguinal anatomy, allowed him to devise a legendary surgical technique for repairing femoral hernias, influencing renowned surgeons of his time and thereafter.

3.13 Sweden

3.13.1 Emanuel Swedenborg

Emanuel Swedenborg (1688–1772) (Figure 3.133) was born in Stockholm, the third of eight children in a devoutly Lutheran family, and was raised by his influential father, who was a Lutheran bishop. At the age of 11, Swedenborg enrolled in Uppsala University, where he pursued an education in philosophy and delved into the writings of Publilius and Seneca, on which he based his dissertation. Upon completing his education, he served on the Royal Board of Mines and in the Swedish House of Lords, but he frequently traveled to pursue scientific endeavors and to perfect his writing.

On a journey to England in 1710, Swedenborg explored the fields of mathematics and astronomy, meeting with eminent physicists such as Sir Isaac Newton. The knowledge he acquired in London sparked in him a rare scientific creativity, which he used to propose the construction of innovational contraptions such as submarines and flying machines. Enthralled with his scientific endeavors in England, he spearheaded the creation of a journal entitled *Daedalus Hyperboreus* upon his return to Sweden.

Swedenborg assumed an interest in the field of metal mining, and later proposed the construction of a canal between Goteborg and Stockholm. Similarly, he invented a mechanism for lugging boats overland for the Siege of Frederickshall. When Queen Ulrica granted his family nobility in 1719, Swedenborg became overwhelmed with Sweden's monetary issues and attempted to organize a decimal-based system.

Beginning in 1734, he developed an interest in the issues behind physical science, and researched in great detail the theories of anatomy and physiology, the origin of the planets and the sun, and the mathematical aspect of mechanics. From his studies, he hoped to master the study of psychology. He then focused on the relationship of the body and the soul, which in turn sparked his studies on the brain. Later, he stepped down from the Board of Mines to focus entirely on the study of religion. Many of his peers disagreed with his fervent religious beliefs, forcing him to leave Sweden and pursue a less opulent life in London.

The impoverished Swedenborg passed away in London on March 29, 1772, and his body was buried in the Swedish Lutheran church.

Figure 3.133 Drawing of Emanuel Swedenborg (1688–1772).

3.13.1.1 Contributions to Neuroanatomy

Swedenborg's groundbreaking neurological discoveries provided vast insight into the complex anatomy and physiology of the brain. However, he was not credited for his work until after his death. Most of his publications, including *The Brain* and *The Cerebrum*, were not released to the public until the 1880s. It was not until Max Neuburger contacted the Swedish Royal Academy for a cohesive collection of Swedenborg's work that his writings became publicly available. During his travels in Europe, Swedenborg wrote to his brother-in-law, describing the anatomy and physiology of the brain. In his publication *Oeconomia Regni Animalis* ("The Economy of the Animal Kingdom"), Swedenborg references prior scientific discoveries to develop novel insight into the brain. In his writing, he acknowledged scientists including Eustachius, Ruysch, and Leeuwenhoek for their contributions, and used their ideas as a basis for observation.

One of Swedenborg's most astute discoveries lay in his description of the miniscule yet complex structure of the neurons within the cortex. In his paper "Tremulationes," he discussed the oscillation of particles through the neural process and claimed that they moved rhythmically from the brain to the peripheral nerves and back to the brain through the center of the medulla. He further claimed that the movement was spurred by wave-like palpitations of the brain and by stimuli to the sense organs. He coined the term "little brains" for the cerebellula, or independent units of the cerebral matter, to describe the nerve cells of the cerebrum. Upon examining the fibrous, chain-like configuration of the cerebellula that extends through the white matter and medulla to the spinal cord and then travels from the periphery to diverse parts of the body, he came to believe that the cerebellula drove various psychological states and bodily operations, including sensation and movement. He defied Descartes' view that the third ventricle housed the final configuration of nerve fibers, and stated that the fibers originate and terminate within the cortical cerebellula. If the fibers were to extend beyond the cerebellula, he explained, they would travel into the periphery. He described the means by which a stimulus traveled from each sense organ to the cortex. Overall, his discoveries regarding the cerebellula defined many of his other contributions to the study of the brain. Martin Ramstrom (1861–1930) deemed Swedenborg's "neuron theory" an innovative and intuitive breakthrough in the field of neuroanatomy.

Swedenborg also focused on sensation governed by the nervous system. Through thorough observation and anatomical review, he distinguished the cortex as the prominent structure of the brain in which the nerve fibers terminate. He denoted the cortex the primary control center for the sensory organs, as it innervates actions of the nerves and muscles and enables movement, cognition, imagination, judgment, and will. Inside the cortex, he claimed, are cerebellula that are configured in accordance with their role in sensation; thus, the cortex is configured concisely so as to receive visual, auditory, olfactory, and gustatory sensations. Not only did Swedenborg discuss the sensory components of the cortex, but he also explained the role of the cortex in governing motor activity. He claimed that the motor activity was made possible by the fibrous connection of the cerebellula to distinct muscle fibers. Thus, cerebellula have the ability to innervate an entire muscle or a group of adjacent muscles.

Swedenborg observed the cortices of living animals to illustrate their role in motor control. He referred to the frontal lobe as the "anterior province of the cerebrum," and claimed that this "anterior province" controlled voluntary action and intellect, with an emphasis on memory and the subconscious. He suggested that, when inflamed, the cerebrum lost its intellectual capabilities. Although written in approximately 1745, Swedenborg's *Oeconomia Regni Animalis*, describing cortical contributions to muscle control, was not discovered or publicized until 1882. In this work, Swedenborg referred to the cortex as having three distinct sections, each innervating the muscular response of a distinct portion of the body. He claimed that the superior lobe of the cortex innervated the muscles of the hands and feet, while the middle lobe controlled the muscles of the abdomen and thorax. In addition, he claimed that the most inferior lobe controlled the muscles of the head and face. Thus, the muscles of the body, according to Swedenborg, were inversely related to the lobes of the cortex.

Some of Swedenborg's most acclaimed discoveries dealt with the hindbrain and the spinal cord. He noted that although the cortex controls movement, the fibers of the cortex ultimately activate the muscles. Swedenborg is credited for describing the upper and lower motor neurons, and for claiming that habitual actions stem from a distinct motion of the cerebrum by means of the medulla. He based his knowledge of habit formation on first-hand observation of the cerebral cortices of deceased animals, and claimed that the subcortical gray matter, such as the corpora striata, is innervated during habitual action. Once an action becomes habitual, motor control shifts from the cerebrum to the corpora striata.

In addition to his findings on the cortex, Swedenborg observed the movements of decapitated reptiles and birds, claiming that not all spinal nerves originate in the same location. Instead, he stated that some spinal fibers originate in the cerebrum or cerebellum, while others originate in the oblongata or gray matter of the medulla. He claimed that the anterior roots of the spinal nerves were connected to fibers from the anterior portion of the gray matter, while fibers from the front and back portions of the gray matter constitute the posterior roots.

In addition, Swedenborg described various anatomical features of the brain well before they were given names by later scientists. Some of these features include the perivascular spaces, the foramen of Magendie (almost 100 years prior to Magendie's description), and cerebrospinal fluid. Interestingly, his description of cerebrospinal fluid predates Cotugno's by many years. He noted the importance of the pituitary gland or "arch gland" in maintaining normal neurological function. His observations demonstrate his keen understanding of the cerebral cortex. Lastly, Swedenborg hypothesized that the tectum was involved with vision and that the corpus callosum acted as a communicator between the left and right cerebral hemispheres.

3.14 Switzerland

3.14.1 Wilhelm His Senior

Few medical families can claim that more than one member contributed significantly to modern-day clinical and research practices. However, within the Swiss family of His, both Wilhelm His Senior (1831–1904) and Wilhelm His Junior (1863–1934) contributed advances that still affect several areas of modern-day research and clinical practice.

Wilhelm His Senior (Figure 3.134) was born in Basel, Switzerland, on July 9, 1831 to a prominent, rich, and distinguished family. His father's real name was Eward Ochs, but he changed his surname to His after becoming engaged to Wilhelm's mother, Katharina La Roche. Ochs translates to "ox" in German, but in slang means "dumb ox." Katharina apparently wanted to avoid any comments and so convinced him to change it.

Figure 3.134 Photograph of Wilhelm His Senior (1831–1904).

Wilhelm began his medical studies in Basel, continued them in Bern, and traveled to Berlin to study anatomy and embryology with Johannes Müller (1801–1858) and Robert Remak (1815–1865), respectively. According to Wendler, His was greatly influenced by Remak's embryology lectures. He was afforded the opportunity to study at Würzburg and meet Rudolf Virchow (1821–1902), who had a further influence on him. He continued his medical studies in Prague and Vienna before returning to Basel to obtain his MD in 1854. His doctoral thesis was entitled "The Anatomy and Pathology of the Cornea," a topic introduced to him by Virchow. During 1855, His traveled to Paris and collaborated with Johann Friedrich Horner (1831–1886). The next year, he returned to Basel, where he defended his habilitation. In 1856, he traveled to Berlin to collaborate with Theodor Billroth (1829–1894) on research concerning lymphatics and connective tissue.

During 1865, His focused on embryological research, publishing his work "On Tissues and Spaces of the Body" and introducing the term "endothelium" from his observations. In 1857, at the age of 26, he was given the Chair of Anatomy at the University of Basel. During his tenure in Basel, he studied human and comparative embryology and the histology of the lymphatics. In 1866, he invented the microtome, one of his most important contributions to modern-day research, and in 1887 he constructed the embryograph, an instrument that allowed him to reconstruct on paper the histological slides of embryos that he was examining. In addition, he described the anatomy of the female genitalia and the lung, and the position of the stomach and its topographic relationships.

With the support of his friend Carl Ludwig, His left Basel in 1872 and accepted the Chair of Anatomy at Leipzig University. He remained there for the rest of his career. It was in Leipzig that the majority of his groundbreaking embryological research was carried out. He created the science of histogenesis and comparative embryology between species, developing several of the current methods used in the fields of anatomy and embryology. He made the first attempt to study the development of the human embryo in full, planning an investigation so large that he never completed it. In 1876, he became a founding member of the *Journal of Anatomy and Development*, and between 1880 and 1885, he published an article in three parts concerning the anatomy of the human embryo. This paper was the most complete picture of human development at the time; according to Garisson, "the human embryo was studied as a whole for the first time."

His documented the development of the nervous system in great detail, and in 1886 presented overwhelming evidence supporting the neuron theory. He discovered that axons and their appendages arose from the neuroblast, forming a solitary entity; in 1889, he named the axonal appendages "dendrites." The last scientific work His produced was in 1904, and concerned the development of the human brain during the first two months of gestation.

His proposed the development of a committee on embryology research, and it was his friend and admirer Franklin P. Mall (1862–1917), an American anatomist, who fulfilled this dream in 1913 by establishing the Department of Embryology of the Carnegie Institution in Washington, DC.

His did not restrict himself to studying embryology and was very active in medical terminology, archeology, photography, and drawing. He tried to use his own photographic material in his lectures, and his own reconstructions of embryology produced from photographs. In addition, he was one of the founders of the Anatomichen Gesellschaft in 1886. In 1895, he was instrumental in creating the International Commission for Anatomical Terminology, whose goal was to unify anatomical nomenclature and standardize its use throughout the research community. One of his best-known archeological ventures began in 1894, with the search for the remains of Johann Sebastian Bach. A coffin had been found outside the Johanniskirche during the building of a local church and the remains were rumored to be Bach's. His and Adam

Politzer (1835–1920), an otologist, examined the skull, in particular the structures involving the ear, in an attempt to identify the skeleton. Their study demonstrated that the cochlea was uncommonly large and that the skull impressions of the inferior temporal gyri suggested strong development in these areas. His and Politzer correlated these findings to perfect pitch and outstanding musical ability, and identified the skeleton as Bach's.

Wilhelm His Senior died on May 1, 1904, while still working as a professor at Leipzig University. He lived a simple life, and was described as a person who did not talk much. His lifelong motto, "that progress in anatomy is most likely to occur when its problems include the study of growth and function, as well as of structure," is viewed by many embryologists as the unifying goal of their investigations.

3.14.2 Wilhelm His Junior

Wilhelm His Junior (1863–1934) (Figure 3.135) was the third of six children born to His Senior while he was in Basel, Switzerland. In 1872, his father took a position at the University of Leipzig in Germany, and His Junior became a student at the city Gymnasium. In this renowned city, noted for its contributions to cultural, political, and scientific progress, His Junior mastered German and French, and became a proficient violinist and painter. He completed his last two years of high school in Basel, and graduated in 1882. His Senior used his connections in the world of medicine to ensure his son studied at the best universities in Europe, including Geneva, Bern, Strasbourg, and Leipzig. In Leipzig, His Senior taught his son anatomy and his groundbreaking procedures in embryological research. This training proved invaluable to His Junior, contributing to his research concerning the conduction system of the heart. In Strasbourg, he gained extensive knowledge working under Oswald Schmiedeberg (1838–1921), the executive director of the world's first pharmacological laboratory. During his time in Strasbourg, after performing experiments on dogs dosed with pyridine acetate, His Junior discovered the body's

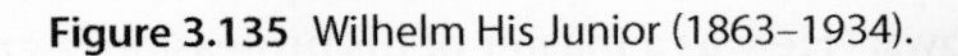

Figure 3.135 Wilhelm His Junior (1863–1934).

capacity to methylate organic compounds. He graduated with a medical doctorate from the University of Leipzig in 1889.

After graduating, His was appointed assistant to Heinrich Curschmann in the Leipzig clinic. While there, he researched the electrical properties of the heart, investigating whether conduction abnormalities arose from the ganglia or muscle. Using his father's techniques, His demonstrated that the heart begins beating before the development of cerebrospinal nerves and ganglia. In 1893, he published a 35-page article describing the activity of the embryonic heart and how this related to the properties of the adult heart. Only one page of this article was devoted to the description of his best-known discovery, the eponymous bundle of His. He experimented on rabbits, producing asynchronous contractions of the atria and ventricles by severing the atrioventricular bundles. The results of his research were presented at the International Physiological Congress in Bern in 1895.

His received a clinical medicine appointment in 1895, and the focus of his studies concerned clinical surveys rather than anatomical research. It was during this time that he researched gout and diseases of the joints. He demonstrated that uric acid produced inflammatory irritation in patients with gout and studied various methods for removing precipitated uric acid from the body, predominantly through dietary control.

After working at a number of hospitals in Germany and Switzerland, His was appointed to the Chair of Internal Medicine in Berlin, and he became Director of the Charité medical clinic in 1907. With the outbreak of World War I, he joined the German army as a consulting internist, taking part in missions to Turkey, Asia Minor, the Western Theater and Russia. It was while in Russia in 1916 that he first described "Volhynia fever" (trench fever). Trench fever became an important aspect of daily life during the war, infecting more than a million soldiers. At the end of the war, His returned to Berlin and was appointed dean of the medical faculty.

His participated in clinical practice and medical education on his return to Berlin, continuing until his retirement due to emphysema in 1932. During his retirement, he wrote a comprehensive account of his discovery of the atrioventricular bundle, including the results of experiments that had never before been published. He died on November 10, 1934, at the age of 71. One of his students described him as follows: "He was a master of his profession, a great physician, investigator and a well cultured gentleman."

Modern researchers and clinicians should consider the importance of the work carried out by Wilhelm His Senior and Wilhelm His Junior. Many advances in embryological, histological, and cardiovascular electrophysiology research can be traced back to this father and son.

Bibliography

Europe

Agutter, P.S. and Wheatley, D.N. (2008). *Thinking About Life.* Dordrecht: Springer.

Getz, F. (1998). *Medicine in the English Middle Ages.* Princeton, NJ: Princeton University Press.

Gordon, B. (1959). *Medieval and Renaissance Medicine.* New York: Philosophical Library.

Infusino, M., Win, D., and O'Neill, Y.V. (1995). Mondino's book and the human body. *Vesalius* 1: 71–76.

Porter, R. (1997). *The Greatest Benefit to Mankind. A Medical History of Humanity from Antiquity to the Present.* London: Harper Collins.

Singer, C. (1957). *A Short History of Anatomy from the Greeks to Harvey.* New York: Dover.

Heinrich Obersteiner

Blom, J.D. (2010). *A Dictionary of Hallucinations*. New York, Dordrecht, Heidelberg, London: Springer.

Freud, S. (1887). Literarische Anzeigen – Anleitung beim Studium des Baues der nervösen Centralorgane im gesunden und kranken Zustande. Von Dr. H. Obersteiner. *Wien. Med. Wochenschr.* 37: 1642–1644.

Hatzigiannakoglou, P.D. and Triarhou, L.C. (2011). A review of Heinrich Obersteiner's 1888 textbook on the central nervous system by the neurologist Sigmund Freud. *Wien. Med. Wochenschr.* 161 (11–12): 315–325.

Jellinger, K.A. (2006). A short history of neurosciences in Austria. *J. Neural Transm.* 113 (3): 271–282.

Marburg, O. (1917). Heinrich Obersteiner zu seinem 70. Geburtstage, miteiner Porträtskizze von Olga Prager. *Wien. Med. Wochenschr.* 67: 2013–2016.

Marburg, O. (1923). Heinrich Obersteiner, obituary. *J. Nerv. Ment. Dis.* 57 (3): 319–321.

Obersteiner, H. (1888). *Anleitung beim Studium des Baues der nervösen Centralorgane im gesunden und kranken Zustande*. Leipzig-Wien: Toeplitz & Deuticke.

Obersteiner, H. (1890). *The Anatomy of the Central Nervous Organs in Health and in Disease*. Translated by Hill A. London: Charles Griffin.

Obersteiner, H. (1892). *Anleitung beim Studium des Baues der nervösen Centralorgane im gesunden und kranken Zustande*, 2e. Leipzig-Wien: Franz Deuticke.

Obersteiner, H. (1893). *Anatomie des centres nerveux: guide pour l.etude de leur structure á l.état normal et pathologique*. Translated by Corenne JX. Paris: Georges Carré.

Obersteiner H, Redlich E. Zur Kenntnis des Stratum (Fasciculus) subcallosum (Fasciculus nuclei caudati) und des fasciculus fronto-occipitalis (reticulirtes coritic-caudales Bündel). *Arbeiten aus dem Neurologischen Institute an der Wiener Universität* 1902;8:286–307.

Pick, A. and Pritchard, P.H. (1890). Hypnotism as a therapeutic agent: its relations to forensic medicine. *N. Eng. Med. Gaz.* 25: 315–328.

Seitelberger, F. (1963). Heinrich Obersteiner (1847–1922). In: *Grosse Nervenärzte, B and 3* (ed. K. Kolle), 21–30. Stuttgart: G. Thieme.

Seitelberger, F. (1992). Heinrich Obersteiner and the Neurological Institute: foundation and history of neuroscience in Vienna. *Brain Pathol.* 2 (2): 163–168.

Shorter, E. (1997). *A History of Psychiatry: From the Era of the Asylum to the Age of Prozac*. New York: Wiley.

Spiegel, E.A. (1953). Heinrich Obersteiner (1847–1922). In: *The Founders of Neurology* (ed. W. Haymaker and K.A. Baer), 199–202. Springfield, IL: Charles C. Thomas.

Emil Zuckerkandl

Adey, W.R. (1951). An experimental study of the hippocampal connexions of the cingulate cortex in the rabbit. *Brain* 74: 233–247.

Bergman RA, Afifi AK, Miyauchi K. Illustrated Encyclopedia of Human Anatomic Variation: Opus V: Skeletal Systems. Wrist and Hand: Part 2 of 5. http://www.anatomyatlases.org/AnatomicVariants/SkeletalSystem/Images/115.shtml. Last accessed July 13, 2018.

Chesbrough, R.M., Burkhard, T.K., Martinez, A.J., and Burks, D.D. (1989). Gerota versus Zuckerkandl: the renal fascia revisited. *Radiology* 173: 845–846.

Isobe, M., Murakami, G., and Kataura, A. (1998). Variations of the uncinate process of the lateral nasal wall with clinical implications. *Clin. Anat.* 11: 295–303.

Lang, J. (1989). *Clinical Anatomy of the Nose, Nasal Cavity and Paranasal Sinuses*. New York: Thieme.

Macapanpan, L.C. and Weinmann, J.P. (1954). The influence of injury to the periodontal membrane on the spread of gingival inflammation. *J. Dent. Res.* 33: 263–272.
Manjunath, K.Y. (2001). Anomalous origin of the middle meningeal artery – a review. *J. Anat. Soc. India* 50: 179–183.
Ober, W.B. (1983). Emil Zuckerkandl and his delightful little organ. *Pathol. Annu.* 18 (1): 103–119.
Ortale, J.R. and Marquez, C.Q. (1998). Anatomy of the intramural venous sinuses of the right atrium and their tributaries. *Surg. Radiol. Anat.* 20: 23–29.
Pernkopf E. Josef Hyrtl, Carl von Langer, Emil Zuckerkandl, Carl Toldt, Julius Tandler und Ferdinand Hochstetter ... Die Wiener Anatomen Schule (The Vienna School of Anatomy). In Österreichische Naturforscher Ärzte und Techniker: Vienna; 1957.
Pina, J.A.E. (1975). Morphological study on the human anterior cardiac veins, venae cordis anteriores. *Acta Anat.* 92: 145–159.
Ravin, J.G. (1999). The statesman, the artist, and the ophthalmologist: Clemenceau, Lautrec, and Meyer. *Arch. Ophthalmol.* 117: 951–954.
Royle, G. and Motson, R. (1973). An anomalous origin of the middle meningeal artery. *J. Neurol. Neurosurg. Psychiatry* 36: 874–876.
Stammberger, H.R. and Kennedy, D.W. (1995). Paranasal sinuses: anatomic terminology and nomenclature. The anatomic terminology group. *Ann. Otol. Rhinol. Laryngol. Suppl.* 167: 7–16.
Tosun, F., Yetiser, S., Akcam, T., and Ozkaptan, Y. (2001). Sphenochoanal polyp: endoscopic surgery. *Int. J. Pediatr. Otorhinolaryngol.* 58: 87–90.
West SF, Correa JD, Germain M, Balsam D, Rosen J. Neuroblastoma imaging. http://www.emedicine.com/radio/topic472.htm. Last accessed July 13, 2018.
Zuckerkandl, E. (1976). Zur anatomie der orbitalarterien. *Med. Jb. (Wien)* 3: 343–350.
Zuckerkandl E. Über ein abnormes verhalten der zungenschlagadern. Wien. Med Wschr. 1881.
Zuckerkandl, E. (1882). *Normale und pathologische anatomie der nasenhöhle und ihrer pneumatischen anhänge.* Wien: Braumüller.
Zuckerkandl, E. (1883). Uber den fixationsapaparat der nieren. *Med. Jahrb.* 59–67.
Zuckerkandl, E. (1884). *Über den zirkulationsapparat in der nasenschleimhaut.* Wien: Kaiserl. Königl. Hof-und staatsdruckerei.
Zuckerkandl, E. (1892a). Die siebbeinmuscheln des menschen. *Anat. Anz.* 7: 13–25.
Zuckerkandl, E. (1892b). *Normale und pathologische anatomie der nasenhöhle und ihrer pneumatischen anhänge. Anatomie der nasenscheidewand, Bd. II.* Wien: Braumüller.
Zuckerkandl, E. (1892c). Die entwicklung des siebbeins. *Anat. Anz.* 6: 261–264.
Zuckerkandl, E. (1893). *Normale und pathologische anatomie der nasenhöhle und ihrer pneumatischen anhänge,* 2e, vol. I. Wien: Braumüller.
Zuckerkandl, E. (1896). Geruchsorgan. In: *Ergebnisse der anatomie und entwicklungsgeschichte* (ed. F. Merkel and R. Bonnet). Wiesbaden: Bergmann.
Zuckerkandl, E. (1901). Ueber Nebenorgane des Sympathicus im Retroperitonealraum des Menschen. *Verh. Anat. Ges.* 15: 85–107.

Vincent Alexander Bochdalek

Haller, J.A. Jr. (1986). Professor Bochdalek and his hernia: then and now. *Prog. Pediatr. Surg.* 20: 252–255.
Harvey, W. (1910). *Exercitatio Anatomica de Motu Cordis et Sanguinis in Animalibus.* Translated by Willis W. Harvard Classics, vol. 38. New York: Collier and Son.
Kops, J. (1975). Doktor Vincenc Bochdalek [Vincenc Bochdalek, MD]. *Zdrav. Prac.* 25: 44.

Locke, J. (1960). *Essay Concerning the Human Understanding*. London: Fontana Library Edition, Collins.
Matousek, O. (1952). Vincenc Alex. Bochdalek, první professor pathologické anatomie v Praze [Vincenc Alex. Bochdalek, first professor of pathological anatomy in Prague]. *Cas. Lek. Cesk.* 91: 407.
Stylopoulos, N. and Rattner, D.W. (2005). The history of hiatal hernia surgery. From Bowditch to laparoscopy. *Ann. Surg.* 241: 185–193.
Urbach, P. (1986). *Francis Bacon's Philosophy of Science: An Account and a Reappraisal*. La Salle, IL: Open Court.
Wondrak, E. (1983). Český anatom a patolog V.A. Bochdalek. 100 let od smrti [The Czech anatomist and pathologist V.A. Bochdalek. 100 year since his death]. *Cas. Lek. Cesk.* 122: 1334–1337.

Richard Lower

Brown, T.M. (1980). *Dictionary of Scientific Biography*, vol. VIII. New York: American Council of Learned Societies.
Donovan, A.J. (2004). Lower, MD, physician and surgeon (1631–1691). *World J. Surg.* 28: 938–945.
Feldman SM. *Thomas Willis, Richard Lower and Cerebri Anatome*. New York University: New York; 1664. http://www.bri.ucla.edu/nha/ishn/abs2005.htm. Last accessed July 13, 2018.
Felts, J.H. (2000). Richard Lower: anatomist and physiologist. *Ann. Intern. Med.* 132: 420–423.
Franklin, K.J. (1932). The work of Richard Lower (1631–1691). *Proc. R. Soc. Med.* 25: 113–118.
Gotch, F. (1908). *Two Oxford Physiologists*. Oxford: Clarendon Press.
Hoff, E.C. and Hoff, P.M. (1936). The life and times of Richard Lower, physiologist and physician (1631–1691). *Bull. Inst. Hist. Med.* 4: 517–535.
Ivins, W.M. (1992). *Prints and Visual Communication*. Cambridge, MA: MIT Press.
JAMA (1967). Richard Lower (1631–1691). Physiologic cardiologist. *JAMA* 199: 146.
Lower, R. (1932). A treatise on the heart. Translated by Franklin KJ. In: *Early Science in Oxford* (ed. R.T. Gunther). London: Dawsons of Pall Mall.
Lower, R. (1963). *De Catarrhis*. Translated by Hunter R, Macalpine I. London: Dawsons of Pall Mall.
Molnár, Z. (2004). Thomas Willis (1621–1675), the founder of clinical neuroscience. *Nat. Rev. Neurosci.* 5: 329–335.
O'Connor, J.P. (2003). Thomas Willis and the background to Cerebri Anatome. *J. R. Soc. Med.* 96: 139–143.
Symonds, C. (1955). The circle of Willis. *Br. Med. J.* 1: 119–124.
Tessman, P.A. and Suarez, J. (2002). Influence of early printmaking on the development of neuroanatomy and neurology. *Arch. Neurol.* 59: 1964–1969.
Willis, T. (1664). *Cerebri Anatome*. London: Cui Accessit Descriptio et Usis.
Willis, T. (1971). *The Anatomy of the Brain*. Translated by Pordage S. New York: USV Pharmaceutical Corp.
Woo H, Fung C. Remarkable physicians associated with pulsus paradoxus, the classic sign – Richard Lower and Adolf Kussmaul. http://www.priory.com/homol/pulsus.htm. Last accessed July 13, 2018.
Zimmer, C. (2004). *Soul Made Flesh. The English Civil War and the Mapping of the Mind*. London: Free Press.

John Browne

Brown, J. (1685). *Phil. Trans.* 15 (78): 1266–1268.
Buelow, G.J. (1990). Originality, genius, plagiarism in English criticism of the eighteenth century. *Int. Rev. Aesthet. Sociol. Music* 21: 117–128.
Forbes, T.R. (1978). The case of the casual surgeon. *Yale J. Biol. Med.* 51: 583–588.
Porter, R. (2006). *The Cambridge History of Medicine.* Cambridge: Cambridge University Press.
Russell, K.F. (1940). John Browne and his treatise on the muscles. *ANZ J. Surg.* 10: 113–116.
Russell, K.F. (1959). John Browne, 1642–1702. A seventeenth century surgeon, anatomist, and plagiarist. *Bull. Hist. Med.* 33: 503–525.
Russell, K.F. (1962). A list of the works of John Browne (1642–1702). *Bull. Med. Libr. Assoc.* 50: 675–683.
Werrett, S. (2000). Healing the nation's wounds: royal ritual and experimental philosophy in Restoration England. *Hist. Sci.* 38: 377–399.

James Drake

Baker, D.E. (1812). *Biographia Dramatica; or a Companion to the Playhouse*, vol. 3. London: Longman, Hurst, Rees, Orme, and Brown.
Drake, J. (1702). A discourse concerning some influence of respiration on the motion of the heart, hitherto unobserved. *Philos. Trans. R. Soc.* 23: 1217–1240.
Drake, J. (1707). *Anthropologia Nova; or A New System of Anatomy Describing the Animal Economy, and a Short Rationale of Many Distempers Incident to Human Bodies.* London: Smith and Walford.
Gysel, C. (1987). The oral medicine of Govert Bidloo (1685), William Cowper (1698) and James Drake (1707). *Rev. Belg. Med. Dent.* 42: 160–165.
Lee, S. (1888). *Dictionary of National Biography*, vol. 15. London: Macmillan.
Loukas, M., Akiyama, M., Shoja, M.M. et al. (2010). John Browne (1642–1702): anatomist and plagiarist. *Clin. Anat.* 23: 1–7.
Munk, W. (1878). *The Roll of the Royal College of Physicians of London*, vol. 2. London: The College, Pall Mall East.
Smith, H. (2001). English "feminist" writings and Judith Drake's "An essay in defence of the female sex.". *Hist. J.* 44: 727–747.

Francis Sibson

Abbondanzo, S. (2003). *Thomas Hodgkin.* Washington, DC: Department of Hematopathology, Armed Forces Institute of Pathology.
(1859). *The Annual Register, or a View of the History and Politics of the Year 1858.* London: Woodfall and Kinder.
Brown, K. (2004). Sibson, Francis (1814–1876). In: *Oxford Dictionary of National Biography* (ed. H.C.G. Matthew and B. Harrison). Oxford: Oxford University Press.
Davy, H. (ed.) (1800). *Researches, Chemical and Pphilosophical; Chiefly Concerning Nitrous Oxide.* London: J. Johnson.
Hughes, J.E. (1977). Francis Sibson. *Med. J. Aust.* 1: 545–546.
Irwin, R.A. (ed.) (1955). *Letters of Charles Waterton.* London: Rockliff.
Lee, S. (ed.) (1897). *Dictionary of National Biography*, vol. LII. London: Oxford University Press.
Maltby, J.R. (1977). Francis Sibson, 1814–1876: pioneer and prophet in anaesthesia. *Anaesthesia* 32 (1): 53–62.

Ord, W.M. (1881). Sibson, F (1814–1876). In: *Collected Works of Francis Sibson*, vol. IV. London: Macmillan.

Royal College of Physicians. Francis Sibson archives. 2003. https://aim25.com/cgi-bin/vcdf/detail?coll_id=7157&inst_id=8&nv1=search&nv2=basic. Last accessed July 13, 2018.

Sibson, F. (ed.) (1848). *On the Movements of Respiration in Disease and on the Use of a Chest-Measurer. Medico-Chirurgical Transactions*. London: R. Kinder.

Sibson, F. (ed.) (1869a). *Medical Anatomy: Or, Illustrations of the Relative Position and Movements of the Internal Organs*. London: John Churchill & Sons.

Sibson, F. (ed.) (1869b). *Iconographic, The Body of a Man Lying down*. London: John Churchill, Hullmandel and Walton.

Toghill, P. (2001). *The Royal College of Physicians and its Collections: An Illustrated History* (ed. G. Davenport, I. McDonald and C. Moss-Gibbons). London: James and James.

William Turner

Davis, H.W. and Weaver, J.R. (1927). *Dictionary of National Biography*. London: Oxford University Press.

Macalister, A. (1916). Principal Sir William Turner, K.C.B., F.R.S., etc., 1832–1916. *J. Anat. Physiol.* 50: 281–284.

Macintyre, I.M. (2002). Sir William Turner: 1832–1916. *Surgeons' News* 1: 105–106.

Magee, R. (2003). Sir William Turner and his studies on the mammalian placenta. *ANZ J. Surg.* 73: 449–452.

Robinson, A. (1916). Principal Sir William Turner, K.C.B., F.R.S., etc. *Edinb. Med. J.* 16: 218–223.

Turner, W. (1863). On the existence of a system of anastomosing arteries between and connecting the visceral and parietal branches of the abdominal aorta. *Brit. Foreign Medico-Chir. Rev.* 32: 222–227.

Turner, W. and Goodsir, J. (1857). *Handbook to Atlas of Human Anatomy and Physiology*. Edinburgh: W. & A.K. Johnston.

Jacobus Sylvius

Adeloye, A. (1977). The aqueduct of Sylvius. *Surg. Neurol.* 8: 458–460.

Baker, F. (1909). The two Sylviuses: an historical study. *Bull. Johns Hopkins Hosp.* 20: 329–339.

Ball, J.M. (1910). *Andreas Vesalius: The Reformer of Anatomy*. St. Louis, MO: Medical Science Press.

Conrad, L., Neve, M., Nutton, V. et al. (2006). *The Western Medical Tradition: 800 BC to AD 1200*. Cambridge: Cambridge University Press.

Dobson, J. (1962). *Anatomical Eponyms: Being a Biographical Dictionary of those Anatomists whose Names Have Become Incorporated into Anatomical Nomenclature, with Definitions of the Structures to which their Names Have Been Attached and References to the Works in which they Are Described*. Edinburgh and London: E & S Livingstone.

JAMA (1966). Jacobus Sylvius (Jacques Dubois) 1478–1555 – preceptor of Vesalius. *JAMA* 195: 169.

Kompanje, E.J. (2005). An historical homage from Denmark: the aqueduct of Sylvius. *Neurosurg. Rev.* 28: 77–78.

Lassek, A.M. (1958). *Human Dissection: Its Drama and Struggle*. Springfield, IL: Charles C. Thomas.

Leite dos Santos, A.R., Fratzoglou, M., and Perneczky, A. (2004). A historical mistake: the aqueduct of Sylvius. *Neurosurg. Rev.* 27: 224–225.

Nutton, C. and Nutton, V. (2004). Noël du Fail, Cardano, and the Paris medical faculty. *Med. Hist.* 48 (3): 367–372.
Saunders, J.B. and O'Malley, C.D. (1950). *The Illustrations from the Works of Andreas Vesalius of Brussles: With Annotations and Translations, a Discussion of the Plates and their Background, Authorship and Influence, and a Biographical Sketch of Vesalius.* Cleveland, OH: The World Publishing Company 1950.
Scultetus, A.H., Villavicencia, J.L., and Rich, N.M. (2001). Facts and fiction surrounding the discovery of the venous valves. *J. Vasc. Surg.* 33: 435–441.
van Gijn, J. (2001). Franciscus Sylvius (1614–1672). *J. Neurol.* 248: 915–916.

Charles Estienne

Choulant, L. (1852). *Geschichte und Bibliographie der Anatomischen Abbildung Nach Ihrer Beziehung auf Anatomische Wissenschaft und Bildende Kunst.* Leipzig: Rudolph Weigel.
Choulant, L. (1920). *History and Bibliography of Anatomic Illustration in its Relation to Anatomic Science and the Graphic Arts.* Translated by Frank M. Chicago: University Chicago Press.
Kellett, C.E. (1964). Two anatomies. *Med. Hist.* 8: 342–353.
Mumey, N. (1944). *Quarter-Centenary of the Publication of Scientific Anatomy (1543–1943),* 1944. Denver, CO: Range Press.
Rath, G. (1964). Charles Estienne: contemporary of Vesalius. *Med. Hist.* 8: 354–359.
Scultetus, A.H., Villavicencio, J.L., and Rich, N.M. (2001). Facts and fiction surrounding the discovery of the venous valves. *J. Vasc. Surg.* 33: 435–441.

Raymond de Vieussens

Amalric P. La naissance de la neuro-ophtalmologie a Montpellier au XVIIME siecle a travers l'oeuvre de Raymond Vieussens et de Guilhem Briggs. Bull. Soc. Opht. France 1983.
Bowman, I.A. (1987). Jean-Baptiste Sénac and his treatise on the heart. *Tex. Heart Inst. J.* 14: 5–11.
Dalla-Volta, S. (1993). A short historical survey of heart failure. *Clin. Investig.* 71: S167–S176.
Debus, A.G. (1971). The history of chemistry and the history of science. *Ambix* 18: 169–177.
Gailloud, P.H. (2003). Histoire de l'anatomie du thalamus de l'antiquité a la fin du XIX siecle. *Schweiz. Arch. Neurol. Psychiatr.* 154: 2–15.
JAMA (1968). Raymond de Vieussens (1641–1715). French neuroanatomist and physician. *JAMA* 18: 206.
Kellet, C.E. (1959). Raymond de Vieussens on mitral stenosis. *Br. Heart J.* 21: 440–444.
Kocica, M.J., Vranes, M.R., Djukic, P.L. et al. (2004). Giant pseudoaneurysm from Vieussens' arterial ring. *Ann. Thorac. Surg.* 78: 1833–1836.
Millar, B.C. and Moore, J.E. (2004). Emerging issues in infective endocarditis. *Emerg. Infect. Dis.* 10: 1110–1116.
Mounier-Kuhn, P. and Guerrier, Y. (1982). Vieussens et l'anatonie de l'oreille. *Acta Otorhinolaringol. Belg.* 36: 1029–1038.
O'Leary, E.L., Garza, L., Williams, M., and McCall, D. (1998). Vieussens' ring. *Circulation* 98: 487–488.
Podolsky, E. (1952). Raymond Vieussens and the affairs of the heart. *Med. Womans J.* 1: 29–31.
Robinson, B. (1907). *The Abdominal and Pelvic Brain with Autonomic Visceral Ganglia.* Hammond, IN: Frank S. Betz.
Swanson, L.W. (2000). What is the brain? *Trends Neurosci.* 23: 519–527.
Thebesius, A.C. (1708). *De circulo Sanguinis in Corde.* Leiden: Lugduni Batavorum.

Tubbs, R.S., Wellons, J.C. III, Salter, G., and Oakes, W.J. (2004). Fenestration of the superior medullary velum as treatment for a trapped fourth ventricle: a feasibility study. *Clin. Anat.* 17: 82–87.

Vieussens, R. (1706). *Nouvelles découvertes sur le coeur, expliquées dans une lettre écrite à Monsieur Boudin, conseiller d'etat, premier medecin de Monseigneur*. Paris: Chez Laurent D'Houry.

Vieussens, R. (1709). *Nouvelles découvertes sur le Coeur*. Toûlouse: Jean Guillmette.

Wearn, J.T. (1927). The role of the Thebesian vessels in the circulation of the heart. *J. Exp. Med.* 47: 293–315.

Zawadzki, M., Pietrasik, A., Pietrasik, K. et al. (2004). Endoscopic study of the morphology of Vieussens valve. *Clin. Anat.* 17: 318–321.

Pierre Dionis

Boutaric, J.J. (2001). Pierre Dionis, a surgeon in Molièie's time. *Rev. Prat.* 51: 1285–1288.

Brockliss, L. and Jones, C. (1997). *The Medical World of Early Modern France*. Oxford: Clarendon Press.

Cotlar, A. (2000). Extrauterine pregnancy: a historical review. *Curr. Surg.* 57: 484–492.

Dionis, P. (1705). *Anatomie de l'homme, suivant la circulation du sang, & les derniéres découvertes*. Paris: d'Houry.

Dionis, P. (1710). *A Course of Chirurgical Operations Demonstrated in the Royal Garden at Paris*. London: Tonson.

Dionis, D. (1719). *A General Treatise on Midwifery*. London: Bell.

Laissus, Y. (2001). Jardin du Roi. French Royal Botanical Garden. In: *Encyclopedia of the Enlightenment Vol. 1*, 736–739. Chicago: Fitzroy Dearborn.

Simpson, D. (2003). Pierre Dionis and the Franco-British dialogue in surgery. *ANZ J. Surg.* 73: 336–340.

Félix Vicq d'Azyr

Adams, C.F. (1853). *The Works of John Adams, Second President of the United States*, vol. VIII. Boston: Little, Brown and Company.

Clarac, F. and Boller, F. (2010). History of neurology in France. In: *History of Neurology* (ed. M.J. Aminoff, F. Boller and D.F. Swaab), 629–656. Edinburgh: Elsevier.

Farrell, P.S. and McHenry, L.C. (1987). Fragments of neurologic history: Felix Vicq d'Azyr and neuroanatomy. *Neurology* 37 (8): 1349–1350.

Goldblatt, D. (1986). The key to the brain Felix Vicq d'Azyr (1748–1794). *Semin. Neurol.* 6 (2): 231–237.

Goodrich, T. (2000). A millennium review of skull base surgery. *Childs Nerv. Syst.* 16 (10–11): 669–685.

Hannaway, C. (1994). Vicq d'Azyr, anatomy and a vision of medicine. *Clin. Med.* 25: 280–290.

Mandressi R. Félix Vicq d'Azyr: l'anatomie, l'État, la médecine. 2005. http://www.bium.univ-paris5.fr/histmed/medica/vicq.htm. Last accessed July 13, 2018.

Mandressi, R. (2008). The past, education and science. Félix Vicq d'Azyr and the history of medicine in the 18th century. *Medicina nei secoli* 20 (1): 183–212.

Parent, A. (2007). Felix Vicq d'Azyr: anatomy, medicine and revolution. *Can. J. Neurol. Sci.* 34: 30–37.

Peumery, J.J. (2001). Vicq d'Azyr et la Revolution Française. *Hist. Sci. Med.* 35: 263–270.

Sournia, J.C. (1994). Felix Vicq d'Azyr, founder of the academy of medicine (1748–1794). *Bull. Acad. Natl Med.* 178 (7): 1237–1243. disc. 1243–1244.
Sparks, J. (1838). *The Works of Benjamin Franklin*, vol. VI. Boston: Hilliard, Gray, and Company.
Spillman, R. (1941). Felix Vicq D'Azyr and Benjamin Franklin. *J. Nerv. Men. Dis.* 94 (4): 428–444.
Vicq d'Azyr, F. (1786). *Traité d'anatomie et de physiologie – avec des planches colorës représentant au naturel les divers organes de l'Homme et des Animaux. F.A.* Paris: Didot.

Marie-François Xavier Bichat

Andral, G. (1844). *An Essay on the Blood in Disease.* Translated by Meigs J, Stillé A. Philadelphia: Lea and Blanchard.
Anonymous (1815). Biographical sketch of Bichat. In: *Eclectic Repertory and Analytical Review, Medical and Philosophical*, vol. 5, 507–518. Philadelphia: Thomas Dobson.
Avila, V.L. (1995). *Biology: Investigating Life on Earth*, 2e. Boston: Jones and Bartlett.
Bartolucci, S. and Forbis, P. (2005). *Stedman's Medical Eponyms*, 2e. Baltimore: Lippincott Williams & Wilkins.
Bichat, X. (1809). *Physiological Researches upon Life and Death.* Translated by Watkins T. Philadelphia: Smith & Maxwell.
Bichat, X. (1824). *General Anatomy, Applied to Physiology and to the Practice of Medicine* Translated by Coffyn C. London: Shackell, Arrowsmith.
Bichat, X. (1827). *Physiological Researches on Life and Death.* Translated by Gold F, with notes by Magendie F (notes translated by Hayward G). Boston: Richardson and Lord.
Bos, J.D. (1997). A history of immunodermatology. In: *Skin Immune System (SIS): Cutaneous Immunology and Clinical Immunodermatology*, 2e (ed. J.D. Bos), 4. Baton Rouge: CRC Press.
Burns, W.E. (2003). *Science in the Enlightenment: An Encyclopedia.* Santa Barbara, CA: ABC-CLIO.
Byers, P.K. and Bourgoin, S.M. (eds.) (1998). *Encyclopedia of World Biography*, 2e. Gale Publishing: Detroit.
Castaglioni, A. (1941). *A History of Medicine.* Translated by Krumbhaar EB. New York: Alfred A. Knoff.
Chai, L. (2006). *Romantic Theory: Forms of Reflexivity in the Revolutionary Era.* Baltimore: Johns Hopkins University Press.
Chisholm H (ed.). Encyclopedia Britannica, 11. 1911. https://www.studylight.org/encyclopedias/bri.html. Last accessed July 13, 2018.
Conrad, L.I., Neve, M., Nutton, V. et al. (1995). *The Western Medical Tradition: 800 BC to 1800 AD.* Cambridge: Cambridge University Press.
Cook, H. (1997). From the scientific revolution to the germ theory. In: *Western Medicine: An Illustrated History* (ed. I. Loudon), 80–101. New York: Oxford University Press.
Debré, P. (2000). *Louis Pasteur.* Translated by Forster E. Baltimore/London: Johns Hopkins University Press.
Djordjevic, S.P. (2004). *Dictionary of Medicine: French-English with English-French Glossary*, 2e. Schreiber: Rockville.
Dobo, N. and Role, A. (1989). *Bichat. La vie fulgurante d'un génie.* Paris: Perrin.
Drake, D. and Wood, W.M. (eds.) (1836). Lesions of the blood [Review of G. Andral. A Treatise on Pathological Anatomy (translated by R. Townsend and W. West). New York. 1832]. In: *The Western Journal of the Medical and Physical Sciences*, vol. 9, 57–76. Cincinnati, OH: N.S. Johnson.
Ewald, P.W. (1994). *Evolution of Infectious Disease.* New York/Oxford: Oxford University Press.

Forget, E.L. (1999). *The Social Economics of Jean-Baptiste Say: Markets and Virtue*. London/New York: Routledge.

Foucault, M. (2003). *The Birth of the Clinic, an Archaeology of Medical Perception*. Translated by Sheridan AM. London/New York: Routledge Classics.

Gemmill, C.L. (1973). Medical numismatic notes XI: commemorative medals issued at the international physiological congresses. *Bull. N. Y. Acad. Med.* 49: 556–566.

Griffiths, R. (ed.) (1822). *The Monthly Review or Literary Journal, England*, vol. XCIX. London: A. and R. Spottswoode.

Halevi, L. (2006). Death and dying. In: *Medieval Islamic Civilization: An Encyclopedia*, vol. 1 (ed. J.W. Meri), 198. New York: Routledge.

Haller, J.S. (1981). *American Medicine in Transition, 1840–1910*. Chicago: University of Illinois Press.

Hamadani Z. *Javahe-o-tashrih [Pearls of Anatomy]* (Persian). National Library of Tabriz; Tabriz: 1860.

Harrison, R. (1977). *National Tuberculosis Association, 1904–1954: A Study of the Voluntary Health Movement in the United States*. New York: Arno Press.

Hegel, G.W.F. (2002). *Philosophy of Nature*, vol. 3. Abingdon: Routledge.

Hoblyn, R.D. (1844). *A Dictionary of Terms Used in Medicine and the Collateral Sciences*, 2e. London: Sherwood, Gilbert & Piper.

King, L.S. and Meehan, M.C. (1973). A history of the autopsy. A review. *Am. J. Pathol.* 73: 514–544.

Konstam, A. (2003). *Historical Atlas of the Napoleonic Era*. Guilford, Connecticut: Lyons Press.

La Berge, A.F. (1992). *Mission and Method: The Early Nineteenth-Century French Public Health Movement*. Cambridge/New York: Cambridge University Press.

Lawlor, L. (2006). *The Implications of Immanence: Toward a New Concept of Life*. New York: Fordham University Press.

Levinson, A. (1943). Medical Medallions. *Bull. Med. Libr. Assoc.* 31: 5–34.

Lieber, F., Wigglesworth, E., and Bradford, T.G. (1835). *Encyclopedia Americana: A Popular Dictionary of Arts, Sciences, Literature, History, Politics and Biography, Brought Down to the Present Time; Including A Copious Collection of Original Articles in American Biography; on the Basis of the Seventh Edition of the German Conversations-Lexicon*, vol. 13. Philadelphia: Desilver, Thomas, & Co.

Long, J.C. (1992). Foucault's clinic. *J. Med. Humanit.* 13: 119–138.

Marsh, J. (ed.) (1822). Life and writings of Bichat. Article VII. In: *Present Literature of Italy: The North American Review, July*, 132–162. Boston: Oliver Everett.

Maulitz, R.C. (1987). *Morbid Appearances: The Anatomy of Pathology in the Early Nineteenth Century*. Cambridge/New York: Cambridge University Press.

Mayr, E. (1997). *This Is Biology: The Science of the Living World*. Massachusetts/London: Harvard University Press.

Morgan, M., Calnan, M., and Manning, N. (1985). *Sociological Approaches to Health and Medicine*. Sydney: Croom Helm Ltd.

Mostofi, S.B. (2005). *Who's Who in Orthopedics*. London: Springer.

Myers, J.A. (1969). *Fighters of Fate, a Story of Men and Women Who Have Achieved Greatly despite the Handicaps of the Great White Plague*. New York: Ayer Publishing.

Osborne, T. (1995). Medicine and epistemology: Michel Foucault and liberality of clinical reason. In: *Michel Foucault's Critical Assessments*, vol. IV (ed. B. Smart). New York/London: Routledge.

Otis, L. (2002). *Literature and Science in the Nineteenth Century: An Anthology*. New York: Oxford University Press.

Paine, M. (1840). *Medical and Physiological Commentaries*, vol. 2. London: John Churchill.

Pattison, G.S. and Hagan, J. (eds.) (1835). *The Register and Library of Medical and Chirurgical Science*, vol. 1. Washington, DC: Duff Green.
Porter, R. (ed.) (2006). *The Cambridge History of Medicine*. Cambridge: Cambridge University Press.
Prichard, R. (1979). Selected items from the history of pathology: Marie-François-Xavier Bichat (1771–1802). *Am. J. Pathol.* 96: 256.
Reuter, P. (2000). *Worterbuch der humanbiologie*. Berlin: Birkhauser Verlag.
Ripley, G. and Dana, C.A. (1859). *The New American Cyclopaedia: A Popular Dictionary of General Knowledge*, vol. 3 and 6. New York: D. Appleton and Company.
Rose, H.J. (1857). *A New General Biographical Dictionary*, vol. 4. London: T. Fellowes etc.
Rosenblum, R. (1967). *Transformations in Late Eighteenth Century Art*, 39. Princeton, NJ: Princeton University Press.
Shryock, R.H. (1947). *The Development of Modern Medicine: An Interpretation of the Social and Scientific Factors Involved*. New York: Alfred A. Knopf.
Simmons, J.G. (2002). *Doctors and Discoveries: Lives that Created Today's Medicine*. Boston/New York: Houghton Mifflin Co.
Solly, S. (1847). *The Human Brain: Its Structure, Physiology and Diseases*, 2e. London: Longman, Brown, Green and Longmans.
Sterne, J. (2001). Mediate auscultation, the stethoscope, and the "autopsy of the living": medicine's acoustic culture. *J. Med. Humanit.* 22: 115–136.
Ten, C.L. (ed.) (1994). *The Nineteenth Century*. London/New York: Routledge.
Thomas, J. (1871). *Universal Pronouncing Dictionary of Biography and Mythology, Part One. J.B.* Philadelphia: Lippincott and Co.
Tubbs, R.S., Loukas, M., Shoja, M.M. et al. (2008). François Magendie (1783–1855) and his contributions to the foundations of neuroscience and neurosurgery. *J. Neurosurg.* 108: 1038–1042.
Weber, A.S. (ed.) (2000). *Nineteenth Century Science: An Anthology*. Ontario/New York: Broadview Press.
Whewell, W. (1847). *The Philosophy of the Inductive Sciences, Founded upon their History*, vol. 1. London: JW Parker.

François Magendie

Berkowitz, C. (2006). Disputed discovery: vivisection and experiment in the 19th century. *Endeavour* 30: 98–102.
Dawson, P.M. (1908). *A Biography of François Magendie*. New York: A.T. Huntington.
Fenton, P.F. (1951). François Magendie. *J. Nutr.* 43: 3–15.
Haas, L.F. (1994). François Magendie (1783–1855). *J. Neurol. Neurosurg. Psychiatry* 57: 692.
Jørgensen, C.B. (2003). Aspects of the history of the nerves: Bell's theory, the bell-Magendie law and controversy, and two forgotten works by P.W. Lund and D.F. Eschricht. *J. Hist. Neurosci.* 12: 229–249.
Koehler, P. (2003). Magendie, François. In: *Encyclopedia of the Neurological Sciences*, vol. 2 (ed. M.J. Aminoff and R.B. Daroff), 3–4. Cambridge, MA: Academic Press.
Magendie, F. (1824). *An Elementary Compendium of Physiology; for the Use of Students*. Translated by Milligan E. Philadelphia: J. Webster.
Magendie, F. (1844). *An Elementary Treatise on Human Physiology, on the Basis of the précis elementaire de physiologie*, 5. Translated by Revere Je. New York: Harper & Brothers.
Olmsted, J.M.D. (1944). *François Magendie. Pioneer in Experimental Physiology and Scientific Medicine in XIX Century France*. New York: Schuman's.

Poynter, F.N.L. (1968). Doctors in the human comedy. *JAMA* 204: 105–108.
Shampo, M.A. and Kyle, R.A. (1987). François Magendie: early French physiologist. *Mayo Clin. Proc.* 62: 412.
Sourkes, T. (2002). Magendie and the chemists: the earliest chemical analyses of the cerebrospinal fluid. *J. Hist. Neurosci.* 11: 2–10.
West, C.M. (1931). The foramen of Magendie. *J. Anat.* 65: 457–467.

Jules Germain Cloquet

Bishop, W.I. (1954). Jules Germain Cloquet; 1790–1883. *Med. Bilo. Illus.* 4: 141–143.
Cloquet, J.G. (1817). *Recherches Anatomiques sur les Hernies de l'Abdomen*. Paris: Méquignon-Marvis.
Cloquet, J.G. (1819). *Recherches sur les Causes et l'Anatomie des Hernies Abdominales*. Paris: Méquignon-Marvis.
Cloquet, J.G. and Ameline, A. (1963). L'acupuncture de grand-papa. Jules Cloquet (1790–1883). *La Presse Med.* 71: 207–208.
Dumont, M. (1988). Les vagabondages de Jules Cloquet et de Gustave Flaubert. *J. Gynecol. Obstet. Biol. Reprod.* 17: 151–154.
Loukas, M., El-Sedfy, A., Tubbs, R.S., and Wartman, C. (2007). Jules Germain Cloquet (1790–1883) – drawing master and anatomist. *Am. Surg.* 73: 1169–1172.
Pariente, L. (1998). Avis au lecteur sur la vie et l'oeuvre de Jules Germain Cloquet. In: *Manuel d'anatomie descriptive du corps humain* (ed. J.G. Cloquet). Paris: Louis Pariente.
Rutkow, I.M. (1998). A selective history of groin hernia surgery in the early 19th century. The anatomic atlases of Astley Cooper, Franz Hesselbach, Antonio Scarpa, and Jules-Germain Cloquet. *Surg. Clin. North Am.* 78: 921–940.

Augusta Déjerine-Klumpke

Bogousslavsky, J. (2005). The Klumpke family – memories by doctor Dejerine, born Augusta Klumpke. *Eur. Neurol.* 53: 113–120.
Dejerine-Klumpke, A. (1885). Contribution a l étude des paralysies radiculaires du plexus brachial. Paralysies radiculaires totales. Paralysies radiculaires inférieures. De la participation des filets sympathiques oculo-pupillaires dans ces paralysies. *Rev. Med.* 5: 591–616. 739–790.
Lecours, A.R. and Caplan, D. (1984). Augusta Dejerine-Klumpke or "the lesson in anatomy.". *Brain Cogn.* 3: 166–197.
Pearce, J.M. (2006). Dejerine-Sottas disease (progressive hypertrophic polyneuropathy). *Eur. Neurol.* 55: 115–117.
Satran, R. (1974). Augusta Dejerine-Klumpke. First woman intern in Paris hospitals. *Ann. Intern. Med.* 80: 260–264.
Schurch, B. and Dollfus, P. (1998). The "Dejerines": a historical review and homage to two pioneers in the field of neurology and their contribution to the understanding of spinal cord pathology. *Spinal Cord* 36: 78–86.
Tubbs, R.S., Salter, E.G., and Oakes, W.J. (2007). Wilhelm Erb and Erb's point. *Clin. Anat.* 20 (5): 486–488.

Adam Christian Thebesius

Frazier, O.H., Bergheim, M., and Kadipasaoglu, K.A. (2006). *Myocardial Laser Revascularization: Scientific and Historical Precedents*. Boston: Blackwell Publishing.

Klotz, O. and Lloyd, W. (1930). Sclerosis and occlusion of the coronary arteries. *Can. Med. Assoc. J.* 23: 359–365.
Mettenleiter, A. (2001). Adam Christian Thebesius (1686–1732) und die Entdeckung der Vasa Cordis Minima. *Sudhoffs Arch. Z. Wissenschaftsgesch. Beih.* 47: 3–580.
Skalski, J.H. and Kuch, J. (2006). Polish thread in the history of circulatory physiology. *J. Physiol. Pharmacol.* 57: 5–41.
Thebesius, A.C. (1708). *De circulo sanguinis in corde.* Leiden: Lugduni Batavorum.
Vieussens, R. (1706). *Nouvelles découvertes sur le Coeur.* Toûlouse: Jean Guillmette.
Vilallonga, J.R. (2004). Anatomical variations in the coronary arteries. *Eur. J. Anat.* 8: 39–53.
Wearn, J.T. (1927). The role of the Thebesian vessels in the circulation of the heart. *Exp. Med. Surg.* 57: 293–317.

Franz Kaspar Hesselbach

Hesselbach HK. Vollständige Anleitung zur Zergliederungskunde des menschlichen Körpers, 3 Rudolfstadt: Hefte, 1805–1808.
Hesselbach, H.K. (1806). *Anatomisch-chirurgische Abhandlung über den Urspurng der Leistenbrüche.* Würzburg: Baumgärtner.
Hesselbach, H.K. (1814). *Neueste anatomisch-pathologische Untersuchungen über den Ursprung und das Fortschreiten der Keisten- und Schenkelbrüche.* Würzburg: Stahel.
Hesselbach, H.K. (1815). *Beschreibung und Abbildung eines neuen Instrumentes zur sicheren Entdeckung und Stillung einer bei dem Bruchschnitte entstandenen gefährlichen Blutung.* Würzburg: Baumgärtner.
Keynes, M. (1954). Battle of eponymy. *Br. Med. J.* 1: 96.
Lermann, H. (1962). *Die Prosektoren Hesselbach Franz Caspar Hesselbach und Adam Kaspar Hesselbach als Prosektoren der Würzburger Anatomischen Anstalt.* Würzburg: Inaugural Dissertation.
Quain, J. (1828). *Elements of Descriptive and Practical Anatomy.* London: Simpkin and Marshall.
Rutkow, I.M. (1998). A selective history of groin hernia surgery in the early 19th century. The anatomic atlases of Astley Cooper, Franz Hesselbach, Antonio Scarpa, and Jules-Germain Cloquet. *Surg. Clin. North Am.* 78: 921–940.
Skandalakis, P.N., Skandalakis, J.E., Colborn, G.L. et al. (2004). *Abdominal wall and hernias in Skandalakis' Surgical Anatomy. T he Embryologic and Anatomic Basis of Modern Surgery.* Athens: Paschalidis Medical Publication.

Hubert von Luschka

Dvorak, J. and Sandler, A. (1994). Historical perspective Hubert von Luschka. *Spine* 19: 2478–2482.
Fisher, R.G. (1967). Surgery of the congenital anomalies. In: *A History of Neurological Surgery* (ed. A.E. Walker), 334–361. New York: Hafner.
Haubrich, W. (2007). Luschka of the crypts of Luschka. *Gastroenterology* 132: 1216.
Luschka, H. (1850a). *Die Nerven in der harten Himhaut.* Tübingen: H. Laupp:.
Luschka, H. (1850b). *Die Nerven des menschlichen Wirbelcanals.* Tübingen: H. Laupp.
Luschka, H. (1853). *Der Nervus phrenicus des Menschen.* Tübingen: H. Laupp.
Luschka, H. (1855). *Die Adergeflechte des menschlichen Gehirns.* Berlin: Georg Reimer.
Luschka, H. (1860). *Der Hirnanhang und die Steissdrüse des menschlichen Herzens.* Berlin: Georg Reimer.
Luschka, H. (1862–1867). *Die Anatomie des Menschen in Rücksicht auf die Bedürfnisse der praktischen Heilkunde bearbeitet,* vol. 3. Tübingen: H. Laupp.

Luschka H. Anatomie des Menschen in Rücksicht auf das Bedürfnis der praktischen Heilkunde. 1863–1869.

Luschka, H. (1871). *Über Maß- und Zahlenverhältnisse des menschlichen Körpers*. Tübingen: H. Laupp.

Noble, E.R., Smoker, W.R.K., and Ghatak, N.R. (1997). Atypical skull base paragangliomas. *Am. J. Neuroradiol.* 18: 986–990.

Swedenborg, E. (1887). *The Brain Considered Anatomically, Physiologically and Philosophically*, vol. II. London: James Speirs.

Tubbs, R.S., Loukas, M., Shoja, M.M. et al. (2008). Francois Magendie (1783–1855) and his contributions to the foundations of neuroscience and neurosurgery. *J. Neurosurg.* 108: 1038–1042.

Walker, A.E. (1998). *The Genesis of Neuroscience*. Park Ridge, IL: American Association of Neurological Surgeons.

Adolf Wallenberg

Hydén, D. and Norrving, B. (2005). Adolf Wallenberg (1862–1949). *J. Neurol.* 252: 1135–1136.

Okun, M.S. (2003). Neurological eponyms – who gets the credit? Essay review. *J. Hist. Neurosci.* 12: 91–103.

Pearce, J.M. (2000). Wallenberg's syndrome. *J. Neurol. Neurosurg. Psychiatry* 68: 570.

Tatu, L., Moulin, T., and Monnier, G. (2005). The discovery of encephalic arteries. From Johann Jacob Wepfer to Charles Foix. *Cerebrovasc. Dis.* 20: 427–432.

Wallenberg, A. (1888). Veränderungen der nervösen Centralorgane in einem Falle von cerebraler Kinderlähmung. *Eur. Arch. Psychiatry Clin. Neurosci.* 19: 297–313.

Wallenberg, A. (1895). Acute Bulbäraffection (Embolie der Art. cerebellar. post. inf. sinistr.?). *Arch. Psychiatr. Nervenkr* 27: 504–540.

Wallenberg, A. (1918). Neue BeitrÄge zur Diagnostik der Hirnstammerkrankungen. *J. Neurol.* 58: 105–114.

Wallenberg, A. (1931a). Beiträge zur vergleichenden Anatomie des Hirnstammes. *J. Neurol.* 117: 677–698.

Wallenberg, A. (1931b). Über basale Verbindungen des Vorder- und Zwischenhirns mit sensiblen Trigeminuskernen bei Vögeln und Knochenfischen. *Eur. Arch. Psychiatry Clin. Neurosci.* 94: 246–267.

Wallenberg-Chermak, M. (1963). Adolf Wallenberg. In: *Grosse Nervenärzte*, vol. 3 (ed. K. Kolle), 1–13. Stuttgart: New York George Thieme.

Korbinian Brodmann

Bentivoglio, M. (1998). Cortical structure and mental skills: Oskar Vogt and the legacy of Lenin's brain. *Brain Res. Bull.* 47 (4): 291–296.

Brodmann, K. (1908). Beitrage zur histologischen Lokalisation der Grosshimrinde, VI: Mitteilung: Die Cortexgliederung des Menschen. *J. Psychologie Neurologie* 10: 231–246.

Campbell, A. (1905). *Histological Studies on the Localisation of Cerebral Function*. Cambridge: Cambridge University Press.

Danek A, Rettig J. Korbinan Brodmann (1868–1918) (German). *Schweiz Arch. Neurol. Psychiatr.* 1989;140(6):555–566.

Garey, L.J. (1999). *Brodmann's "Localization in the Cerebral Cortex."*. London: Imperial College Press.

Garey LJ. Prof. Dr. Korbinian Brodmann. http://www.korbinian-brodmann.de. Last accessed July 13, 2018.

Haymaker, W. and Rose, J.E. (1953). *Korbinian Brodmann (1868–1918).* The Founders of Neurology: One Hundred and Thirty-Three Biographical Sketches Prepared for the Fourth International Neurological Conference in Paris. Springfield, MO: C.C. Thomas.
Pearce, J.M. (2005). Historical note: Brodmann's cortical maps. *J. Neurol. Neurosurg. Psychiatry* 76: 259.
Rose, J.E. (1970). *Korbinian Brodmann.* Springfield, MO: Charles C. Thomas.
Vogt, O. (1959). Korbinian Brodmann, Lebenslauf. In: *Grosse Nervenarzte,* vol. 2, 40–44. Stuttgart: Thieme.

Aristotle

Barnes, J. (1984). *The Complete Work of Aristotle, the Revised Oxford Translation.* Princeton, NJ: Princeton University Press.
Crivellato, E. and Ribatti, D. (2006). A portrait of Aristotle as an anatomist: historical article. *Clin. Anat.* 20: 447–485.
Dunn, P.M. (2006). Aristotle (384–322 BC): philosopher and scientist of ancient Greece. *Arch. Dis. Child Fetal Neonatal Ed.* 91: F75–F77.
Durant, W. (1926). *The Story of Philosophy.* New York: Simon & Schuster.
French, R.K. (1978). The thorax in history 1. From ancient times to Aristotle. *Thorax* 33: 10–18.
Harris, C.R.S. (1973). *The Heart and the Vascular System in Ancient Greek Medicine, from Alcmaeon to Calen.* Oxford: Clarendon Press.
Jori, A. (2005). Aristotle on the function of blood in the processes of life. *Med. Secoli* 17: 603–625.
Lonie, I.M. (1973a). The paradoxical test "On the heart." I. *Med. Hist.* 17: 1–15.
Lonie, I.M. (1973b). The paradoxical text "On the heart." II. *Med. Hist.* 17: 136–153.
Loukas, M., Tubbs, R.S., Louis, R.G. Jr. et al. (2007). The cardiovascular system in the pre-Hippocratic era. *Int. J. Cardiol.* 120: 145–149.
Peck, A.L. (1965). *Historia Animalium,* vol. 1–3. London: Heinemann.
Shoja, M.M. and Tubbs, R.S. (2007). The history of anatomy in Persia. *J. Anat.* 210: 359–378.
Van Praagh, R. and Van Praagh, S. (1983). Aristotle's "triventricular" heart and the relevant early history of the cardiovascular system. *Chest* 84: 462–468.

Galen

Bartolucci, S.L. and Forbis, P. (2005). *Stedman's Medical Eponyms,* 2e. Philadelphia: Lippincott Williams & Wilkins.
Cooper, S. (ed.) (1818). Bandage. In: *A Dictionary of Practical Surgery: Comprehending all the most Interesting Improvements, from the Earliest Times Down to the Present Period; an Account of the Instruments and Remedies Employed in Surgery; the Etymology and Signification of the Principal Terms; a Copious Bibliotheca Chirugica; a Variety of Original Facts and Observations,* 3e, 143–148. London: Longman.
Derenne, J., Debru, A., Grassino, A.E., and Whitelaw, W.A. (1995). History of diaphragm physiology: the achievements of Galen. *Eur. Respir. J.* 8: 154–160.
Djordjević, S.P. (2004). *Dictionary of Medicine: French–English with English–French Glossary,* 2e. Rockville, MD: Schreiber.
Dunn, P.M. (2003). Galen (AD 129–200) of Pergamun: anatomist and experimental physiologist. *Arch. Dis. Child Fetal Neonatal Ed.* 88: F441–F443.
French, R. (ed.) (2003). Galen. In: *Medicine before Science: The Rational and Learned Doctor from the Middle Ages to the Enlightenment,* 1e, 34–58. Cambridge: Cambridge University Press.

Gleason MW. Shock and awe: the performance dimension of Galen's anatomy demonstrations. Princeton/Stanford Working Papers in Classics. Version 5 January 2007. http://www.princeton.edu/~pswpc/pdfs/gleason/010702.pdf. Last accessed July 13, 2018.

Grmek MD, Gourevitch D. Aux sources de la doctrine médicale De Galien: l'enseignement de Marinus, Quintus et Numisianus. ANRW 1994;II.37.2:1491–1528.

Kaplan, E.L., Salti, G.I., Roncella, M. et al. (2009). History of the recurrent laryngeal nerve: from Galen to Lahey. *World J. Surg.* 33: 386–393.

Lardner, N. (1838). *The Works of Nathaniel Lardner D.D., with a Life by Dr. Kippis*, vol. 7. London: William Ball.

Lehoux, D. (2007). Observers, objects, and the embedded eye; or, seeing and knowing in Ptolemy and Galen. *Isis* 98: 447–467.

MerckMedicus. Dorland's Medical Dictionary. https://www.dorlandsonline.com/dorland/home. Last accessed July 13, 2018.

Mijnwoordenboek. http://www.mijnwoordenboek.nl/thema/ME/DE/NL/G/1.html. Last accessed July 13, 2018.

Missios, S. (2007). Hippocrates, Galen, and the uses of trepanation in the ancient classical world. *Neurosurg. Focus.* 23: E11.

Nutton, V. (2005). The life and career of Galen. In: *Ancient Medicine*, 1e, 216–229. New York: Routledge.

Reilly, W.J. (2005). Pharmaceutical necessities. In: *Remington: The Science and Practice of Pharmacy*, 21e (ed. D.B. Troy), 1078–1079. Philadelphia: Lippincott Williams & Wilkins.

Singe, P.N. (1997). Levels of explanation in Galen. *Class. Q. (New Series)* 47 (02): 525–542.

Sternbach, G.L., Varon, J., Fromm, R.E. et al. (2001). Galen and the origins of artificial ventilation, the arteries and the pulse. *Resuscitation* 49: 119–122.

Temkin, O. (ed.) (1991). Hippocratism encounters Christianity. In: *Hippocrates in a World of Pagans and Christians*, 2e, 179–196. Baltimore: Johns Hopkins University Press.

Todman, D. (2007). Galen (129–199). *J. Neurol.* 254: 975–976.

Ustun, C. (2004). Galen and his anatomic eponym: vein of Galen. *Clin. Anat.* 17: 454–457.

van Hoof, H. (1993). *Dictionnaire des éponymes médicaux: français-anglais*. Leuven: Peeters Publishers.

Vasiliadis, E.S., Grivas, T.B., and Kaspiris, A. (2009). Historical overview of spinal deformities in ancient Greece. *Scoliosis* 4: 6.

Xarchas, K.C. (2003). Galen on the anatomy and treatment of the injured hand. *Clin. Anat.* 16: 105–107.

Polybus

Angeletti, L.R. (1989). Polybus and the heritage of Hippocrates. *Med. Secoli* 1: 23–38.

Barnes, J. (1984). *The Complete Work of Aristotle, the Revised Oxford Translation*. Princeton, NJ: Princeton University Press.

Buffa, G.F. (1999). Traces of Hippocratic medicine in two late roman tracts: Polybus of Kos. *Koinonia* 23: 39–54.

Bujalkova, M. (2001). Hippocrates and his principles of medical ethics. *Bratisl. Lek. Listy* 102: 117–120.

Kempf, E.J. (1904). From Hippocrates to Galen. *Med. Library Hist. J.* 2: 282–307.

Kuhn, S.T. (1970). *The Structure of Scientific Revolutions*. Chicago: University of Chicago Press.

Lewis WJ, Beach JA. Galen on Hippocrates' On the Nature of Man. https://www.stmarys-ca.edu/sites/default/files/attachments/files/On_Hippocrates.pdf. Last accessed July 13, 2018.

Loukas, M., Tubbs, R.S., Louis, R.G. Jr. et al. (2007). The cardiovascular system in the pre-Hippocratic era. *Int. J. Cardiol.* 120: 145–149.
Nutton, V. (2005). The fatal embrace: Galen and the history of ancient medicine. *Sci. Context.* 18: 111–121.
Shoja, M.M. and Tubbs, R.S. (2007). The history of anatomy in Persia. *J. Anat.* 210: 359–378.
Shoja, M.M., Tubbs, R.S., Loukas, M., and Ardalan, M.R. (2007). The Aristotelian account of "heart and veins.". *Int. J. Cardiol.* 125: 304–310.
Smith, W. (1867). *Dictionary of Greek and Roman Biography and Mythology*, vol. 3. Boston: Little, Brown, and Company.

Franciscus Sylvius

Dobson, J. (1962). *Anatomical Eponyms: Being a Biographical Dictionary of those Anatomists whose Names Have Become Incorporated into Anatomical Nomenclature, with Definitions of the Structures to which their Names Have Been Attached and References to the Works in which they Are Described.* Edinburgh and London: E & S Livingstone.
Kompanje, E.J. (2005). An historical homage from Denmark: the aqueduct of Sylvius. *Neurosurg. Rev.* 28: 77–78.
Lassek, A.M. (1958). *Human Dissection: Its Drama and Struggle.* Springfield, IL: Charles C. Thomas.
Leite dos Santos, A.R., Fratzoglou, M., and Perneczky, A. (2004). A historical mistake: the aqueduct of Sylvius. *Neurosurg. Rev.* 27: 224–225.
Preul, M.C. (1997). A history of neuroscience from Galen to Gall. In: *A History of Neurosurgery in its Scientific and Professional Contexts* (ed. Greeblatt), 99–130. Park Ridge, IL: American Association of Neurological Surgeons.
van Gijn, J. (2001). Franciscus Sylvius (1614–1672). *J. Neurol.* 248: 915–916.

Niels Stensen

Cioni, R. (1962). *Niels Stensen. Scientist-Bishop.* New York: PJ Kennedy & Sons.
Cutler, A. (2003). *The Seashell on the Mountaintop: A Story of Science, Sainthood, and the Humble Genius Who Discovered a New History of the Earth.* Boston: Dutton Adult.
Kermit, H. (2003). *Niels Stensen, 1638–1686: The Scientist Who Was Beatified.* Leominster: Gracewing.
Maar, V. (1948). Nicholai stenosis opera philosophica. *Proc. Staff Meet. Mayo Clin.* 2: 317–320.
Schafer, E.A. and Thane, G.D. (1890). *Quain's Anatomy. Vol II, Part I, Osteology.* London: Longmans, Green and Co.
Sobiech, F. (2005). Blessed Nicholas Steno (1638–1686). Natural-history research and science of the Cross. *Austral. E. J. Theol.* 5: 1–8.
Stensen N. De musculis et glandulis observationem specimen. Hafnia, lit. M. Godiechenii; 1664.
Stensen, N. (1667). *Elementorum Myologicae Specimen.* Florentiae: Stellae.
Stensen, N. (1950). *A Dissertation on the Anatomy of the Brain with a Preface and Notes by Edv Gotfredsen.* Copenhagen: NYT Nordisk Forlag Arnold Busch.

Pieter Camper

Balget, B. and Oostra, R.J. (1998). Historical aspects of the study of malformations in the Netherlands. *Am. J. Med. Genet.* 77: 91–99.

Bindman, D. (2002). *Ape to Apollo: Aesthetics and the Idea of Race in the 18th Century*. Ithaca, NY: Cornell University Press.

Camper, P. (1760). *Demonstratum Anatomico-Pathologicarum Liber Primus Brachii Humani Fabricum et Morbos*. Amsterdam.

Camper, P. (1786). Conjectures to the petrifactions found in St. Peter's mountain near Maefricht. *Philos. Trans. R. Soc.* 76: 443–456.

Cunningham, D.J. (1908). Presidential address: anthropology in the 18th century. *J. R. Anthropol. Inst.* 38: 16–20.

Garrison, F.H. (1921). *An Introduction to the History of Medicine. W.B.* Orlando, FL: Saunders Co.

Haas, L.F. (1993). Neurological Stamp: Pieter Camper (1721–89). *J. Neurol. Neurosurg. Psychiatry* 69: 844.

Ijpma, F.F.A., van de Graaf, R.C., van Geldere, D., and van Gulik, T.M. (2009). An early observation on the anatomy of the Inguinal Canal and the etiology of inguinal hernias by Petrus Camper in the 18th century. *World J. Surg.* 33: 1318–1324.

Ijpma, F.F.A., van de Graaf, R.C., and van Gulik, T.M. (2010). Petrus Camper's work on the anatomy and pathology of the arm and hand in the 18th century. *J. Hell. Stud.* 35A: 1382–1387.

JAMA (1964). Editorial. *JAMA* 189: 852–853.

Meijer, M.C. (1997). Petrus camper on the origin and color of blacks. *Hist. Anthropol. Newsl.* 24 (2): 3–9.

Mostofi, S.B. (2005). *Who's Who in Orthopedics*. London: Springer.

Rutkow, I.M. (2003). A selective history of hernia surgery in the late 18th century: the treatises of Percivall Pott, Jean Louis Petit, D. Augest Gottlieb Pichter, Don Antonio de Gimbernat, and Pieter camper. *Surg. Clin. N. Am.* 83: 1021–1044.

Shampo, M.A. and Kyle, R.A. (1975). Dutch anatomist. *J. Am. Med. Assoc.* 233: 803.

Spencer, F. (1997). *History of Physical Anthropology*. New York: Garland.

Trenmouth, M.J. (2003). Petrus Camper: originator of cephalometrics. *Dent. Hist.* 40: 3–14.

Turner, G.T. (1939). Peter Camper. *Br. Med. J.*. 1939 2: 976.

Van Nouhuys, C.E. (1988). The lacrimal surgery of Petrus camper and his contemporaries. *Doc. Ophthalmol.* 68: 125–133.

Wolpoff, M.H. and Caspari, R. (1997). *Race and Human Evolution: A Fatal Attraction*. New York: Simon and Schuster.

Abraham Colles

Brian, V.A. (1976). The man behind the name. Abraham Colles. *Nurs. Times* 72: 1064.

Buxton, J.D. (1966). Colles and Carr: some history of the wrist fracture. *Ann. R. Coll. Surg. Engl.* 38: 253–257.

Colles, A. (1814). On the fracture of the carpal extremity of the radius. *Edinburgh Med. Surg. J.* 10: 182. Reprinted as Colles' classic description of fractures of the lower end of the radius. J. Bone Joint Surg. 1973;55:454–456.

Colles, A. (1815). Three different situations of scirrhous lymphatic glands. *Edinb. Med. J.*

Colles, A. (1818). On the distortion termed varus, or club feet. *Dublin Hosp. Rep.* 1: 175.

Fallon, M. (1976). Abraham Colles of Dublin and Edinburgh. *J. R. Coll. Surg. Edinb.* 21: 378–382.

McDonnell, R. (1881). *The Works of Abraham Colles Consisting Chiefly of his Practical Observations on the Venereal Disease, and on the Use of Mercury*. London: The New Sydenham Society.

Mostofi, S.B. (2008). Abraham Colles: 1773–1843. In: *Who's Who in Orthopedics?* 73–77. New York: Springer-Verlag.
Peltier, L.F. (1960). Six eponymic fractures. *Bull. Med. Libr. Assoc.* 48: 345–351.
Petit, J.L. (1723). *Traite des maladies des os*, vol. 2, 144. Paris: L d'Houry.
Pouteau, C. (1783). *Oeuvres posthumes*, vol. 2, 251. Paris: Ph. D. Pierres.

Leonardo da Vinci

Belt, E. (1955). *Leonardo the Anatomist*. Lawrence, KS: University of Kansas Press.
Bortolon, L. (1967). *The Life and Times of Leonardo*. London: Paul Hamlyn.
Capra, F. (2007). *The Science of Leonardo; inside the Mind of the Genius of the Renaissance*. New York: Doubleday.
Coghlan, C. and Hoffman, J. (2006). Leonardo da Vinci's flights of the mind must continue: cardiac architecture and the fundamental relation of form and function revisited. *Eur. J. Cardiothorac. Surg.* 29: S4–S17.
Gharib, M., Kremers, D., Koochesfahani, M.M., and Kemp, M. (2002). Leonardo's vision of flow visualization. *Exp. Fluids* 33: 219–223.
Hajar R. Da Vinci's anatomical drawings the heart. Heart Views. 2010. http://www.heartviews.org/article.asp?issn=1995-705X;year=2002;volume=3;issue=2;spage=12;epage=12;aulast=Hajar. Last accessed July 13, 2018.
Jose, A.M. (2001). Anatomy and Leonardo da Vinci. *Yale J. Biol. Med.* 74: 185–195.
Keele, K.D. (1951). Leonardo da Vinci, and the movement of the heart. *Proc. R. Soc. Med.* 44: 209–213.
Keele, K.D. (1952). *Leonardo da Vinci on Movement of the Heart and Blood*. Philadelphia: Lippincott.
Keele, K.D. (1964). Leonardo da Vinci's influence on Renaissance anatomy. *Med. Hist.* 8: 360–370.
Keele, K.D. (1973). Leonardo da Vinci's views on arteriosclerosis. *Med. Hist.* 7: 304–308.
Keele, K.D. (1983). *Leonardo da Vinci's Elements of the Science of Man*. New York: Academic Press.
Keele, K.D. (1997). *Leonardo da Vinci and the Art of Science*. River Forest, IL: Priory Press Ltd.
Kemp, M. (2006). *Leonardo da Vinci: Experience, Experiment and Design*. London: V&A Publications.
O'Malley, C.D. and Saunder, J.B. (1983). *Leonardo on the Human Body*. New York: Dover Publications.
Perace, J.M. (2007). Malpighi and the discovery of capillaries. *Eur. Neurol.* 58: 253–255.
Pormann, P.E. and Smith, E.S. (2007). *Medieval Islamic Medicine*. Washington, DC: Georgetown University.
Richter, J.-P. (1970). *The Notebooks of Leonardo da Vinci, Compiled and Edited from the Original Manuscripts in 1883*, vol. 2. New York: Dover.
Robicsek, F. (1991). Leonardo da Vinci and the sinuses of Valsalva. *Ann. Thorac. Surg.* 52: 328–335.
Schleich, J.M., Dillenseger, J.L., Houyel, L. et al. (2009). A new dynamic 3D virtual methodology for teaching the mechanics of atrial septation as seen in the human heart. *Anat. Sci. Educ.* 2: 69–77.
Vezzosi, A. (1997). *Leonardo da Vinci: Renaissance Man*. London: Thames & Hudson.
Welch, W.H. (1920). *Papers and Addresses, Vol. 3: Medical Education, History and Miscellaneous. Vivisection, and Bibliography*. Baltimore: Johns Hopkins University Press.
Wells, F.C. and Crowe, T. (2004). Leonardo da Vinci as a paradigm for modern clinical research. *J. Thorac. Cardiovasc. Surg.* 127: 929–944.
White, M. (2000). *Leonardo: The First Scientist*. New York: St. Martin's Griffin.

Vidus Vidius (Guido Guidi)

Brockbank, W. (1956). The man who was Vidius. *Ann. R. Coll. Surg. Engl.* 19: 269–295.
Choulant, L. (1852). *Geschichte und Bibliographie der Anatomischen Abbildung Nach Ihrer Beziehung auf Anatomische Wissenschaft und Bildende Kunst.* Leipzig: Rudolph Weigel.
Choulant, L. (1920). *History and Bibliography of Anatomic Illustration in its Relation to Anatomic Science and the Graphic Arts.* Translated and edited by Frank M. Chicago: University of Chicago Press.
Garrison, D. and Hast, M. (2014). *The Fabric of the Human Body: An Annotated Translation of the 1543 and 1555 Editions of "De Humani Corporis Fabrica Libri Septem.".* Basel: Karger.
Lang, J. (1989). *Clinical Anatomy of the Nose, Nasal Cavity and Paranasal Sinuses.* New York: Thieme.
Lang, J. (1995). *Clinical Anatomy of the Masticatory Apparatus and Peripharyngeal Spaces.* New York: Thieme.
Sanders, M. and Zuurmond, W.W. (1997). Efficacy of sphenopalatine ganglion blockade in 66 patients suffering from cluster headache: a 12- to 70-month follow-up evaluation. *J. Neurosurg.* 87: 876–880.
Üstün, Ç. (2003). Guido Guidi's short biography and his eponyms (the vidian artery, nerve and canal). *Inönü Üniversitesi Tip Fakültesi Dergisi* 10: 51–53.

Giambattista Canano

Canano, I.B. and Da Capri, G. (1925). *Musculorum Humani Corporis Picturata Dissectio (Ferrara 1541?).* Facsimile edition annotated by Cushing H, Streeter ES. Florence: R. Lier & Co.

Matteo Realdo Colombo

Eknoyan, G. (2000). Michelangelo: art, anatomy, and the kidney. *Kidney Int.* 57: 1190–1201.
Eknoyan, G. and De Santo, N.G. (1997). Realdo Colombo (1516–1559): a reappraisal. *Am. J. Nephrol.* 17: 261–268.
Fye, W.B. (2002). Realdo Colombo. *Clin. Cardiol.* 25: 135–137.
Lassek, A.M. (1958). *Human Dissection: Its Drama and Struggle. Charles C.* Springfield, IL: Thomas.
Moes, R.J. and O'Malley, C.D. (1960). Realdo Colombo: on those things rarely found in anatomy, an annotated translation from the De Re Anatomica (1559). *Bull. Hist. Med.* 34: 508–528.
Saunders, J.B. and O'Malley, C.D. (1950). *The Illustrations from the Works of Andreas Vesalius of Brussels with Annotations and Translations, a Discussion of the Plates and their Background, Authorship and Influence, and a Biographical Sketch of Vesalius.* Cleveland, OH: World Publishing Company.

Costanzo Varolio (Constantius Varolius)

Banzi, C. (1958). Unpublished facts on Costanzo Varolio. *Minerva Med.* 49: 4261–4268.
Choulant, L. (1962). *History and Bibliography of Anatomic Illustration.* Translated by Frank M. New York: Hafner.
DiDio, L.J.A. (1980). Highlights of a career in medical science. *Ohio J. Sci.* 80: 195–205.
Di Ieva, A., Tschabitscher, M., and Rodriguez y Baena, R. (2007). Lancisi's nerves and the seat of the soul. *Neurosurgery* 60: 563–568.
Dobson, J. (1962). *Anatomical Eponyms*, 2e. Edinburgh: E. & S. Livingstone.

Frati, P., Frati, A.L., Salvati, A. et al. (2006). Neuroanatomy and cadaver dissection in Italy: history medicolegal issues, and neurosurgical perspectives. *J. Neurosurg.* 105: 789–796.
Sappey, P.C. (1879). *Traité d'Anatomie Descriptive*, 3e. Paris: V. A. Delahaye et Cie.
Schmahmann, J.D., Ko, R., and MacMore, J. (2004). The human basis pontis: motor syndromes and topographic organization. *Brain* 127: 1269–1291.
Varolio C. De Nervis Opticis Nonnullisque Aliis Praeter Communem Opinioneu in Humano Capite Observatis. Ad Hieronymum Mercurialem, Patavii apud Paul et Anton. Meiettos fratres; 1573.
Varolio C. Anatomiae Sive de Resolutione Corporis Humani ad Caesarem Mediovillanum Libri IV. Eiusdem Varolii et Hieron. Mercurialis De nervis Opticis, epistolae, Francofurti, a pud Joh. Wechelum et Petr. Fischerum consortes; 1591.

Giulio Cesare Casseri

Anonymous (1973). *Julius Casserius's "Tabulae Anatomicae & De Formato Foetu Tabulae."*. New York: Medicina Rara, Folio.
Choulant, L. (1852). *Geschichte Und Bibliographie Der Anatomischen Abbildung*. Leipzig: Rudolph Weigel.
Hiatt, J.R. and Hiatt, N. (1997). The conquest of Addison's disease. *Am. J. Surg.* 174: 280–283.
Loukas, M., Akiyama, M., Shoja, M.M. et al. (2010). John Browne (1642–1702): anatomist and plagiarist. *Clin. Anat.* 23: 1–7.
Riva, A., Orru, B., Pirino, A., and Riva, F.T. (2001). Iulius Casserius (1552–1616): the self-made anatomist of Padua's golden age. *Anat. Rec.* 265: 168–175.
Sawday, J. (1996). *The Body Emblazoned: Dissection and the Human Body in Renaissance Culture*. New York: Routledge.
Sterzi G. Giului Casseri, anatomico e chirurgo (1552 c.–1616). Nuovo Arch Veneto, N.S. 1910; XVIII, Pt. II.
Tomasini I. 1630. Illustrium virorum elogia iconibus exornata. D. Pasquardum et Socium: Paravii; 1630.
Wright, J. (1902). The nose and throat in the history of medicine. *Laryngoscope* 12: 9–32.

Giovanni Maria Lancisi

Acierno, L.J. (1994). *The History of Cardiology*. New York: Informa Health Care.
Angeletti LR, Gazzaniga V, Conforti M, Fiorilla M, Marinozzi S. Giovanni Maria Lancisi and the Academy which Does Not Exist. XLII Congress of the Italian National Society of the History of Medicine, Bologna; 2002.
Anonymous (1852). *Edinburgh Medical and Surgical Journal Volume 78*. Edinburgh: Adam and Charles Black Longman.
Arroyo-Guijarro J, Prats-Galino A, Ruano-Gil D, Costa-Llobet C. Comparative study of the medial longitudinal striae of the hippocampus (formally called "nerves of Lancisi") in the fighting bull and domestic bull (French). *Bull. Assoc. Anat.* 1988;72:15–19.
Assalto P. 1720. Vita (of Lancisi) by Assalto. In: Lancisi GM (ed.). *De Aneurysmatibus, Opus Posthumum ... Aneurysms, the Latin Text of Rome, 1745*. Revised with translation and notes by Cave Wright W. Macmillan: New York; 1952, pp. ix–xxi.
Barr, M.L. and Kirnan, J.A. (1995). *Anatomy of the Human Nervous System*. Milan: McGraw-Hill.
Choulant, L. (1852). *Geschichte und bibliographie. Anatomichen Abbildung*. Leipzig: Rudolf Weigel.
Conforti, M. and Fiorilla, M. (2002). The Lancisiana library as a resource for the transmission of medical culture. *Med. Secoli* 14: 499–513.

Corradetti, A. (1987). In the Roman campagna from G. M. Lancisi to G. B. Grassi: two centuries of ideas, hypotheses and misunderstandings about malaria fevers. *Parassitologia* 29: 123–126.

Di Leva, A., Tschabitscher, M., and Rodriguez y Baena, R. (2007). Lancisi's nerves and the seat of the soul. *Neurosurgery* 60: 563–568.

Eycleshymer, A.C. (1917). *Anatomical Names*. New York: William Wood & Co.

Fineschi, V., Baroldi, V., and Silver, M.D. (2006). *Pathology of the Heart and Sudden Death in Forensic Medicine*. New York: Taylor and Francis.

Fleming, P.A. (1997). *Short History of Cardiology*. New York: Rodopi.

Foote, J. (1917). Giovanni Maria Lancisi (1654–1720). *Int. Clin.* 2: 292–308.

Fye, W.B. (1990). Giovanni Maria Lancisi, 1654–1720. *Clin. Cardiol.* 13: 670–671.

Gazzaniga, V. (2003). Giovanni Maria Lancisi and urology in Rome in early modern age. *J. Nephrol.* 16: 939–944.

Gazzaniga, V. and Marinozzi, S. (2006). Nephrology in the Lancisi Medical Dictionary (1672–1720). *J. Nephrol.* 19: S44–S47.

Grondona F. The dissertation of Giovanni Maria Lancisi about the seat of the rational soul (Italian). *Physis* 1965;7:401–430.

JAMA (1964). Giovanni Maria Lancisi (1654–1720) – cardiologist, forensic physician, epidemiologist. *JAMA* 189: 375–376.

Jarcho, S. (1969). The correspondence of Morgagni and Lancisi on the death of Cleopatra. *Bull. Hist. Med.* 43: 299–325.

Lancisi, G.M. (1728). *De Motu Cordis et Aneurysmatibus*. Rome: J. M. Salvioni.

Lancisi, G.M. (1952). *De Aneurysmatibus, Opus Posthumum ... Aneurysms, the Latin Text of Rome, 1745*. Revised with translation and notes by Cave Wright W. New York: Macmillan.

Mantovani, A. and Zanetti, R. (1993). Giovanni Maria Lancisi: De bovilla peste and stamp. *Hist. Med. Vet.* 18: 97–110.

McDougall, J.I. and Michaels, L. (1972). Cardiovascular causes of sudden death in "De Subitaneis Mortibus" by Giovanni Maria Lancisi. A translation from the original Latin. *Bull. Hist. Med.* 46: 486–494.

Michaels, L. (1972). Pain of cardiovascular origin in the writings of Giovanni Maria Lancisi. *Can. Med. Assoc. J.* 106: 371–373.

Ripley, G. and Dana, C.A. (1858). *The New American Cyclopedia*. New York: Appleton and Co. 1858.

Schullian, D.M. (1983). News and events: a survey of variant title page vignettes in Lancisi's "De subitaneis mortibus". *J. Hist. Med.* 38: 336–339.

Singer, C. (1953). A great papal physician. *Br. Med. J.* 1: 93–94.

Weber, G. (1999). Lesser known profiles of morbid anatomists in the XVII and XVIII centuries (and at the beginning of the XIX century). *Med. Secoli* 11: 107–116.

Willius, F.A. and Dry, T.J. (1948). *A History of the Heart and Circulation*. Philadelphia and London: W.B. Saunders Co.

Giovanni Battista Morgagni

Andrioli, G. and Trincia, G. (2004). Padua: the renaissance of human anatomy and medicine. *Neurosurgery* 55: 746–754.

Androutsos, G. (2006). Giovanni-Battista Morgagni (1682–1773): creator of pathological anatomy. *J. Buon.* 11: 95–101.

Antonello, A., Cal, L., Bonfante, L. et al. (1999). Giovani Battista Morgagni, a pioneer of clinical nephrology. *Am. J. Nephrol.* 19: 222–225.

Buck, A.H. (1920). *The Dawn of Modern Medicine*. New Haven, CT: Yale University Press.

Hajdu, S.I. (2010). A note from history: the first printed case reports of cancer. *Cancer* 116: 2493–2498.
Hakan, T. (2009). Neurosurgery and a small section from the Greek myth: the god pan and syrinx. *Childs Nerv. Syst.* 25: 1527–1529.
Kechagiadakis, G. and Antonopoulou, M. (1994). G.B. Morgagni: the father and founder of contemporary pathology. *Forum* 4: 485–487.
Ritter, G. (1979). Giovanni Battista Morgagni (Venedig 1761). *Nervenarzt* 50: 144–146.
Ventura, H.O. (2000). Giovanni Battista Morgagni and the foundation of modern medicine. *Clin. Cardiol.* 23: 792–794.

Ruggero Ferdinando Antonio Guiseppe Vincenzo Oddi

Flati, G. and Andren-Sandberg, A. (2002). Wirsung and Santorini: the men behind the ducts. *Pancreatology* 2: 4–11.
Modlin, I. and Ahlman, H. (1994). Oddi: the paradox of the man and the sphincter. *Arch. Surg.* 129: 549–556.
Morelli, A. and Sorcetti, F. (1988). Rugerro Oddi's life. *Z. Gastroenterol. Verh.* 23: 203–207.
Oddi R. Di una special disposizione q sfintere allo sbocco del coledoco. Annali dell' Università Libera di Perugia, 2, Vol. 1, Facoltà medico-chirurgica; 1886–87.
Oddi, R. (1887). D'une disposition a sphincter speciale de l'ouverture du canal choledoque. *Arch. Ital. Biol.* 8: 317–322.
Oddi, R. (1888). Sulla tonicità dello sfintere del coledoco. *Archivio par le scienze mediche* 12: 333–339.
Oddi, R. (1894a). Sul centro spinale dello sfintere del coledoco. Lo Sperimentale, sec. biologica. *Florence* 48: 180–191.
Oddi, R. (1894b). Sulla esistenza di speciali gangli nervosi in prosssimità dello sfintere del coledoco. *Monitore Zool. Ital.* 5: 216.
Oddi R. Sulla fisio-patologia delle vie biliari. Conferenze cliniche italiane dirette dal Prof. Achille de Giovanni, 1, Folge, 1 bd Malland o J.; 1894c, pp. 77–124.
Oddi R. Gli alimenti e laloro finzione neil' economia dell'organismo individuale e sociale. Torino; 1902.
Oddi R. Nuova terapla delle malletti d'infezizone e delle cachessie maligne: Modo di prevenirle e curarle: La vitalina Gatchkowsk: Citta di Castello; 1906.
Ono, K. and Hada, R. (1988). Ruggero Oddi: To Commemorate the Centennial of his Original Article – "Di una Speciale Disposizione a Sfintere allo Sbocco del Coledoco". *Jpn. J. Surg.* 18: 373–375.
Ritter, U. (1988). One hundred years of research on the sphincter of Oddi. *Z. Gastroenterol. Verh.* 23: 208–209.
Woods, C.M., Mawe, G.M., Toouli, J., and Saccone, G.T.P. (2005). The sphincter of Oddi: understanding its control and function. *Neurogastroenterol. Motil.* 17: 31–40.

Johan Georg Ræder

Boniuk, M. and Schlezinger, N.S. (1962). Raeder's paratrigeminal syndrome. *Am J. Ophthalmol.* 54: 1074–1084.
Coughlan, T., Lawson, S., and O'Neill, D. (2004). French without tears? Foreign accent syndrome. *J. R. Soc. Med.* 97: 242–243.
Goadsby, P.J. (2002). "Paratrigeminal" paralysis of the oculopupillary sympathetic system. *J. Neurol. Neurosurg. Psychiatry* 72: 297–299.

Hem, E. (2006). A peculiar speech disorder due to bomb injury of the brain. *Tidsskr. Nor. Laegeforen.* 126: 3311–3313.
Kazakova, D., Mermoud, A., and Krieglstein, G. (2005). Mechanisms of filtration after deep sclerectomy with collagen implant. In: *Focus on Glaucoma Research* (ed. S.M. Reece), 197–216. New York: Nova Biomedical.
Kerty, E. (1999). Johan Georg Raeder and Raeder's syndrome. *Tidsskr. Nor. Laegeforen.* 119: 3117.
Larsen, O. (1996). *Norges Leger*, vol. 4. Oslo: Den norske lageforening.
Lowe, R.F. (1977). Primary angle closure glaucoma: a review of ocular biometry. *Aust. N. Z. J. Ophthalmol.* 5: 9–17.
Marmor, M.F. (2000). A brief history of macular grids: from Thomas Reid to Edvard Munch and Marc Amsler. *Surv. Ophthalmol.* 44: 343–353.
Munson, P.D. and Heilman, B. (2005). Foreign accent syndrome: anatomic, pathophysiologic and psychosocial considerations. *S. D. J. Med.* 58: 187–189.
Nogaki, H., Nakamura, M., and Tatsumi, S. (1989). A case of Raeder's syndrome caused by metastatic malignant lymphoma. *No To Shinkei* 41: 259–262.
Raeder, J.G. (1918a). Et nyt instrument til maaling av camera anteriors dybde (thalamometer). *Norsk Magazin for Lægevidenskaben* 79: 862–871.
Raeder, J.G. (1918b). Et tilfælde av intrakraniel sympaticuslammelse. *Norsk Magazin for Lægevidenskaben* 79: 999–1016.
Raeder, J.G. (1924). Paratrigeminal paralysis of oculopupillary sympathetic. *Brain* 47: 149–158.
Ræder, J.G. and Harbitz, F.G. (1926). Ansigts – og øienatrofi (presenil cataract og "glaukom") foraarsaket av symmetrisk carotisaffektion. *Norsk Magazin for Lægevidenskaben* 87: 529–548.
Schück, H., Sohlman, R., Österling, A. et al. (1951). *Nobel – The Man and his Prizes*. Stockholm: University of Oklahoma Press.
Solomon, S. (2001). Raeder syndrome. *Arch. Neurol.* 58: 661–662.
Solomon, S. and Lustig, J.P. (2001). Benign Raeder's syndrome is probably a manifestation of carotid artery disease. *Cephalalgia* 21: 1–11.
Spaeth, G.L. (2000). The normal development of the human anterior chamber angle. A new system of descriptive grading. In: *Classic Papers in Glaucoma* (ed. R. Ritch and R.M. Caronia), 245–260. The Hague: Kugler Publications.
Strachan, R.W. (1964). The natural history of Takayasu's arteriopathy. *Q. J. Med.* 33: 57–69.

Andrew Fyfe the Elder

Anonymous (1824). Obituary, the late Andrew Fyfe, ESQ. *Edinburgh Mag. Lit. Miscellany* 93: 510.
Anonymous (1825). *Edinb. Med. Surg. J.* 24: 8–9.
Burch, D. (2007). *Digging Up the Dead: Uncovering the Life and Times of an Extraordinary Surgeon*, 2007. London: Chatto and Windus.
Cooper, B.B. (1843). *The Life of Sir Astley Cooper*. London: John W. Parker.
Currie, J. (2004). Fyfe, Andrew (1752–1824), anatomist. In: *Dictionary of National Biography*, vol. 21, 217–218. Oxford: Oxford University Press.
Fyfe, A. (1784). *A System of Anatomy and Physiology*. Edinburgh.
Fyfe, A. (1800a). *A Compendium of the Anatomy of the Human Body*. Edinburgh.
Fyfe, A. (1800b). *View of the Bones, Muscles, Viscera and Organs of the Senses*. Edinburgh.
Fyfe, A. (1803). *Views of the Bones and Muscles with Concise Explanations*. Edinburgh.
Fyfe, A. (1813). *Outlines of Comparative Anatomy, Intended Principally for the Use of Students*. Edinburgh: Adam Black.

Fyfe, A. (1814). *A System of Anatomy of the Human Body Illustrated by Upwards of two hundred Tables, Taken Partly from the most Celebrated Authors and Partly from Nature,* 1814. Edinburgh: Adam Black.
Fyfe, A. (1818a). *A Probationary Essay on Crural Hernia.* Edinburgh.
Fyfe, A. (1818b). *A Probationary Essay on Alvine Concretions.* Edinburgh.
Horrocks, W.H. (1900). *The Life of Sir Astley Cooper.* London: Leadenhall.
Kaufman, M.H. (1999). Observations on some of the plates used to illustrate the lymphatics section of Andrew Fyfe's compendium of the anatomy of the human body, published in 1800. *Clin. Anat.* 12: 27–34.
Kaufman, M.H. (2006). John Barclay (1758–1826) extra-mural teacher of anatomy in Edinburgh: honorary fellow of the Royal College of Surgeons of Edinburgh. *Surgeon* 4: 93–100.
Kemp, M. (2010). Style and non-style in anatomical illustration: from renaissance humanism to Henry Gray. *J. Anat.* 216: 192–208.
Liggett, L.M.A. (1904). Extracts from the journal of a Scotch medical student of the eighteenth century. *Med. Lib. Hist. J.* 7: 107.
Matthew, H.C.G. and Harrison, B.H. (2004). *Oxford Dictionary of National Biography: In Association with the British Academy: From the Earliest Times to the Year 2000,* vol. 21. Oxford: Oxford University Press.
Monro, A., Fyfe, A., Archibald, W. et al. (1797). *Three Treatises. On the Brain, the Eye and the Ear Illustrated by Tables.* Edinburgh: Bell and Bradfute.
Risse, G.B. (1986). *Hospital Life in Enlightenment Scotland: Care and Teaching at the Royal Infirmary of Edinburgh.* Cambridge: Cambridge University Press 1986.
Rock, J. (2000). An important Scottish anatomical publication rediscovered. *Book Collect.* 49: 27–60.
Rock J. Richard Cooper and anatomical illustration. 2011. http://sites.google.com/site/richardcooperengraver/home/richard-cooper-and-anatomical-illustration. Last accessed July 13, 2018.
Russell, K.F. (1963). *British Anatomy: 1525–1800, a Bibliography.* Victoria: Melbourne University Press.
Stephen, L. and Lee, S. (1889). *Dictionary of National Biography.* London: Smith, Elder, & Co.
Struthers, J. (1867). *Historical Sketch of the Edinburgh Anatomical School.* Edinburgh: MacLachlan and Stewart.

Sir John Struthers

Association of American Medical Colleges (AAMC). Learning Objectives for Medical Student Education: Guidelines for Medical Schools. Washington, DC; 1998. https://members.aamc.org/eweb/upload/Learning%20Objectives%20for%20Medical%20Student%20Educ%20Report%20I.pdf. Last accessed July 13, 2018.
Darwin, C. (2000). *The Descent of Man, and Selection Relation to Sex.* Boston: Adamant Media Corporation.
De Jesus, R. and Dellon, A.L. (2003). Historic origin of the "arcade of Struthers.". *J. Hand Surg.* 28: 528–531.
Doyle, J.R. and Botte, M.J. (2002). *Surgical Anatomy of the Hand and Upper Extremity.* Philadelphia: Lippincott Williams & Wilkins.
Gillespie, A.L. (1889). Obituary: Sir John Struthers MD LLD. *Edinb. Med. J.* 5: 433–434.
Keith, A. (1911). *Anatomy in Scotland during the Lifetime of Sir John Struthers (1823–1899).* Sir John Struthers Anatomical Lecture. Edinburgh: Royal College of Surgeons of Edinburgh.

Lordan, J., Rauh, P., and Spinner, R.J. (2005). The clinical anatomy of the supracondylar spur and the ligament of Struthers. *Clin. Anat.* 18: 548–551.

McGonagall, W. (2002). *The Famous Tay Whale: The World's Worst Poet.* Springfield, IL: Templegate Publishers.

Struthers, J. (1848). On a peculiarity of the humerus and humeral artery. *Month J. Med. Sci.* 28: 264–267.

Struthers, J. (1854). On some points in the abnormal anatomy of the arm. *Br. Foreign Med. Chir. Rev.* 14: 170–179.

Struthers, J. (1856). Hints to students on the prosecution of their studies: being extracts from the introductory address at Surgeons' Hall, session 1855–56. *Edinb. Med. J.* 2: 353–360.

Struthers, J. (1869). *Notes on Medical Education: Being Replies to the Inquiries Addressed to Teachers by the General Medical Council.* Aberdeen: D. Chalmers and Company.

Struthers, J. (1883). On a method of promoting maceration for anatomical museums by artificial temperature. *J. Anat. Physiol.* 18: 49–53.

Struthers, J. (1891). *Notes on the Progress of Aberdeen University during the Last thirty Years.* Aberdeen: Aberdeen Free Press.

Struthers, J. (1893a). On the rudimentary hind-limb of a great fin-whale (*Balaenopterae musculus*) in comparison with those of the humpback whale and the Greenland right-whale. *J. Anat. Physiol.* XXVII: 291–335.

Struthers, J. (1893b). On the development of the bones of the foot of the horse, and of digital bones generally; and on a case of polydactyly in the horse. *J. Anat. Physiol.* 28: 51–62.

Struthers, J. (1893c). The new five-year course of study: remarks on the position of anatomy among the earlier studies, and on the relative value of practical work and of lectures in modern medical education. *Edinb. Med. J.*

Waterston, S.W., Laing, M.R., and Hutchison, J.D. (2007). Nineteenth century medical education for tomorrow's doctors. *Scott. Med. J.* 52: 45–49.

Waterston, S.W. and Hutchison, J.D. (2004). Sir John Struthers MD FRCS Edin LLD Glasg: anatomist, zoologist, and pioneer in medical education. *Surgeon* 2: 6.

Williams, M.J. (1996). Professor Struthers and the Tay Whale. *Scott. Med. J.* 41: 92.

Charles Bell

Bell, C. (1812). *A System of Operative Surgery Founded on the Basis of Anatomy.* Hartford, CT: Hale & Hosmer.

Bell, C. (1821). *Illustrations of the Great Operations of Surgery: Trepan, Hernia, Amputation, Aneurism, and Lithotomy.* London: Longman, Hurst, Rees, Orme and Brown.

Grzybowski, A. and Kaurman, M.H. (2007). Sir Charles Bell (1774–1842): contributions to neuro-ophthalmology. *Acta Ophthalmol. Scand.* 85 (8): 897–901.

Howe, A.J. (1894). *Miscellaneous Papers.* Cincinnati, OH: Robert Clarke & Co.

Jay, V. (1999). A portrait in history: Sir Charles Bell – artist extraordinaire. *Arch. Pathol. Lab. Med.* 123: 463–465.

Kazi, R.A. and Rhys-Evans, P. (2004). Sir Charles Bell: the artist who went to the roots! *Postgrad. Med. J.* 50: 158–159.

Major, R.H. and Schullian, D.M. (1953). Notes and queries. *J. Hist. Med. Allied Sci.* 8 (2): 215–217.

Pearce, J.M. (1993). Sir Charles Bell. *J. R. Soc. Med.* 86 (6): 353–354.

Shaw, A. (1839). *Discoveries of Sir Charles Bell in the Nervous System.* London: Longman.

Wilkins, R.H. and Brody, I.A. (1969). Bell's palsy and Bell's phenomenon. *Arch. Neurol.* 21: 661–699.

Antonio de Gimbernat y Arbós

Albiol R. Pere Virgili i el seu temps. Thesis, University of Tarragona; 1992.
Beddoes, T. (1795). *A New Method of Operating for the Femoral Hernia, Translated from the Spanish of Don Antonio de Gimbernat, with Plates to Which are Added by the Translator, Queries Respecting a Safer Method of Performing Inoculation*. London: J. Johnson.
Cardoner-Planas A. Creación y historia del Real Colegio de Cirugía de Barcelona. In: Proceedings of the IX Congress of Medicine and Biology of Cataluna. Academia de Medicina: Barcelona; 1936.
Febrer JLF. Epónimos médicos. Ligamento de Gimbernat; 1999. http://www.historiadelamedicina.org/gimbernat.html. Last accessed July 13, 2018.
Ferrer, D. (1963). *Pedro Virgili*. Barcelona: Asociacion de Cirugia de Barcelona, Editorial Emporium.
Gimbernat, A. (1793). *Nuevo Método de operar en la Hernia Crural*. Madrid: Ibarra.
Llagostera-Sala F. Reseña bibliográfica del Dr. Antonio de Gimbernat. In: Proceedings of the Third Congress on the History of Catalonian Medicine. Real Academia de Medicina: Lleida; 1881.
Puig-La Calle, J. and Marti-Pujol, R. (1995). Antonio de Gimbernat (1734–1816). Anatomist and surgeon. *Arch. Surg.* 130: 1017–1020.
Rutkow, I.M. (2003). A selective history of hernia surgery in the late eighteenth century: the treatises of Percivall Pott, Jean Louis Petit, D. August Gottlieb Richter, Don Antonio de Gimbernat, and Pieter Camper. *Surg. Clin. N. Am.* 83: 1021–1044.

Emanuel Swedenborg

Akert, K. and Hammond, M.P. (1962). Emanuel Swedenborg (1688–1772) and his contributions to neurology. *Med. Hist.* 6: 254.2–266.1.
Anonymous (1968). Emanuel Swedenborg (1688–1772) natural scientist, neurophysiologist, theologian. *JAMA* 206: 887–888.
Gross, C.G. (1997). Emanuel Swedenborg: a neuroscientist before his time. *Neuroscientist* 3: 142–147.
Gross, C.G. (2009). Three before their time: neuroscientists whose ideas were ignored by their contemporaries. *Exp. Brain Res.* 192: 321–334.
Haas, L.F. (1993). Emanuel Swedenborg (1688–1772). *J. Neurol. Neurosurg. Psychiatry* 56: 343.
Ramstrom, M. (1911). Emanuel Swedenborg's investigations in natural science and the basis for his statements concerning the functions of the brain. *J. Am. Med. Assoc.* LVI: 614.
Rodgers, R.R. (1910). Swedenborg, the philosopher and theologian. *Trans. Int. Swedenborg Congr.* 269–284.
Swedenborg, E. (1789). *A Brief Exposition of the Doctrine of the New Church (L)*. London: R. Hindmarsh.
Swedenborg, E. (1882). *The Brain*. Translated and edited by Tafel RL., vol. 1. London: J. Spiers Vol. 2, 1887.
Swedenborg, E. (1932). Oeconomia Regni Animalis. Translated by A. Clissold, Boston, 1868. In Pratt, F.H.: Swedenborg on the Thebesian blood flow of the heart. *Ann. Med. Hist.* 4: 434–439.

Wilhelm His Senior and Wilhelm His Junior

Anonymous (1964a). Wilhelm His Sr. (1831–1904) – embryologist and anatomist. *JAMA* 187: 58.
Anonymous (1964b). Wilhelm His, Jr. (1863–1934). *JAMA* 187: 453–454.
Berry, D. (2006). History of cardiology: Wilhelm His, Jr, MD. *Circulation* 113: f72.
Garrison, F.H. (1929). *An Introduction to the History of Medicine*, 4e. Philadelphia: Saunders.

Hildebrand, R. (2005). "...that progress in anatomy is most likely to occur when its problems include the study of growth and function, as well as of structure." Über den Anatomen und Physiologen Ernst Heinrich Weber (1795–1878) und über Wilhelm His (1831–1904), seinen Nachfolger auf dem Lehrstuhl für Anatomie an der Universität Leipzig. *Ann. Anat.* 187: 439–459.

His, W. (1870). Description of a microtome. *Arch. Microskop. Anat.* 6: 229–232.

His, W. (1880–1885). *Anatomy of the Human Embryo. Three Parts and Atlas.* Leipzig: FCW, Vogel.

His, W. (1895). *Anatomical Nomenclature.* Leipzig: Veit & Co.

His, W. (1965). *On Tissues and Spaces of the Body.* Basel: Schweighauser.

His, W. Jr. (1887). Ueber das Stoffwechselproduct des Pyrdins. *Arch. Exp. Pathol. Pharmakol.* 22: 253–260.

His, W. Jr. (1933). Zur Geschichte des Atrioventrikular-bundles nebst Bemerken uber die embyonale Herztatigkeit. *Klin. Wochenschr.* 12: 569–574.

Mall, F.P. (1905). Wilhelm His. His relation to institutions of learning. *Am. J. Anat.* 4: 139–161.

Mall, F.P. (1914). A plea for an institute of human embryology. *J. Am. Med. Assoc.* 60: 1599–1601.

Müller, F. and O'Rahilly, R. (1986). Wilhelm His und 100 Jahre Embryologie des Menschen. *Acta Anat.* 125: 73–75.

Peipert, J.F. and Roberts, C.S. (1986). Wilhelm His, Sr.'s finding of Johann Sebastian Bach. *Am. J. Cardiol.* 57: 1002.

Picken, L. (1956). The fate of Wilhelm His. *Nature* 178: 1162–1165.

Roguin, A. (2006). Wilhelm His Jr. (1863–1934) – the man behind the bundle. *Heart Rhythm.* 3: 480–483.

Silverman, M.E., Grove, D., and Upshaw, C.B. Jr. (2006). Why does the heart beat? The discovery of the electrical system of the heart. *Circulation* 113: 2775–2781.

Wendler, D. and Rother, P. (1982). Wilhelm His sen. – Leben und wirken des bedeutenden Leipziger Morphologen. *Z. Gesamte Inn. Med.* 37: 810–813.

4

Middle East

Although not medical books, the early religious texts of the Islamic and non-Islamic cultures of the Middle East often discussed anatomical structures and demonstrated an understanding of the human body that was sophisticated for their day. Medieval Islamic medicine was, in general, Galenic medicine. The original Greek writing of Galen is lost to history, but it was saved in part by the diligence of its Arabic translators. The Golden Age of Arabian learning peaked between 750 and 850 CE, the century after the establishment of the Abbasid Caliphate, with its center in Baghdad. Although taboo, some dissection did occur in this culture. For example, it is said that Yuhanna ibn Masawayh dissected apes in a specially built room built on the banks of the Tigris. Anatomic knowledge, in all Middle Eastern cultures, was investigated most often during surgical procedures.

4.1 Anatomy in the Bible and Talmud

4.1.1 Neuroanatomy

The writings referred to as "the Bible," a term derived from the Greek word for "book," were written in antiquity in Hebrew, Aramaic, and Greek over a period exceeding a millennium. The number of books regarded as canonical and the order in which they are incorporated differs among the religions and denominations that hold these writings as sacred. The Jewish sacred writings are referred to as the Tanakh. Christian writers subsequently referred to these writings as the Old Testament. The Christian Bible (Figure 4.1) incorporates the Old and New Testaments.

The Torah, the Hebrew Bible, is the written component of Jewish Law. This written law is meant to be studied with an accompanying Oral Law component, originally passed only from teacher to disciple. The Mishneh, an annotated version of the Oral Law, was first put into writing around 200 CE, followed by the Gemarah, a further expansion on the Mishneh, at around 500 CE. Together, the Mishneh and the Gemarah are referred to as the Talmud (Figure 4.2). While the Talmud primarily expounds Jewish Law, it contains several references to medical practices of the day. It is important to remember that the Torah and the Talmud are not inherently medical texts, and as such often provide concise descriptions, illustrating primarily those facets concerned with law. Nevertheless, they provide valuable insight into the surgical practices and anatomic knowledge of the time.

History of Anatomy: An International Perspective, First Edition. R. Shane Tubbs, Mohammadali M. Shoja, Marios Loukas and Paul Agutter.

IESV CHRISTI
DOMINI NOSTRI NOVVM
TESTAMENTVM, SIVE NOVVM
fœdus è Græco archetypo Latino
sermone redditum.

Theodoro Beza interprete, & iam vltimò ab eo recognitum.

Cui additur ex aduerso eiusdem Noui Testamenti ex vetustissima Syriaca translatione, translatio Latina IMMANVELIS TREMELLII *Theologiæ Doctoris, & Professoris celeberrimi.*

LONDINI,
Excudebat Henricus Midletonus Typographus.
1585.

Figure 4.1 Title page from a 1585 Latin translation of the Greek New Testament.

4.1.1.1 Spinal Cord

The spinal cord is found mentioned only in the Talmud. It is referred to as the "silver cord" in Talmudic poetic allegory, an expression preserved by anatomists of the Middle Ages as the *funis argenteus*. Interestingly, the common Hebrew name for the spinal cord is *khut ha-shedra*, meaning "the string of the vertebral column," with the lower part of the canal called the *parashoth* or "partings" (cauda equina). Notably, the Talmud also reports the story of a ewe owned by Rabbi Habiba in early fifteenth-century Spain that was dragging its hind legs. The rabbi asked, "perhaps its spinal cord is severed?" Such comments suggest an early understanding of the motor fibers carried by this neuroanatomical structure.

4.1.1.2 Peripheral Nerves

Mention of peripheral nerves, and specifically the sciatic nerve, are scantly found in the Talmud and Old Testament. Interestingly, the Hebrew *gid* refers to nerves, but also to ligaments and tendons. The sciatic nerve was called the *gid ha-nashe*, which was by law not to be eaten. In fact, the Talmud gives precise instructions as to its removal from slaughtered animals. Jacob, as found in Genesis, is thought to have had injury to his sciatic nerve, and it is in deference to him that this law is followed. Jacob may have lost his wrestling contest with God (angel or man) due to this injury (Figure 4.3). When the angel saw that he could not overpower him, he touched the

מאימתי · פרק ראשון ברכות ב

מאימתי

Figure 4.2 The first page of the Babylonian Talmud (Talmud Bavli, Vilna edition).

Figure 4.3 "Jacob Wrestles the Angel" (detail), painted in 1861 by Eugène Delacroix (1798–1863) (Church of St. Sulpice, Paris).

socket of Jacob's hip so that it was wrenched as he wrestled. Jacob called the place of his contest Peniel, saying, "It is because I saw God face to face, and yet my life was spared." The sun rose above him as he passed Peniel, and he was limping because of his hip. The term *shigrona,* found in the Talmud, is generally accepted as referring to sciatica. For the treatment of sciatica, the Talmud recommends that fresh brine be rubbed 60 times over painful areas.

4.1.1.3 Brain and Meninges

References specifically to the brain are found only in the Talmud. Many rabbis believed that the spinal cord, bone marrow, and brain were all essentially similar in their constitution, as seen in the Hebrew word *moach,* which refers to both the brain and bone marrow. Rabbi Abaye (c. third century) thought sperm originated in the brain, as was also taught by Aristotle. Interestingly, Plato thought sperm arose from the spinal cord. The area or hair overlying the anterior fontanelle is called the *kodkod* ("bregma" or "βρέγμα" in Greek). An example of this term is found in reference to Absalom, the third son of Kind David in II Samuel 14 : 25: "Absalom was very handsome: from the sole of his foot to the crown [*kodkod*] of his head there was no blemish on him." The school of Rabbi Yannai indicated that *tefillin* (phylacteries) should only be worn over this location, understanding that the anterior fontanelle was once the shortest route to the cerebrum. Talmudic sages called it "the place where the brain of a child is soft." As a result, Aramaean translators interchanged *kodkod* with the word *moach.* Rosner has recalled a story set in Jerusalem where a rooster that thought the pulsatile anterior fontanelle of an infant was an insect, pecked out the child's brain. Two coverings of the brain are discussed, one hard and one soft; it is said that these coverings resemble those of the testes. The text further mentions that the "skin" (arachnoid, dura?) of the brain is rich in blood vessels.

4.1.1.4 Skull

Although the exact term "skull" is not found in these ancient writings, it is interesting to mention that some modern-day terms are derived from Biblical events. Germane to the skull is that "Calvary," a derivative of "calvaria," is the purported site of Jesus' crucifixion in Jerusalem. Calvary is found only in the King James Version of the Bible; the Hebrew equivalent is Golgotha (Aramaic Gûgaltâ, Chaldean Gulgalta, Greek κρανίου τόπος), the "place of the skull" or a plateau containing a pile of skulls or a geographic feature resembling a skull. The traditional identification of the site of Calvary was made by St. Helena, the mother of Emperor Constantine, when she found (c. 326 CE) what was believed to be a relic of the actual cross on which Jesus was crucified. This spot, represented by a pile of rocks approximately 5 m high, is today to be found inside the Church of the Holy Sepulchre, built by the Roman emperor Constantine the Great.

4.1.1.5 Cranial Malformation

References to misshapen crania can be found in the Talmud and, perhaps, the Bible. The Mishnah refers to several cranial abnormalities that would render priests unfit to serve in the Temple. The term *kilon,* used in the Gemara, describes a head that is in the shape of a "keg cover" or is egg-shaped. *Liftan* head shapes resemble a turnip, being wide at the top and narrower as one moves down. *Makban* refers to the shape of a hammer, with a markedly protruding forehead and occiput, which may represent scaphocephaly, attributable to sagittal suture synostosis. Although a round head would not cause a person to be prohibited from serving as a priest, it was still thought of as unsightly; the round head was typical of the Babylonians, who were thought to have unskilled midwives.

Some Biblical and Old Testament scholars, including Moskopp, have hypothesized that Goliath, who was slain by David, had acromegaly with resultant macrocephaly. On the other hand, Feinsod suggests that David may have been able to approach the giant unnoticed because of a visual defect resulting from a pituitary adenoma. Notably, references to gigantism are also found in II Samuel and I Chronicles. Although not related to cranial malformations, the term *sefikas* was used to disqualify potential priests; it was described as a "receding neck," and may have represented the Klippel–Feil syndrome.

4.1.1.6 Headache

Interestingly, headache is not mentioned in the Bible, and is mentioned only infrequently in the Talmud. The Talmud considers hypersensitivity of the nerve that "grows from the mouth of the stomach," possibly the vagus nerve, as a cause of headache (abdominal migraine), and treats such headache with dietary restrictions. Yabez (Temurah 16a) prays to the Lord that "I not suffer from headache." One Talmudic treatment for a headache is to rub the head with wine, vinegar, or oil.

4.1.2 General Anatomy

There are several surgical accounts in the Talmud, many surprising given their early date. Strict Jewish law forbade dissection. The only known dissection was of a cadaver of a prostitute, who had been condemned to be burned to death, by the disciples of Ishmael Ben Elisha (90–135 CE). An account of the dissection of a human scalp for teaching purposes is also known. Another dissection is described by Mar Bar Rav Ashi (?–467 CE), who performed plastic surgery on a penis.

The rabbis were also well versed in veterinary anatomy and embryology. Because it is considered impure to eat an animal that was ill before its death, both live and slaughtered animals were inspected to ensure their health. This provided the rabbis with extensive knowledge of animal anatomy. In fact, an early account of intubation of a lamb with a cut trachea is to be found in the Talmud (Chullin, 57b). Although the extent that this can be extrapolated to humans is limited, the surgical accounts indicate significant knowledge of anatomy.

Perhaps the most detailed account of a surgical procedure is that of cranial surgery performed to remove a disease called *ra'atan*. The exact nature of this disease is uncertain to us today, but the Talmud clearly describes a growth or organism present upon the lining membranes of the brain. Three hundred cups of a presumable anesthetic solution (although it may have been an antiseptic or anticoagulant) were poured over the skull. A physician's drill was then used to open it. Four myrtle leaves, which are both sharp and flexible, were used to separate the organism, which could then be removed with tongs. The Talmud clearly states that the surgeon must be sure to completely remove and burn the organism, lest it return to the victim. The use of leaves and tongs illustrates the knowledge that damage to the brain itself would likely result in death, so that care was taken to leave both the brain and meninges intact.

Trephination, while remote from surgery practiced today, was often used. There are several accounts of trephination being used as a treatment for epilepsy. Rosner describes cases of opened skulls where the wounds were closed with pumpkin peel to prevent infection. Splenectomy is also described, and confirmation that a healthy life can be lived without a spleen is given. There is some debate as to when the first successful splenectomy in history occurred. As described by Oren et al. (1998), the Chinese surgeon Hua Tuo (115–205 CE) is thought to have been the first to successfully perform this surgery. However, they also mention that Rav Yehuda (219–299 CE) commented on a section of the Bible from the time of King David (approximately 960 years before the advent of Christianity) in which at least 50 men are described as being "without spleens." There is debate as to whether the saying "without spleens" is referring to a surgical splenectomy, as we know it today, or to some other interpretation of the words. Rav Yehuda also mentioned that these men "had the soles of their feet carved out." Oren et al. (1998) go on to describe this as a method to elevate the longitudinal plantar arches to make the men run faster, since they were members of the elite guard of Adonijah (a pretender to David's throne). Rabbi Jacob Khouly maintained that glowing-hot copper trays were applied to the soles of the men until the skin was dead; it could then be chiseled down to the bone. Rosner and Preuss refer to the belief, again, that the two interventions were undertaken in order to improve the speed and endurance of the members of this elite team of guards.

These ancient Jewish texts make reference to plastic surgery and catheterization of the throat. It is stated that when a surgical wound is sewn up, the lips of the wound are renewed. Procedures to open an imperforate anus of a newborn (a crosswise incision is made using a barley twig, so as to avoid the irritation that would result from the use of metal), remove splinters using hot water and a needle, and treat several skin ailments and burns are all described. The insertion of artificial teeth is also mentioned. Mar Bar Rav Ashi performed plastic surgery on a penis. Joint dislocations and broken bones and their casting are mentioned, although the details are not described. An advanced knowledge of obstetrics is demonstrated by the use of the cesarean section. In general, the rule was that the mother's life took precedence over the life of her unborn child, even it desecrated the Sabbath. One account tells of a caesarean section being performed on a woman who had just died in order to maintain the life of her child. Another mentions a caesarean section in which both the mother and child survived. The Talmud also describes several birth defects; according to Rosner, these were probably cases of myelomeningocele.

Although not strictly within the surgical realm, venesection was performed. In keeping with medical practices of the time, bloodletting was valued as health-preserving and was used both as a prophylaxis and as treatment. Several guidelines are elucidated, including the nutritional state of the patient, the regularity of the pulse, the location of the bloodletting, and the recovery period. The Talmud indicates that one should wait for a short period of time before walking following the bloodletting. The tools used to induce blood flow included a small lancet and a nail-like instrument.

It is interesting to note that a cleft lip, the most common facial birth defect, is not mentioned a single time in the Talmud, while many other less common congenital anomalies are described many times. Westreich provides three plausible explanations for this. First, it is possible that the deformity did not exist in the region during the time that the Talmud was being written. Second, it is possible that the congenital anomaly existed but was not recorded. Third, and most likely according to Westreich, it is possible that the cleft lip was seen and recorded but that it was included as part of a larger spectrum of deformities. This spectrum is known as the Sandal, and it is the most common birth defect mentioned in the Talmud. Westreich describes the Sandal as including "cleft lip along with aborted tissue and products of gestation." It is called the Sandal due to the resemblance of the sandal fish to a cleft lip. According to Westreich, if the cleft lip were in fact included in the Sandal spectrum of deformities, then this would be the earliest known historical reference to this defect.

4.2 Anatomy in the Qur'an and Hadeeth

The Qur'an and Hadeeth contain accurate descriptions of anatomical structures, surgical procedures, physiological characteristics, and medical remedies. Notably, both texts place an emphasis on the heart and blood as vehicles for life and as central to emotion and attitude. Furthermore, the lifestyle prescribed by these Islamic traditions promotes longevity of life, prevention of cardiovascular diseases, and avoidance of risk factors associated with such diseases.

4.2.1 History of the Qur'an and Hadeeth

The Qur'an is believed by Muslims to be the direct word of God, revealed to Mohammad through the Angel Gabriel over a span of 23 years (610–632 CE). Although revealed during these years, the transmission of the verses was conducted orally until the text was compiled and canonized the year after Mohammad's death. The exegesis of the Qur'an was carried out by scholars in later centuries, the most popular being Ibn Kathir in the fourteenth century. The Hadeeth are the sayings, rulings, advice, actions, and habits of Mohammed, which are distinct from the direct words of God, and which were also transmitted orally until they were organized into a comprehensive permanent record in the ninth century. Scholars of the time were meticulous in their work and employed stringent rules as to which sayings of Mohammad would be included in the compilation, to ensure accuracy and authenticity. Both the Qur'an and the Hadeeth were used when creating the Islamic law Shariah, "path."

4.2.2 The Cardiovascular System

Various aspects of the cardiovascular system are mentioned in both the Qur'an and the Hadeeth. The two texts discuss the importance of the heart, the blood, and the circulation, and how these are vital to the maintenance of life.

4.2.2.1 Blood and Circulation

Blood is mentioned in several passages of the Qur'an and Hadeeth, generally in relation to lineage and identity, menstruation, the slaughtering of animals for consumption, and embryology.

The relationship between God and man is illustrated in the following verse: "We created man – We know what his soul whispers to him: We are closer to him than his jugular vein." By noting the importance of the internal jugular vein and its connection with the heart, the author of the Qur'an was well aware of the centrality of blood and the heart to the maintenance of life.

Another great vessel mentioned in the Qur'an is the *Al-Aatín* or aorta: "We would certainly have seized his right hand and cut off his *Al-Watín*." *Al-Watín* has been translated into a number of different words, including "aorta," "life artery," and simply "artery." This verse confirms that (i) blood was viewed as a vehicle for life and (ii) the artery directly leading from the heart was known to be vital to survival. By analyzing the different translations of *Al-Watín*, it can be safely assumed that it is the aorta that the author of the Qur'an is referring to in this verse.

Blood is also mentioned numerous times in verses discussing food. For instance, the drinking of blood is forbidden, and it is stated that all of the blood of a slaughtered animal must be drained when the carotid arteries and jugular veins are severed. There seems to be an acknowledgement in the Qur'an that some blood is impure and can contain and transmit pathogens leading to disease. In addition, during menstruation, women are to abstain from sexual intercourse and the ritual prayer, because menstrual blood is considered impure. However, not all blood is impure, as Mohammad distinguishes between menses and blood "from a blood vessel": if a woman's uterine vessels rupture, causing bleeding, the restrictions placed on her during menstruation do not apply. Blood is also mentioned when the Qur'an describes the early stages of the embryo as "congealed blood" or "blood clot."

4.2.2.2 The Heart

The heart is mentioned numerous times in both the Qur'an and the Hadeeth, in many different contexts (e.g., "in the heart," "from the heart"). The repetitive use of the concept illustrates the heart's centrality to the core of every individual. First, the importance of the heart is demonstrated in the fact that we find different states of the heart in the three groups of people that the Qur'an describes: the *mu'minun* (believers) have hearts that are alive; the *kafirun* (the rejecters of faith) have hearts that are dead; and the *munafiqun* (the hypocrites) have hearts that are diseased. The two general types of heart that are described are the spiritual heart and the physical heart. Although there are multiple Qur'anic verses and prophetic traditions regarding the spiritual heart, only a few references are made to the anatomy and physiology of the physical heart as a vulnerable organ vital to the human being. We first see the heart referred to as a muscle in prophetic tradition, where it is stated, "Beware! There is a piece of flesh in the body if it remains healthy the whole body becomes healthy, and if it is diseased, the whole body becomes diseased. Beware, it is the heart."

4.2.3 Contributions to Medicine

The Qur'an and Hadeeth provide detailed, accurate descriptions of the major events that occur during embryological development. For example:

> We [God] created man from a quintessence of clay. We then placed him as a *nutfah* [drop] in a place of settlement, firmly fixed, then We made the drop into an *'alaqah* (leech-like structure), and then We changed the *'alaqah* into a *mudhah* (chewed-like substance, somite stage), then We clothed the bones with *lahm* (muscles, flesh), then We caused him to grow and come into being and attain the definitive (human) form. So, blessed be God, the best to create.
>
> When forty-two nights have passed over the conceptus, God sends an angel to it, who shapes it (into human form), makes its hearing, sight, skin, muscles and bones

Shortly after Mohammad's death, not only did his followers vastly expand the Islamic Empire, but they also became scientific and medical innovators and educators. The Islamic Empire, for more than 1000 years, was the most advanced and civilized empire in the world. The inspiration for all its scientific and medical discoveries and practices stemmed from the teachings of the Qur'an and the Hadeeth – teachings that strongly encouraged and supported the drive to seek knowledge and to make scientific achievements and discoveries. A few centuries after the death of Mohammad, the medical education that developed closely resembled that which we have today. The curriculum consisted of training in the basic sciences, including anatomy (taught by dissecting apes), skeletal studies, and didactics, and clinical training, including therapeutics, pathology, surgery, and orthopedics.

An example of a notable contributor to cardiovascular anatomy during this time was Ibn al-Nafis, a thirteenth-century Syrian physician, who boldly rejected Galen's assertion that there was a direct (but invisible) passage through the interventricular septum between the right and left ventricles. Ibn al-Nafis, who wrote medical, theological, and philosophical works, made his greatest contribution in *Sharh tashrih ibn Sina* ("Explanation of the Dissection of Avicenna"), where he asserted that there was no direct interventricular opening, and outlined, for the first time in history, the pulmonary circulation:

> The blood, after it has been refined in this cavity [i.e., the right ventricle], must be transmitted to the left cavity where the [vital] spirit is generated. But there is no passage between these two cavities; for the substance of the heart is solid in this region and has neither a visible passage, as was thought by some persons, nor an invisible one which could have permitted the transmission of blood, as was alleged by Galen. The pores of the heart there are closed and its substance is thick. Therefore, the blood after having been refined, must rise in the arterial vein [i.e., pulmonary artery] to the lung in order to expand and be mixed with air so that its finest part may be clarified and may reach the

venous artery [i.e., pulmonary vein] in which it is transmitted to the left cavity of the heart after having been mixed with the air and having attained the aptitude to generate the [vital] spirit. That part of the blood which is less refined is used by the lung for its nutrition.

Ibn al-Nafis is but one of numerous examples of scholars working in the tradition of the Qur'an and Hadeeth who made a large contribution to modern medicine. The works of many Islamic physicians and scientists greatly influenced the scientific discoveries and advancements made during the Renaissance.

4.3 Persia

Persia, officially renamed Iran ("the Land of the Aryans") in 1949, was founded by Cyrus II (Cyrus the Great, 599–530 BCE) and his successors, the Achaemenids, in the middle of the sixth century BCE. The early Persians were one of several Aryan tribes that settled in the Iranian plateau in approximately 1700 BCE. These people probably originated from the shores of the Caspian Sea. Later, Persian leaders allied with the Medes, where Cyrus the Great made himself ruler of the so-called Medo-Persian Empire. Extending from Egypt to the Punjab and from the Dardanelles to Samarkand, this empire was the largest the world had seen until the conquests of Alexander the Great.

The practice of surgery became widespread during this period, in which the Code of Hammurabi (the sixth king of the first dynasty of Babylon, 1955–1912 BCE) was established. According to this code, surgeons were subjected to serious penalties for making surgical errors. Such strict laws may have led to more realistic approaches to the practice of medicine and surgery in order to decrease potential iatrogenic errors and subsequent punishment (Figure 4.4). Little is known about the practice of anatomy and dissection during this period, but a model of a sheep's liver with surface segmentations was used for divination and instruction during the nineteenth century BCE (Figure 4.5). However, it is not clear whether this model with cuneiform writing actually belonged to the Babylonians or had been acquired from the Elamites. The earliest use of animal models for the study of human disease is attributed to the Babylonian school of rabbis. Moreover, early information on anatomy, physiology, and pathology using animal models is found in the Babylonian Talmud (Figure 4.2). The number of human bones was counted as 248, and the muscles (flesh) were described. The psoas muscle was mentioned as having two parts, whose respective fibers ran longitudinally and transversely. Rust has stated, "The medical knowledge of the Talmudist was based upon tradition, the dissection of human bodies, observation of disease and experiments upon animals." Hence, this represents the first use of human dissection for medical education in Babylon. Persaud states that notwithstanding their devotion to spiritual pursuits, the Aryans (Persians) developed a rational and secular approach to the practice of healing. He further writes that without the knowledge of anatomy, the diagnosis and treatment of diseases might have been inconceivable in this era.

4.3.1 Following the Establishment of the Persian Empire (Sixth to Seventh Century CE)

4.3.1.1 The Achaemenian (Achaemenid) Dynasty (559/558–330 BCE)

Cyrus the Great (ruled c. 559/558–530 BCE) was the first Achaemenid emperor. Having united the Medes and Persians, he expanded his empire to the west, conquering Babylon in 539 BCE. His rule was characterized by tolerance and respect, and the Achaemenian Dynasty was the first major Persian dynasty to promote the development of science. During this period, records

Figure 4.4 Ancient Babylonian cuneiform script dealing with the Code of Hammurabi. As the first written code of laws, this covered extensive medico-legal issues.

Figure 4.5 Babylonian anatomical model of a sheep's liver from the nineteenth century BCE.

indicate that the bodies of condemned criminals were used for dissection and medical research. According to the Avesta, the holy book of Zoroastrianism, every individual is a microcosm of the vast world, with any of his or her constituents having a counterpart in the universe (macrocosm). The so-called Bundahishn of the Avesta states, "The skin is like the sky, the flesh is like the earth, the bones are like the mountains, the veins are like the rivers, the blood in the body is like the water in the sea, the hair is like the plants, the hairier parts are like the forests." Elgood states that through their development of the microcosm theory, the Persians had established the fundamentals of anatomy, physiology, and pathology long before the Greeks.

4.3.1.2 The Islamic Conquest of Persia (637–651 CE) and the Ascendency of Baghdad (762–1259 CE): The Islamic Golden Age

The Islamic Conquest of Persia took place in the seventh century CE. With the fall of the Sassanid Dynasty, the Academy of Gondishapur persisted for several more centuries. Arabic became the universal language of the empire. The destruction of major libraries by Muslims created intense apprehension among Persian scholars, as significant knowledge of the sciences was lost. For approximately two centuries, a major effort was made to retain the surviving non-Arabic literature by translating it into Arabic, which also indirectly served to preserve the continuity of Persian history. Begun in the Umayyad Period, such conversion flourished and was supported by the Abbasids (see later).

Opinions differ on the effects of Islam on the practice of anatomy. Savage-Smith states that Muslim medical practitioners disavowed human anatomical dissection, due to general cultural and specific religious prohibitions. Wakim opines that while Islamic scholars did little dissection of the human body, they did dissect animals and separate human organs, such as the eye. Abdel-Halim and Abdel-Maguid believe that the practice of dissection was not prohibited in either the religion of Islam or in the Islamic world. These authors write that the knowledge of anatomy was thought by Muslims to deepen the appreciation of God's wisdom and omniscience. Abul Waleed ibn Rushd, a Muslim philosopher-physician, even stated, "Anyone who undertakes dissection increases their faith in God." Anatomy had been so intertwined with theology that Imam Fakhr al-Din al-Razi (1149–1209 CE), a well-known Persian theologian and philosopher, devoted a significant part of his theological treatise to the descriptions of the human body. Moreover, a lack of knowledge regarding the human body was a criterion for the illiterateness of laypersons. Batirel postulates that in the Islamic era, anatomical knowledge progressed through animal dissections, evaluation of human corpses, observations of bones that had been found accidentally, and surgical intervention on patients. However, evidence exists of the direct involvement of these scholars in human dissection.

4.3.2 Abu Zakariya Yuhanna ibn Masawaih, Mesue

The Ibn Masawaih (777–857 CE), known in the West as Masuya, Mesue Major, Mesue Senior, Mesue the Elder, or Jean Mesue, was Persian. This Christian physician was the son of a pharmacist from the Academy of Gondishapur. Mesue left Gondishapur for Baghdad to study under Jibrail ibn Bakhtyashu. He soon became a knowledgeable physician and anatomist, and was appointed director of a hospital in Baghdad. Mesue dissected apes for his anatomical studies and teaching, because of their similarity to humans. Reportedly, he was supported by Calif al-Mutasim, the eighth Abbasid caliph. It has been said that Mesue wished to autopsy his mentally delated son in order to discover the cause of his disability so that physicians could find a treatment for similar patients. Several anatomical and medical treatises are attributed to him, the most important of which are the *Kitab al-Kankash le-Mashajer al-Kabir* ("Textbook of Medical Consultation"), comprising 80 sections, and *Daghal al-Ain* ("Disorders of the Eye"), both written in Arabic. In sections 23–26 of the *Kitab al-Kankash,* Mesue eloquently describes facial nerve paralysis and differentiates between its treatable and untreatable forms. He wrote extensively on embryology and provided the first treatise on diet, *Kitab al-Hawass al-agdiyah.*

Mesue's student Hunayn Ibn Ishaq continued his ophthalmologic studies in Rome, Alexandria, and Persia, and wrote the book *Al-Ashr Maqalat fi al-Ayn* ("Ten Treatises on the Eye"), in which the first detailed anatomical illustration of the eye appears. Hunayn also translated the anatomical writings of Galen.

Wakim states that Mesue and his follower Johannitus were representative of a transition period in Arabic medicine when physicians no longer limited themselves to translations but began to develop original medical studies and practices.

4.3.3 Ali ibn Sahl Rabban al-Tabari

Tabbari (807–870 CE), another major physician of the ninth century CE, wrote *Firdous al-Hikmat* ("Paradise of Wisdom"), which contained extensive information from Greek, Syrian, Persian, and Indian sources, including anatomical information. Al-Tabari came from a well-known Jewish family in Merv of Tabaristan (Old Persia). Due to its contiguity with India, Merv was a strategic city and served as a critical passage between the two countries. Tabbari's father was a clergyman in the court of the king of Tabaristan. Following the anarchy of Merv and murder of its king, Tabbari went to Ray and then to Samarrah (in present-day Iraq). *Firdous al-Hikmat* was written in Arabic in seven sections, and was translated into Syriac. The second section dealt mostly with embryology, explanations of the brain, nerves, heart, vessels, liver, and stomach, and the mechanisms of voluntary and involuntary movements.

4.3.4 Abubakr Muhammad Ibn Zakaria Razi, Rhazes

Rhazes (865–925 CE) (Figure 4.6), also known as Ibn Zakariya, ar-Razi, or Razi, was born in Ray, a city a few miles south of modern Tehran. He studied under a pupil of al-Tabari. He proudly stated, "I never wrote about things unless I first examined them myself." His two major contributions to medicine were the *Kitab al-Mansuri* ("Liber Al Mansuri"), which was dedicated to the Samanid ruler of Ray, and the *Kitab al-Hawi* ("Liber Continens," meaning comprehensive book or encyclopedia). His smaller work was a treatise on smallpox and measles that

Figure 4.6 Painting of Rhazes (Abubakr Muhammad Ibn Zakaria Razi, Rhazes, 865–925 CE).

was considered a masterpiece of medicine for centuries. The anatomy section of *Kitab al-Mansuri* was organized into chapters on (i) the simple organs (bones, nerves, muscles, veins, and arteries) and (ii) the compound organs (eyes, nose, heart, and intestines). This format was later used by Joveini and Jorjani (see later). Rhazes was the first to utilize neuroanatomy in the localization of lesions of the nervous system and to correlate them with clinical signs. He described nerves as having motor and sensory functions, while enumerating seven cranial nerves from the optic to the hypoglossal and 31 spinal nerves. Some of his anatomical descriptions were original, having never been reported before him. For example, he excluded Galen's assumption of the existence of a bone at the base of the heart and was the first to describe the recurrent laryngeal nerve as a mixed sensory and motor nerve.

A pioneer of applied neuroanatomy, Rhazes used the differential diagnosis approach for the evaluation of his patients; an approach that continues to be used in modern medicine. He noted that different diseases might have similar signs and symptoms. He used music to promote healing (music therapy) and was one of the first to appreciate the influence of diet on the function of the body and predisposition to disease. He described tracheostomy and facial plastic surgery in his *Kitab al-Hawi*. He was an alchemist, mathematician, philosopher, and astronomer as well as physician and anatomist. August 27, his birth date, is now proclaimed the Day of the Pharmacist in Iran.

Rhazes rejected several claims made by Greek physicians regarding the alleged superiority of Greek language and medicine. Although an avid follower of Galenic thought, he disputed many of his teachings. For example, he disagreed that the brain, spinal cord, and cerebral ventricles were formed in pairs. He also disagreed that hemiplegia had a cerebral ventricular origin, rather than being related to hemispheric dysfunction. However, Rhazes praised Galen for his anatomical contributions. Interestingly, when Rhazes tested his students, he began with questions on anatomy. If the students failed to respond correctly, they were dismissed, even if they had already passed their clinical examinations, as he felt failure in this subject made them unworthy to practice as a physician.

4.3.5 Abubakr Rabi ibn Ahmad Joveini Bukhari

Joveini (or Akhawaini) (?–983 CE), also known as Abu Hakim, was from Bukhara in Old Persia; he studied under a pupil of Rhazes. He wrote a comprehensive medical book, *Hidayat al-Mutaallimin fi al-Tibb* ("Guideline for Scholars in Medicine," c. 975 CE) in Persian, and this became the first book on medicine in Farsi. Joveini primarily used Persian anatomical terms, and infrequently Arabic translations. Newman wrote that with the *Hidayat*, the Persian language proved itself capable of incorporating or otherwise providing appropriate anatomical concepts and terms. Extraordinary discussions of the eye and nervous system, a symptom-based approach, and its clear style made this book a masterpiece of anatomy and medicine for its time. However, the book obtained little attention from Arab governors.

Joveini wrote in the *Hidayat*, "where these branches [internal carotid arteries] reach the brain, a strange structure is formed, a rete [arteriosus], which supplies the whole brain." Such a description of the arterial circle of Willis could not be found in previous Greek literature. Joveini also mentioned the cardiac innervation by the same cranial nerve that provides the recurrent laryngeal nerve and supplies the alimentary tract. Some have postulated that Joveini had himself practiced the dissection of the human body.

Although some similarities are to be found between the *Kitab al-Mansuri* of Rhazes and the *Hidayat* of Joveini, the latter includes more details on the anatomy of the organs and their functions. Dr. Jalal Matini, a modern Iranian writer, produced a literary edition of the *Hidayat* in 1965, for which he won the annual Iranian Royal Award.

4.3.6 Ali ibn Abbas al-Majusi, Hally Abbas

Ali ibn Abbas al-Majusi (930–994 CE) ("the Magian," meaning that he or his father was of the Zoroastrian faith), was born in Ahwaz, southwestern Persia; he flourished under the patronage of Buwayhid Amir Adud al-Dawla, a ruler of Persia and Iraq who died in 983 CE. Ali ibn Abbas is now regarded as one of the greatest physicians of the Eastern Caliphate. He dedicated to the Adud al-Dawla a medical encyclopedia (Figure 4.7) called "The Royal Book" (*Kitab al-Maliki*, Liber Regius, Regalis Dispositio, Liber Totius Medicinae; also called *Kamil al-Sana al-Tibbiya*) in c. 965 CE, which is more systematic and concise than Razhes' *Kitab al-Hawi*, but more practical than the *Canon of Avicenna*, by which it was superseded.

Written in Arabic, the *Maliki* is divided into 20 discourses. It was translated by Constantine the African and Stephen of Antioch, and was used as a textbook of surgery in schools across Europe. It describes the concept of a capillary system and uterine contractions during birth. Its anatomical descriptions were separately translated by Christians, and were the main source of such knowledge for centuries, even at Salerno. In 1921, Castiglioni stated that the *Maliki* represented the first attempt at the renewal of the art of medicine in the Middle Ages. Ali ibn Abbas introduced several novel surgical techniques, including ones for the removal of spinal tumors and goiters. He was one of the first to disprove that a passage existed between the right and left ventricles of the heart, as had been erroneously stated by Galen and Avicenna (see later). This notion facilitated the emergence of the concept of the pulmonary circulation in the

Figure 4.7 Page from the *Kitab al-Maliki* by Ali ibn Abbas, c. 965 CE. (*Source:* With permission of the Beinecke Rare Book and Manuscript Library, Yale University.)

thirteenth century CE. Ali ibn Abbas wrote that the pulmonary arteries are composed of two layers – an inner layer of oblique fibers and an outer layer of longitudinal fibers – which allow the relaxation and contraction of these vessels.

4.3.7 Abu Ali al-Hussain ibn Abdallah ibn Sina, Avicenna

Ibn Sina (980–1037 CE) (Figure 4.8), also known as Hakim Bu Ali, Abu Ali, Pur Sina, Ibn Sina, Bin Sina, Sheikh or-Raeis, and Sheikh Ali in the Arab world, and as the Latinized "Avicenna" in the West, was born in 980 CE. As a physician, philosopher, mathematician, astronomer, politician, governor, and administrator, he was a towering figure. His contributions to anatomy represent the most important advances in the subject between those of Claudius Galen (120–200 CE) and Andreas Vesalius (1514–1564 CE). Although it is controversial whether or not he directly undertook human dissection, most of his descriptions of the structures of the body were novel for his time. Some authorities believe that he performed dissection in secret. As the writer of the *Al-Qanun fi al-Tibb* ("The Canon of Medicine," c. 1020 CE), he introduced anatomy in a systems-based approach at the beginning of each chapter, followed by a discussion on the diseases of that system. This approach, pioneered by Avicenna, was the template for modern clinically oriented anatomy. He stated, "Pupils should learn the general principles of medicine, analyze the diseases that affect different organs and to do so, must first study the anatomy of these organs."

Avicenna was born in the village of Afshaneh near Bukhara in Old Persia (modern-day Uzbekistan). His family moved to Bukhara when he was five years old. Although his native language was Persian (Farsi), he educated himself in Arabic, which was the formal language of Islamic territories at that time. Avicenna became versed in Islamic laws and various sciences by the age of ten. He continued to study philosophy and medicine, and at the age of 17 cured Nooh Ibn Mansoor, the King of Bukhhara, of an unknown illness that other physicians had failed to cure. At the age of 21, he finished a 20-volume encyclopedia, *Utility of Utilities* (*Al-Hasel va Al-Mahsoul*), which included all the knowledge of the time, except in mathematics.

Figure 4.8 Drawing of Galen (left), Avicenna (Abu Ali al-Hussain ibn Abdallah ibn Sina, 980–1073 CE) (middle), and Hippocrates (right).

After serving as a librarian and then a court physician, Bu Ali left Bukhara for Jurjan and then Ray, Hamadan, and Isfehan, where he wrote his masterpiece book of medicine, the *Al-Qanun fi al-Tibb*, c. 1020 CE. This text was divided into five parts, the first of which describes the anatomy of simple organs. Gerard of Cremona translated the *Al-Qanun fi al-Tibb* into Latin in the twelfth century, and it was translated into Hebrew in 1279. It was the main medical text in Western medical schools until as late as 1650.

Avicenna always recommended that physicians and surgeons base their knowledge on a close observation of the human body. He observed that the aorta at its origin contains three valves, which open when blood rushes out of the heart during contraction and close during relaxation, thus keeping blood from returning to the ventricle. This observation predated William Harvey's demonstration of the circulation of the blood by six centuries. He asserted that muscular movements may occur as a result of the nerves that supply them and that the perception of pain from the muscles may also be due to these nerves. Furthermore, he observed that the liver, spleen, and kidney do not contain any nerves but that nerves do cover these organs. He also provided precise descriptions of six extraocular muscles and the trigeminal nerve. Avicenna distinguished nerves and tendons as different anatomical structures, in contrast to previous interpretations. His study of the eye produced major advances regarding the anatomy and care of this organ. He was also the first physician to popularize the practice of tendon repair. He stated that if an excretory duct of the body becomes occluded, an adjacent gland will then swell. He described in detail the vertebrae and their parts. He also provided correct anatomical commentary on the cerebellum and caudate nucleus.

Avicenna died in 1037 at the age of 58 and was buried in Hamadan, where his grave may still be visited today. He authored more than 400 treatises and books on various subjects, including medicine, philosophy, theology, and logic. August 23, his birth date, is celebrated as the Day of the Physician in Iran. For his leading role in the sciences and in politics, he received the title *Sheikh al-Rais* or Master the Elder. A crater on the moon is named in his honor.

4.3.7.1 Avicenna Traveling in Persia

Avicenna's life is highlighted by his dangerous travels to a number of Persian cities. Although most of these travels were made for political reasons, they also provided him with an opportunity to access libraries, such as the Khwarizmi Library of Gorganji, the Buwayhid Libraries of Ray and Hamadan, and the Kakuyid Library of Isfahan.

With the fall of the Samanid Dynasty, Avicenna fled Bukhara for Gorganji of the Khwarizmi Kingdom, where he met Abu Reyhan Biruni, a great pharmacist of the day, and Abu Sahl Masihi, a knowledgeable physician. In 1017 CE, the Ghazan ruler, Sultan Mahmood, overthrew the Khawarizmi Dynasty. Although Sultan Mahmood had called for him, Avicenna was unwilling to joint his court and fled to Jurjan; from there, he went to Ray, Qazvin, Hamadan, and Isfahan. Avicenna was given a home by Prince Abu Mohammad Shirazi in Jurjan, a city located in the southeastern part of the Caspian Sea, where he began to write his masterpiece *Al-Qanun*. While residing in Ray, near the modern city of Tehran, Avicenna treated the Buwayhid ruler Sultan Majd al-Douleh, the son of Fakhr al-Douleh, for depression, and reportedly served the court as a business manager. There, Avicenna was appointed a minister by the young sultan. Displeased with this appointment, Sayyida, a very influential and powerful figure of the period and the mother of the sultan, rejected Avicenna. Following this, Avicenna arrived in Hamadan, another Buwayhid state, where he treated Sultan Shams al-Douleh, the older son of Sayyida, for severe colic. It has been said that he spent 40 days and nights beside Sultan Shams al-Douleh in order to cure him. Sultan Shams al-Douleh soon appointed him his minister. In 1021 CE, following the sultan's death, Avicenna refused to continue his court duties under his son, Taj al-Mulk. Charged with betraying the Buwayhids and secretly corresponding with the Kakuyid ruler of Isfahan, Sultan Aala al-Douleh, Avicenna was sent to prison. However, he escaped in

1024 CE and traveled to Isfahan, where he prospered under Sultan Aala al-Douleh. Later, when the sultan moved to Hamadan, Avicenna accompanied him.

Ultimately, at the age of 58, Avicenna died of colic a few days after arriving in Hamadan in 1037 CE. His student and biographer, Juzjani, illustrated his master's illness, which included protracted and recurrent attacks of colic, seizures, and intestinal bleeding. Avicenna treated himself with enemas and used celery seeds to eliminate his bloating. Elgood cites Ducastel as attributing Avicenna's death to gastric cancer based on his course of illness.

Avicenna was described as very intelligent, so much so that a popular Persian legend describes his arguing against Satan and outwitting him. A story also tells that Avicenna fell in love with the daughter of a king, who exiled him from the city once the romance was discovered. Such legends are proof of Avicenna's influence over the public as a brilliant, ingenious, and extraordinary personality. In his will, and in a letter to Abu Said Abu Al-Khayr, a Persian poet and Gnostic, Avicenna reflected his philosophy of life: to bear adversities and difficulties, help mankind, and seek a proper and genteel science in order to satisfy God.

4.3.7.2 Al-Qanun fi al-Tibb (The Canon of Medicine)

Al-Qanun fi al-Tibb is regarded as "the most famous single book in the history of medicine both East and West," according to the *Encyclopedia Britannica*. The Latin translation was made available by Gerard of Cremona in the twelfth century and remained an essential teaching book in European universities up until the seventeenth. *Al-Qanun*, which has also been translated into Hebrew, German, French, and English, consists of five books: Book I is dedicated to general anatomy and principles of medicine; Book II, matrica medica; Book III, diseases of the special organs; Book IV, general medical conditions; and Book V, formulary. In Book III, each chapter begins with a brief account of anatomy, followed by a list of signs and symptoms related to the diseases of a specific organ. *Al-Qanun* is such an influential treasure in the history of medicine that Sir William Osler credited it as a "medical bible" and "the most famous medical textbook ever written."

Avicenna had a vision of blood circulation, but he erroneously accepted the Greek notion regarding the existence of a hole in the ventricular septum by which the blood traveled from the left to the right ventricle. He ignored the pulmonary circulation, which was later described by an Arab physician, Ibn Nafis, in the thirteenth century CE. Interestingly, in describing cardiac morphology, he essentially followed the teachings of Aristotle, rather than Galen, on the three-chambered nature of the heart. Beginning in the sixteenth century, the anatomy of *Al-Qanun* became increasingly criticized by some Western scholars, such as da Vinci and Paracelsus. At about the same time, Lorenz Fries of Colmar, himself a physician, wrote a treatise in defense of Avicenna, highlighting the important influence that he had on the progression and preservation of medicine. Andreas Vesalius and William Harvey both read *Al-Qanun* and Harvey mentioned Avicenna is his treatise, *An Anatomical Disquisition on the Motion of the Heart and Blood in Animals*. Although some of Avicenna's anatomical descriptions were erroneous, he correctly wrote on the cardiac cycles and valvular function. It may be worth mentioning a quotation from Michelangelo, the Italian sculptor, who also studied anatomy: "It is better to be mistaken following Avicenna than to be true following others."

4.3.8 Zinn-ol-Abedin Seyed Esmail ibn al-Hussain ibn Mohammad ibn Ahmad al-Jorjani, Hakim Jorjani

Esmail Jorjani (1042–1137 CE), born in Jorjan, northeastern Persia, was a physician who served in the court of Khwarazm Shah Qutb al-Din Muhammad ibn Nushtikin, the governor of the Persian province of Khwarazm, and his successor, Atsoz. He flourished under Abd al-Rahman ibn Ali ibn Abi Sadegh, a former pupil of Avicenna and an influential physician of the Khwarazmi

Dynasty (1077–1231 CE) in Nishapur, before going to Khwarazm. His comprehensive textbook of medicine, *Zakhireyei Khwarazmshahi* ("The Treasure of Khwarazm Shah," c. 1112 CE), is considered the oldest medical encyclopedia written in Persian; it was later translated into Hebrew and Turkish.

Zakhireyei Khwarazmshahi is composed of 10 books, the first of which contains extensive anatomical descriptions containing almost all the knowledge of human body structures of its time. Jorjani criticized certain details offered by Galen and Avicenna. On the anatomy of the optic nerve, Jorjani, unlike Avicenna, agreed with Galen that it goes at its periphery to the ipsilateral rather than the contralateral eye. In the fifth book of *Zakhireyei Khwarazmshahi*, Jorjani elegantly described three body fluid compartments: the intravascular and the interstitial, and a third within the material of the tissues (intracorporeal). He was also the first to mention the association between a goiter and exophthalmia.

Al-Aghraz-o-Tebbieh and *Khofieh Alaii* are two synopses of *Zakhireyei Khwarazmshahi* created by Jorjani himself for use as manuals or handbooks by medical students. *Zobdat al-Tibb* ("Master of Medicine"), written by Jorjani at an uncertain date, is also a treatise on anatomy and medicine. Jorjani later moved to Merv, the capital of the Seljug sultan Sanjar ibn Malikshah, where he died.

4.4 From the Mongol Invasion of Persia to the Foundations of Modern Anatomy

In the centuries following the Mongolian conquest (c. 1220 CE), the conquest of Baghdad (1258 CE), and the end of the Abbasid Caliphate, three power centers emerged in the Middle East: Persia, Turkey, and Egypt. A line of Mongols called the Ilkhans ruled Persia. The Ilkhanid Dynasty was based in three capitals: Tabriz and Maragha (both in Azerbaijan, presently northwestern Iran), and Baghdad. The Ilkhanids brought about a new political and social tolerance that allowed the arts and sciences to flourish and led to the assertion of prophetic medicine. Elgood postulates that this change facilitated the emergence of the first color-illustrated anatomy text in Persia, which previously, because of religious restrictions, would have had no opportunity for dissemination. Furthermore, Persian–Chinese trade and scientific exchanges, which occurred via the Silk Road and lasted for more than a millennium, increased in this era. Unfortunately, astrology and magic became sources for the interpretation of some anatomical studies. Chinese anatomical ideas were made available through such works as the *Tansuq-nama-yi Ilkhani dar Funun-i Ulum-i Hata* (1314 CE) under the auspices of Rashid al-Din Fazlallah Hamadani. Medical writings were often as much theological and political as scientific.

4.4.1 Nasir al-Din Tusi

Jafar Mohammad ibn Mohammad ibn al-Hasan al-Tusi (1201–1274 CE), also known as Muhaqqiq-i Tusi, Khwaja-i Tusi, and Khwaja Nasir (Figure 4.9), stated hundreds of years before Darwin that "organisms that can gain new features faster are more variable. As a result, they gain advantages over other creatures." Being a great theorist of his time, Tusi developed primary ideas regarding human evolution, hereditary variability, and the conservation of mass. He believed that humans were derived from advanced animals. Tusi was born in Tous in Khurasan province, currently in northwestern Iran, to a family of Shiite jurists. He initially studied religious law, and under the influence of his uncle went to Nishapur to learn logic, physics, mathematics, and metaphysics. His youth was concurrent with the Mongol invasion of the Islamic area (c. 1220 CE), which is now regarded as one of the most devastating events

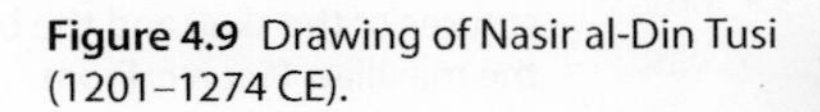

Figure 4.9 Drawing of Nasir al-Din Tusi (1201–1274 CE).

in the history of Persia. Under the rule of Genghis Khan, the leader of the invading Mongols, thousands of innocent people were killed and many libraries were burned. Later, in the middle of the thirteenth century, Tusi advised Hulegu Khan, the grandson of Genghis Khan. Hulegu was deeply impressed by Tusi's knowledge and appointed him as a minister. At least 64 treatises have been recorded for Tusi, including on astronomy, algebra, arithmetic, trigonometry, medicine, metaphysics, logic, ethics, and theology. Tusi died in Kazemain, near Baghdad, in 1274 CE.

4.4.2 Mansur ibn Muhammad ibn Ahmad ibn Yusuf ibn Ilyas, Ibn Ilyas

The color-illustrated anatomy text of Mansur (c. 1380–1422 CE), *Tashrih-i Mansuri* ("Mansur's Anatomy"), also known as the *Tashrih-i Badan-i Insan* ("Human Anatomy"), was compiled in the fourteenth century CE (Figures 4.10–4.15). This Persian book was dedicated to Pir Muhammad ibn Umar ibn Timur, the ruler of the province of Fars, who was the grandson of Timur (Tamerlane). It consisted of seven sections: an introduction, five main chapters, and an appendix on the formation of the fetus and organs. Mansur used both Arabic and informal Persian terms. The main chapters systematically reviewed human bones, nerves, muscles, veins, and arteries, and each contained a diagram. Controversy exists regarding the origin of these illustrations. Some copies of *Tashrih-i Mansuri* include one or two additional illustrations: one showing a pregnant woman, with the fetus lying in a breech position within the uterus, and one illustrating a naked female body. Some authorities believe that these later illustrations may indeed be Persian (belonging to Mansur), as the body is of a rather oblique orientation not seen in previous Greek and Roman works. Moreover, earlier Latin illustrations

و تاج یا بر شکل لام است و درز دوم را سهمی گویند و آن درزی است مستقیم که منصف سر است در طول
او را از آن جهت سهمی گویند که شبیه تیر است در کمان و هرگاه که انضمام سهمی با اکلیلی ملاحظه کنند از اسکودی
گویند منسوب به سفود که آن آلت گوشت بریان کردنست و درز سوم را لامی خوانند بنابر آنکه
شبیه لام است در خط یونان و بر شکل دال است در خط عربی و درزهای کاذبان که ایشان را قشرین
گویند دو درزند که در میروند در جانب طول موازی درز سهمی از یمین
و یسار و درین شکل مجموع ظاهر میگردد و فک اعلی
مرکب است از چهارده استخوان و درزها میان این استخوانها بر آنکه
است هر جیشی سه عدد و کناره دو و عدد که محیط است باصول دندان بغیر ثنایا و رباعیات علیا
و دو عدد کوچک که در ایشان سوراخی نافذ هست از منخرین بدهن و انف دو عدد و اجزای او مختلف
افتاده در صلابت و لین بنابر آنکه عظام و چنین ضرورت صلابت او جهت محافظت و عظام
انف ضرورت که رخو و متخلل باشد تا رسیدن روایح بدو و در شدن فضلات دماغی از و آسان
بود و شکل اجزای او از مثلث و مربع و منحرف الاضلاع ازین صورت معلوم میشود که بعد ازین گفته
می آید و چون فک اعلی محاذی قم دماغست و فضلات دماغی بسیار از و منصب میگردد ضرورت بود
که درزها و مفاصل بسیار باشد بخلاف فک اسفل و دیگر آنکه در فک اعلی
حرکت مطلوب بود تا بر حرکت بسیار رود در فک اسفل همین بس باید که فک اعلی
اخف از اسفل باشد فک اسفل مرکب است از دو عظم و بعضی گویند یکیست
و در میان ایشان مفصل موثق خفیست که از [illegible] خوانند و این دو استخوان
از طرف اعلی دو شعبه دارند یکی باریک و در نشیب هر دو عظم [illegible] واقعست و
متصل بدوست و تری از عضله صدغ و شعبه دوم غلیظ و مستدیر الراس
افتاده و در میرود در نقره که در تحت دائره [illegible] است و در فکین سی و دو دندان مرکوز است از طرف
بالا دو ثنیه و دو رباعی و دو ناب و پنج اضراس از هر طرف و از نشیب همچنین ثنیتین و رباعیتین جهت قطع و
دو ناب جهت کسر و از هر طرف پنج اضراس جهت طحن و نواجذ بعضی را نباشد و اسنان را شعبی چند هست
که در میرود در سوراخ فکین و زائده محیط است بر ایشان و غیر از اضراس یک شعبه دارند اما اضراس
آنچه فوقانیست از ایشان سه شعبه دارد یا چهار و آنچه تحتانیست دو و بعضی بر آنند که دندان عظام نیست
و پیش جالینوس از عظامست و مخصوص است بحس فقرات عنق فقره استخوانیست که در میان آن
سوراخی بود که نخاع در آن گذرد و هر یک از ایشان زوائد متصل دارند که بدان منتظم و مرتبط میگردد و در بعضی
فقره و در بعضی [illegible] موجود یعنی بیکدیگر متصل گردد بهم تا بسبب استحکام شود و این در بعضی چهار باشد و در

Figure 4.10 Excerpt from Mansur's *Tashrih-i Badan-i Insan* describing the sutures of the skull and the bones of the maxillae. (*Source:* From Choulant 1852.)

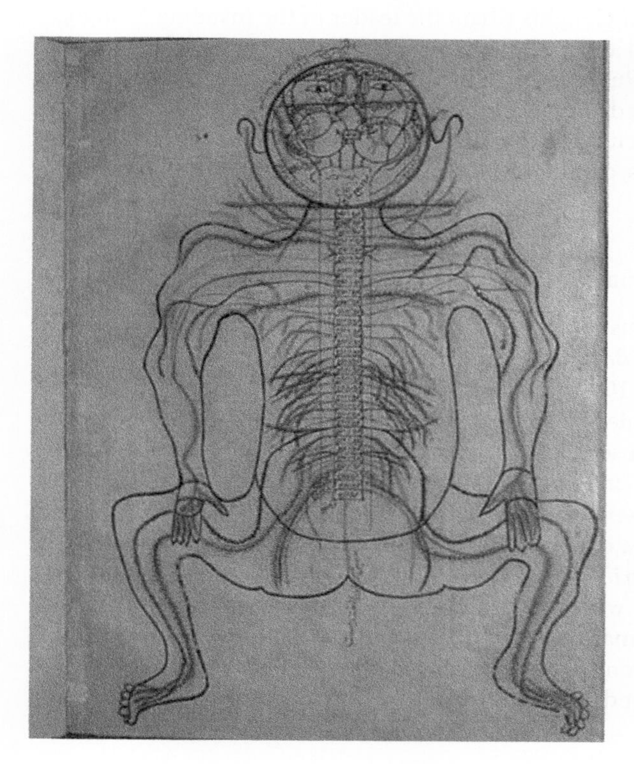

Figure 4.11 Drawing of the nervous system from Mansur's *Tashrih-i Badan-i Insan*. (*Source:* From Choulant 1852.)

Figure 4.12 Drawing of the venous system from Mansur's *Tashrih-i Badan-i Insan*. (*Source:* From Choulant 1852.)

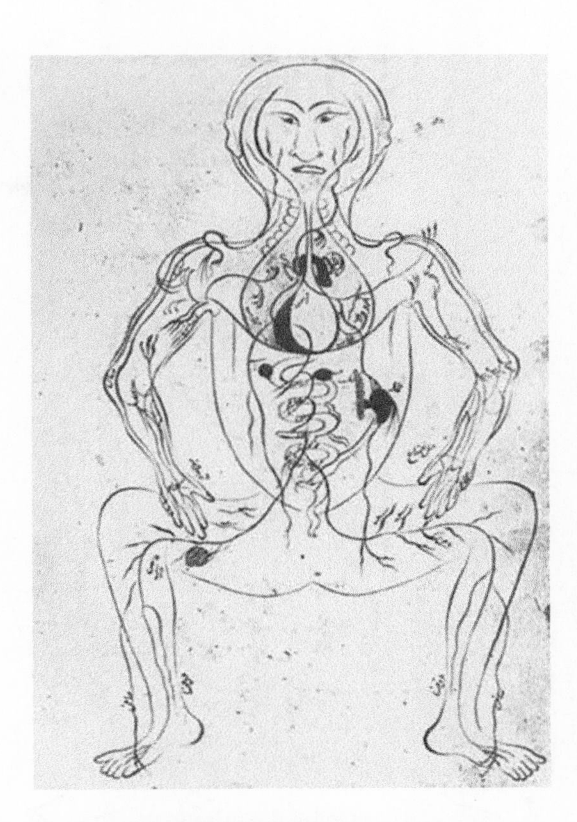

Figure 4.13 Drawing of the skeletal system from Mansur's *Tashrih-i Badan-i Insan*. (*Source:* From Choulant 1852.)

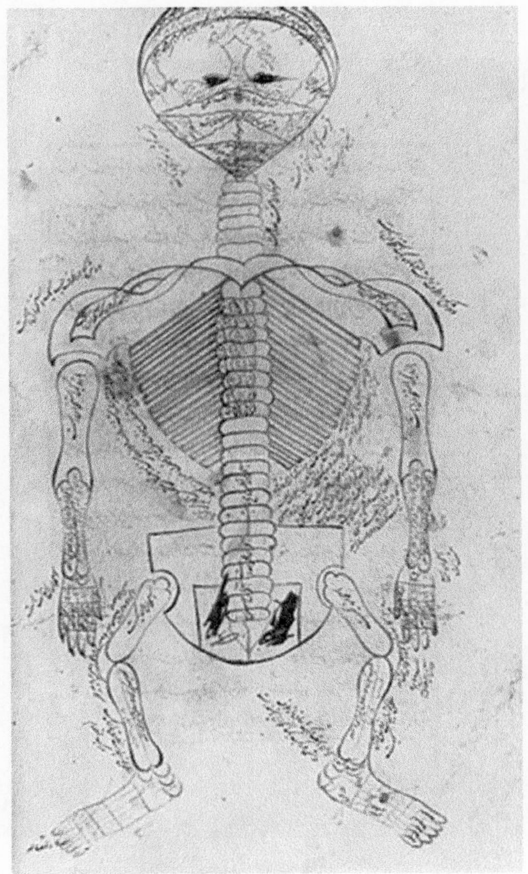

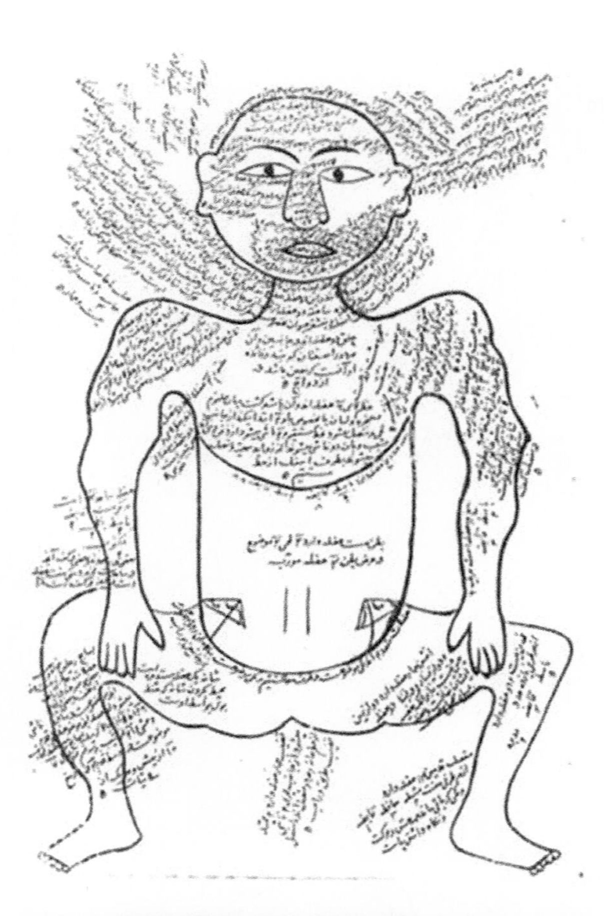

Figure 4.14 Drawing of the named skeletal muscles from Mansur's *Tashrih-i Badan-i Insan*. (*Source:* From Choulant 1852.)

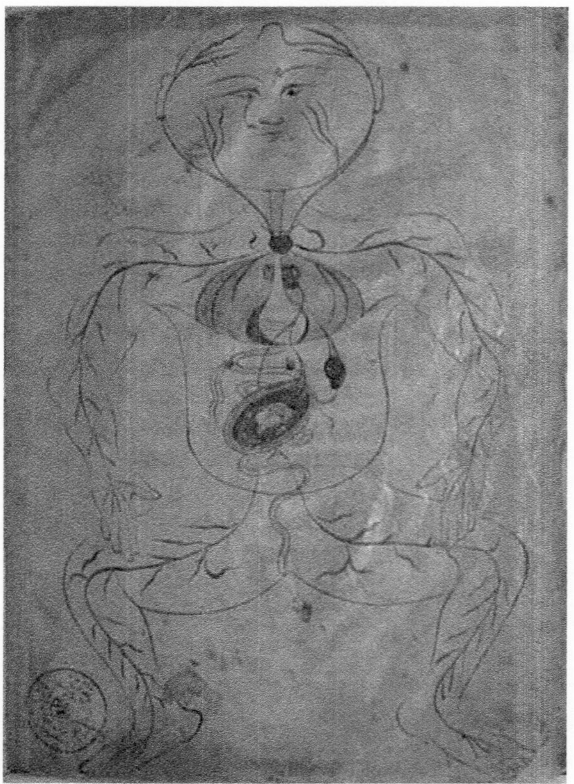

Figure 4.15 Drawing of a pregnant woman from Mansur's *Tashrih-i Badan-i Insan*. (*Source:* From Choulant 1852.)

did not include figures of pregnant women. Mansur organized the chapters on organs in a hierarchical order according to their perceived vital role, with the cardiovascular and reproductive systems as the first and last chapters, respectively.

A unique feature of *Tashrih-i Mansuri* that is rarely found in earlier Persian and Islamic medical writings is its detailed chapters on human embryology. Mansur believed that the shortest gestational age at which a fetus could survive was six months. He rejected the Hippocratic assertion that the brain is the first part of the body formed in the fetus. He agreed with Avicenna that the most important organ of the body is the heart and that this is the source of natural heat and life forces. Occasionally, the book employs poetic descriptions of body structures.

Some authorities believe that *Tashrih-i Mansuri* was a transformation of Galenic dissection to a prophetic tradition aimed at influencing the attitudes of the general public through extensive references to the Qur'an and Islamic leaders. Newman opines that the importance given to the prophetic tradition in *Tashrih-i Mansuri* can be partly explained as an effort to demonstrate the strengths of Islam, and thereby to influence governors to religious tolerance. Newman hence divides the anatomy of Persia into two eras, before and after Mansur, corresponding to the tradition of Galenic dissection and prophetic knowledge. *Tashrih-i Mansuri* was popular in Persia for at least two centuries, and was quoted in the Persian anatomical treatises of Abi al-Majd al-Tabib al-Bayzavi, and Abd al-Razzaq (Khulasat al-Tashrih).

Mansur came from a family of scholars and physicians active for several generations in the city of Shiraz, in the central province of Persia, Fars. Before writing his anatomy text, Mansur wrote a general medical compendium, *Kifaya-yi Mujahidiya* or *Kifaya-yi Mansuri*, which contained a short anatomical section based on Galenic dissection and was dedicated to Sultan Mujahid al-Din Zyn al-Abedin, the last Muzaffarid ruler of Fars before the conquest of the Timurids. Shah Shoja ruled in Fars between 1384 and 1391, and from the dedication of *Kifaya*, it is not unlikely that Mansur was a court physician to this king. Mansur gained his experience in Tabriz but also traveled to many different Eastern countries.

Mansur wrote:

> Though Aristotle's assertion that the heart is the first to develop is from observation and deduction, for those who do dissection, the observation is more superior and better than deduction. And as deduced, semen contains many airy elements and corporeal heat too, and the first thing derived from it is the essence of spirit because its development is easier, and its demands are high. Hence, the essence of spirit is formed and collected first. As the spirit is a flowing existence, from the viewpoint of physicians, it does not stand on itself, but requires a container ... Hence, the first organ, which is developed, is the container of spirit; that is the heart.

For the arteries, he wrote:

> The arteries, the moving vessels, originate from the left ventricle of heart and follow it in contraction and expansion. The arteries are composed of two layers: the inner layer is dense as it is the true receptacle of spirit. Its fibers are circumferential. Hence, the contraction, which expels the exhaled residue, takes place by this layer. The outer layer is composed of oblique and longitudinal fibers and the absorption of air takes place here. However, the venous artery has only one layer and goes to the lung. Know that two moving vessels originate from the left heart ventricle: one is small and composed of one layer and this is the venous artery. The venous artery goes to the lung. The other vessel, which also originates from this side, is large and is called the aorta. After arising from the heart,

> this vessel divides into two branches: the small one goes to the right-sided ventricle and the other branch encircles the heart and supplies its parts. The remaining segment of this vessel divides into two branches. A descending artery is larger because the organs located below the heart are numerous and also larger than those located above the heart. The ascending artery divides into two branches, much of which go to the liver.

On the complex organs, he wrote:

> The heart, the noblest organ, is composed of multi-positioned fibers, hard flesh, and a thick membrane. The heart has two cavities: one on the right side and the other on the left side, both moving continuously. The origin of the arteries is from the left side. The right cavity has two membranous passages. From one the blood comes to the heart from the liver and from the other, blood goes from the heart to the lung. The left ventricle also has two passages: one is for the entry of air from the lung to the heart and the other is the commencement of pulsating vessels or arteries. The hole between the right and left cavities is wider on the right side and gradually becomes narrower as it goes to the left side. Through this hole, pure blood flows from the right to the left side. Each of the two cavities has two prominences. Surrounding the heart is a membrane for its protection, which is called the cardiac sheath. This sheath is separate from the heart, so that when the sheath is damaged, the heart remains intact.

4.4.3 The Safavid Period (1501–1722 CE) and the Reconciliation of Prophetic and Galenic Traditions

One and a half centuries after the Ilkhanids, the Safavid Dynasty began its influence with the capture of Tabriz from the ruling Sunni Turkmen alliance known as the Ak Koyunlu (the White Sheep Emirate). As the heir of Sufism, the Safavids imposed Shia as the state religion. Sadly, this was a time of gradual scientific withdrawal. As an example of the stagnation of Safavid medicine, Elgood writes, "this period was characterized by writings with no new ideas and in which no ancient theory was contested." Two books of note are *Kholasat al-Tajareb* ("The Quintessence of Experience") by Mohammad Baha al-Dawla (?–1501) and *Bihar al-Anvar* ("Oceans of Light") by Mohammad Baqir al-Majlisi (1626–1698), representing early and late medical theory and practice, respectively, during the Safavid Period. Comparisons between these works highlight the scientific trends of this era. While the former primarily dealt with the author's own experiences and contained novel descriptions of such diseases as whooping cough, febrile skin eruptions, and syphilis, the latter focused on achieving a reconciliation between the Shia variant of prophetic medicine approved by the imams and Galenic tradition.

Mohammad Baqir al-Majlisi (1628–1698) (Baqir) was born in Isfahan into a Shiite family. His influence over Shah Sultan Husayn was substantial, as demonstrated by his appointment as Mulla Bashi (Senior Clergyman). Baqir attempted to eradicate Sufism and Sunni Islam and to persecute Zoroastrians, Jews, and Christians. In the West, he was known as one of the most powerful and fanatical mujtahids (Islamic leaders). He compiled the first encyclopedia of Shiite jurisprudence, the *Bihar al-Anwar* ("Oceans of Light"). Chapter 48 of this text, which is entitled "On What al-Hukama [Philosophers] and al-Atibba [Physicians] Say on the Anatomy of the Body and its Parts," is arranged into seven sections on the simple and the compound organs, followed by the organs of hearing, the neck and spinal cord, the thoracoabdominal system, the reproductive organs, and the number of bones. Interestingly, tendons, ligaments, membranes, and cartilage are added to the organs of *Tashrih-i Mansuri* as additional simple organs. The *Bihar*'s anatomical descriptions are in agreement with those of Hellenic, Greek, and the earlier

Sunni Islamic texts of Avicenna and Mansur on different body structures. Baqir al-Majlisi, although known in the West as a Shia extremist, never stressed a single anatomical tradition, but rather applied a liberal approach, which according to Newman highlighted the points of compatibility between the prophetic and Galenic traditions and provided the basis for an anti-radical discourse. Al-Majlisi stated that the most important organs of the body are the brain, the heart, the liver, and the organs of procreation (or reproduction). He wrote that the heart contains innate heat and is the absolute master of the organs.

Jurists of earlier Islamic times did not directly address the issue of human dissection. Perhaps the first jurist who wrote specifically about dissection and its permissibility was Allamah Ahmad ibn Abd al-Munim al-Damanhuri, a highly educated man who died in 1778 CE. He wrote profusely on jurisprudence, medicine, astronomy, and surveying. His treatise on anatomy, *al-Qawl al-Sareeh fi Ilm al-Tashrih* ("The Explicit Word on Dissection"), which he later expanded into a book entitled *Muntaha al-Tashrih bi-Khulasat al-Qawl al-Sareeh fi Ilm al-Tashih* ("The Pinnacle of Dissection with the Quintessence of the Explicit Word on Dissection"), provided the basis for the future of modern medical schools and dissection classes in the Islamic countries. The term *al-musharrihon* ("dissectors") has frequently been used in the Persian-Arabic medical literature.

In addition to the texts by known authors, several anonymous Persian anatomical treatises and drawings come to us from various periods. A set of six anonymous figures dated to the late eighteenth century is provided at the end of the book *Tibb al-Akbar* ("Akbar's Medicine"), written by Mohammad Akbar Arzani ibn Mir Mohammad Mogim (or Akbar Shah). This book is an amplified translation of an Arabic work by a Persian scholar, Nafis ibn Ivad Kirmani. At least 40 anatomical manuscripts dated between the thirteenth and mid-nineteenth centuries and written by Persian scholars are available in the national libraries of Iran.

4.5 Baghdad

4.5.1 Ibn Jazlah

The eleventh century was culturally and medicinally one of the most exciting periods in the history of Islam. The medicine of this time was influenced by the Greeks, Indians, Persians, Coptics, and Syriacs, and one of its most prolific writers was Ibn Jazlah (c. 1030–1100 CE), who resided in Baghdad in the district of Karkh. Ibn Jazlah made many important observations regarding the pathology of the brain and spinal cord.

During the Abbasid Period (750–1258 CE), Arabs and Muslims across North Africa, the Iberian Peninsula, and the Middle East developed a civilization that rivaled the sophistication of earlier cultures such as those of the Syriacs, Persians, and Greeks. Islamic civilization began its worldwide influence in the city of Baghdad, which was built by Calif Al-Mansur in 762 CE and was proclaimed the capital of the Abbasid Caliphate, with legal authority over all Islamic territories. During this period, much attention was directed to science and to the translation of Greek and non-Greek literature so that the information it contained could become widespread throughout Islamic territories. The Abbasid strategy of encouraging translation led to the flourishing of the so-called Islamic Golden Age. Subsequently, several hospitals were built, primarily in Baghdad.

Abu ali Yahya ibn Isa Ibn Jazla Al Baghdadi or Ibn Jazlah, also known as Bengesla, Buhahylyha, and Byngezla, was an important eleventh-century (c. 1030–1100 CE) Arab physician from Karkh, a district of Baghdad on the west side of the Tigris river. Although little is known of his life, Ibn Jazlah was born to Christian parents but converted to Islam in 1074 CE. He embraced

Islam under the tutelage of Abi Ali bin Al-Walid Al-Maghribi, the chief of the Mur'tazalite sect. Ibn Jazlah trained in medicine under Said ibn Hibat Allah, the physician to the Abbasid caliph al-Muqtadi (reigned 1075–1094 CE), and later became physician to this ruler himself. His most important work was a medical synopsis comprising 44 tables that described and outlined the treatment of over 350 diseases, the *Taqwim al-Abdan fi Tadbir al-Insa; Dispositio Corporum de Constittutione Hominis, Tacuin Agritudinum* ("Tables of the Bodies with Regard to their Constitutions"); it was probably modeled upon a similar work by Ibn Butlan (first half of the eleventh century). Latin (translated in 1280 by the Sicilian Jewish physician Faraj ben Salem) and German (translated by Farragut) versions of the *Taqwim* were published in 1532 and 1533, respectively. Ibn Jazla also wrote another work, *Al-Minhaj fi al-Adwiah al-Murakkabah* ("Methodology of Compound Drugs"), which was translated by Jambolinus and was known in the Latin as the *Cibis et Medicines Simplicibus*. This latter work was what produced his fame as a pharmacist in Baghdad.

Ibn Jazlah was known for his care and generosity toward the poor, and for treating friends and relatives, and providing them with medications, free of charge. He wrote for al-Muqtadi (caliph from 1075 to 1094) an alphabetical list of simple and compound medicines called *Minhaj al-Bayan fi Ma Yasta 'Miluhu al-Insan; Methodica Dispositio Eorum, Quibus Homo Uti Solet* ("The Pathway of Explanation as to that which Man Uses"). Although this work was compiled as a medical text, some have highlighted its culinary contributions. For his writings, Ibn Jazlah stated that he consulted the earlier works of Hippocrates, Dioscorides, Rufus, Oribasius, Paul of Aegina, Ishaq, Hunayn, Rhazes, Galen, and al-Majusi. He was perhaps most influenced by Rhazes (865–925 CE). Interestingly, he regarded music as important in the treatment and prevention of diseases, stating "the effect of music on ailing psyches resembles that of medicines on ailing bodies." He was a contemporary of Esmail Jorjani (1042–1137), who was renowned for his contributions to medicine and anatomy, especially regarding the cranial nerves. Other books by Ibn Jazlah include *Al Ishara fi Talkhis al-i'bara, Rissala fi Madh Tib wa Mouafakataho li Shara* and *Rissala fi Ar-rad ala Annasrania*. In the introduction of his *Taqwim al-Abdan*, he stressed hygienic measures as a means of preventing disease and promoting good health. He emphasized the necessity of medical care for a happy life, as well as for the afterlife:

> The existence of both worlds (this world and hereafter) depends on the desire to correct them, as stated in the tradition (*khabar*). It is best for one not to ignore or abandon either this world for the hereafter, nor the hereafter for this world, but to benefit from both ... If the preservation of health per se is not intended, but rather for the sake of work and knowledge, then man is considered as a tool or means to them [work and knowledge]. If he becomes servant to them, then it would be undesirable to devote time to it [health] more than is required.

Neurologically, Ibn Jazlah, in his *Taqwim al-abdan*, stated that brain diseases can be attributed to headache, vertigo, and disorders of the humors such as in encephalitis, memory loss, dizziness, lethargy, and nightmares (of the 44 sections, three were dedicated to the nervous system, relating to headache, diseases of the brain, and nervous diseases). Headaches he attributed to both local and systemic pathology. Specifically, if the headache was over the entire head, then this was due to disease at remote sites, but if it stemmed from the head itself, then it was due to heat, cold, dryness, or moisture. Ibn Jazlah also found that some cases of encephalitis were associated with phlegmatic issue. Surgically, he recommended cephalic-vein bloodletting (*al-Istifragh*) in cases of vertigo, as he attributed the dizziness to the brain. Specifically, he believed that it was caused by the arteries surrounding the brain. Interestingly, the bloodletting performed on such patients was done after first cooling them.

Ibn Jazlah believed that epilepsy was caused by an obstruction of the cerebral veins and an arrest of proper circulation to the brain. He noted preictal symptoms such as headache, visual obscurity, and weakness. The treatment for such spells was bloodletting from the great saphenous vein, drinking of boiled water laced with anise, mastic, and oxymel or fennel, or resting for three days and then drinking oxymel in hot water with salt and fat. The patients was then to have their head bandaged with a combination of chamomile, thyme, mint, clove, and ginger and to be fed garbanzo bean soup and bread. He also used ointments in the treatment of nervous diseases, including almond oil, myrrh, walnut oil, the fruit and leaves of laurel, and marjoram. He observed that seizures were more common in the elderly and in colder climates. He avoided the use of sedatives in the treatment of diseases of the nervous system.

Ibn Jazlah also concurred with Hippocrates that patients with significant stroke would rarely recuperate. He attributed fainting to bad "humors" of the vessels supplying the brain, although the term "humors" was not defined or explained. Diseases of the central parts of the brain were known to prohibit movements. Pathology of the lateral aspects of the brain was described as affecting the limbs and organs. Loss of speech was observed following blows to the head.

Ibn Jazlah recognized that diseases of the spinal cord often resulted in paralysis and numbness, and stated, "When the disease occurs in the beginning of the spinal cord this results in looseness of the body limbs but not the face"; he noted, again, that this "is more common among elderly people and more prevalent in cold-moist climates during winter." Other symptoms of such pathology were poor movements and tremor. Ibn Jazlah believed certain afflictions were more common during particular seasons. For example, he believed joint pain and sciatica to be more common in the fall, and vertigo to be more common among adults in hot climates during the summer. He distinguished sciatic pain from joint pain in that it traveled to the knee and ankle joints. He treated numbness and paralysis due to spinal cord injury/pathology in the same way, by having the patient drink various medicinal concoctions that often included Chinese herbs. Furthermore, he found that if any part of the facial nerve were affected, then the part of the face it served became paralyzed. He identified injuries to the laryngeal nerves as resulting in loss of voice and that injury to the nerve near the chest muscles (phrenic) resulted in difficulty breathing. He reported traumatic nerve injuries, and found that when "looseness" was appreciated, such injuries were not curable. Additionally, he differentiated joint pain from "sciatic" pain due to disc disease.

Ibn Jazlah bequeathed his entire book collection to the famous mosque of al-Imam Abu Hanifah in Baghdad, where he was buried. This early figure in medical history is remembered for his prolific contributions to Islamic medicine. While some of his conclusions regarding the nervous system were erroneous, the greater part were correct.

Bibliography

Anatomy in the Bible and Talmud

Bernstein, A. and Bernstein, H.C. (1951). Medicine in the Talmud. *Calif. Med.* 74: 267–268.

Budrys, V. (2007). Neurology in holy scripture. *Eur. J. Neurol.* 14: e1–e6.

Dan, B. (2005). Titus's tinnitus. *J. Hist. Neurosci.* 14: 210–213.

Fabbro, F. (1994). Left and right in the Bible from a neuropsychological perspective. *Brain Cogn.* 24: 161–183.

Feinsod, M. (1997). Three head injuries: the Biblical account of the deaths of Sisera, Abimelech and Goliath. *J. Hist. Neurosci.* 6: 320–324.

Hermann, H. (1966). La Cirugia en la Biblia y el Talmud. *Folia Clin. Int. (Barc).* 16: 605–619.
Hurwitz, L.J. (1971). A neurologist's anecdotes and the Bible. *Practitioner* 206: 287–292.
Karampelas, I., Boev, A. III, Fountas, K.N., and Robinson, J.S. (2004). Sciatica: a historical perspective on early views of a distinct medical syndrome. *Neurosurg. Focus.* 16: 1–4.
Kaufmann, J.C.E. (1964). Neuropathology in the Bible. *South African Med. J.* 38: 788–789.
Leon, F.E. (1993). The first reported case of radial nerve palsy. *South. Med. J.* 86: 808–811.
Loukas, M., Bilinsky, E., Bilinsky, S. et al. (2011). Surgery in early Jewish history. *Clin. Anat.* 24: 151–154.
Lurie, S. (2006). Vaginal delivery after caesarean delivery in the days of the Talmud (2nd century BCE – 6th century CE). *Vesalius.* 12: 23–24.
Mansour, A.M., Gold, D., Salti, H.I., and Sbeity, Z.M. (2004). History of ophthalmology. *Surv. Ophthalmol.* 49: 446–453.
Moskopp, D. (1996). Neurosurgery and the Holy Bible. *Neurosurg. Rev.* 19: 97–104.
Ohry, A. (1987). The brain injured patient, some poetic and neurologic aspects. *Orthop. Rev.* 16: 94–96.
Oren, M., Herman, J., and Elbaum, J. (1998). Men with no spleens and carved-out feet: what is the meaning in the words. *Ann. Intern. Med.* 129: 756–758.
Preuss, J. and Rosner, F. (1978). *Julius Preuss' Biblical and Talmudic Medicine.* New York: Sanhedrin Press.
Rosner, F. (1975). Neurology in the Bible and Talmud. *Isr. J. Med. Sci.* 11: 385–397.
Rosner, F. (1977). *Medicine in the Bible and the Talmud.* Jersey City, NJ: KTAV Publishing House.
Rosner, F. (1978). *Julius Preuss' Biblical and Talmudic Medicine.* New York: Hebrew Publishing Company.
Rosner, F. (1986). Bloodletting in Talmudic times. *Bull. N. Y. Acad. Med.* 62: 935–946.
Rosner, F. (1998). *The Medical Legacy of Moses Maimonides.* Jersey City, NJ: KTAV Publishing House.
Rosner, F., McCall, D.D., and Berman, S.J. (1971). Anesthesia in the Bible and Talmud. *Anesth. Analg.* 50: 298–301.
Shapiro, R.M. (1990). Head injuries in the Old Testament. *Radiology* 74: 84–86.
Snowman, J. (1974). *A Short History of Talmudic Medicine.* New York: Hermon Press.
Steinberg, A. (2003). *Encyclopedia of Jewish Medical Ethics,* vol. I–III. Jerusalem, New York: Feldheim Publishers.
Tubbs, R.S., Loukas, M., Shoja, M.M. et al. (2008). The intriguing history of the human calvaria: sinister and religious. *Childs Nerv. Syst.* 24: 417–422.
Weinberg, A. (2006). A case of cranial surgery in the Talmud. *J. Hist. Neurosci.* 15: 102–110.
Westreich, M. and Segal, S. (2000). Cleft lip in the Talmud. *Ann. Plast. Surg.* 45: 229–239.

Anatomy in the Qur'an and Hadeeth

Abdel Haleem MAS (trans.). *The Qur'an: A New Translation by M.A.S. Abdel Haleem.*Oxford University Press Oxford; 2005.
Al-Bukhari MI. The English Translation of Sahih Al Bukhari with the Arabic Text (9-Volume Set). Translated by Muhammad Muhsin Khan. Al-Saadawi Publications; 1996a.
Al-Jauziyah IIQ. Healing with the Medicine of the Prophet (Peace be upon Him). Translated by Jalal Abual Rub. Darussalam Publishers & Distributors, Fordham University; 1999
Coats, A.J. (2009). Ethical authorship and publishing. *Int. J. Cardiol.* 131: 149–150.
Hehmeyer, I. and Khan, A. (2006). Islam's forgotten contributions to medical science. *Can. Med. Assoc. J.* 176: 10.
Imam, Y.O. (1995). Health care services in the contemporary world. *Islam. Q.* 39: 234–255.

Muslim I. Sahih Muslim Volumes I–IV. Translated by Muhammad Muhsin Khan. Al Saadawi Publications; 1996.
Prioreschi, P. (2006a). Anatomy in medieval Islam. *J. Int. Soc. History Islamic Med.* 5.
Syed, I.S. (2002). Islamic Medicine: 1000 years ahead of its times. *J. Int. Soc. History Islamic Med.* 2: 4–8.
Tan, S.Y. (2002). Medicine in Stamps Hippocrates: father of medicine. *Singap. Med. J.* 43: 5–6.
Tan, S.Y. and Yeow, M.E. (2003). Medicine in stamps: Andreas Vesalius (1514–1564): father of modern anatomy. *Singap. Med. J.* 44: 229–230.
University of Southern California: Center for Muslim-Jewish Engagement. http://cmje.usc.edu/. Last accessed July 13, 2018.
Zindani, A.A., Johnson, E.M., Goeringer, G.C. et al. (1994). *Human Development as Described in the Qur'an and Sunnah: Correlation with Modern Embryology*. Bridgeview, Illinois: Islamic Academy for Scientific Research.

Persia

Ahmed, M.M. (ed.) (2005). *Encyclopaedia of Islam. Vol. 13: Science in Islam*. New Delhi: Anmol Publications.
Alakbarli, F. (2006). *Azerbaijan: Medieval Manuscripts, History of Medicine, Medicinal Plants*. Baku: Nurlan.
Ameli, N.O. (1965). Avicenna and trigeminal neuralgia. *J. Neurol. Sci.* 2: 105–107.
Bigdeli ZM. Magam-e Abu Ali Sina dar Jahan [The Position of Avicenna in the World] (Persian). The Collection of Articles from the International Symposium of Avicenna. Bu-Ali Sina Scientific and Cultural Foundation: Hamadan; 2004.
Blumenau, R. (2002). *Philosophy and Living*. Charlottesville, VA: Imprint Academic.
Browne, E.G. (trans.)(1921). *Revised Translation of Chahar Maqala (Four Discourses) – Nizami-I-Arudi of Samarqand Followed by an Abridged Translation of Mirz Muhammad's Notes to the Persian Text*. London: Cambridge University Press.
Burrell, D.B. (2002). Avicenna. In: *A Companion to Philosophy in the Middle Ages* (ed. J.J. Gracia and T.B. Noone), 196–208. Malden, MA: Blackwell Publishing.
Coats, A.J. (2009). Ethical authorship and publishing. *Int. J. Cardiol.* 131: 149–150.
Conrad, L.I., Neve, M., Nutton, V. et al. (1995). *The Western Medical Tradition 800 BC to AD 1800*. Cambridge: Cambridge University Press.
Cronin, T.G. Jr. (1988). Yawning: an early manifestation of vasovagal reflex. *AJR Am. J. Roentgenol.* 150: 209.
Dumper, M.R. and Stanley, B.E. (eds.) (2007). *Cities of the Middle East and North Africa: A Historical Encyclopedia*. Santa Barbara, CA: ABC-CLIO.
(1823). *Encyclopeadia Britannica, or, A Dictionary of Arts, Sciences, and Miscellaneous Literature, Enlarged and Improved*, 6e, vol. III. Edinburgh: Archibald Constable.
Field, C. (1910). *Mystics and Saints of Islam*. London: Francis Griffiths. (Reprinted by Kessinger Publishing, 2004.).
Forghani, B. (trans.)(1982). *Persian Translation of Cyril Elgood's A Medical History of Persia and the Eastern Caliphate: The Development of Persian and Arabic Medical Sciences (Cambridge University Press, 1951)*. Tehran: Amir Kabir Publication.
Gohlman, W.E. (trans.)(1974). *The Life of Ibn Sina: A Critical Edition and Annotated Translation [of Avicenna's autobiography]*. New York: State University of New York Press.
Goodrich, J.T. (1997). Neurosurgery in ancient and medieval worlds. In: *A History of Neurosurgery in Its Scientific and Professional Contexts* (ed. S.H. Greenblatt, T.F. Dagi and M.H. Epstein). Park Ridge, IL: American Association of Neurological Surgeons.

Hashemi, S.A. trans.)(2004). *Persian Translation of Hakim S. Zillurahman's Qanun-I Ibn-I Sina, Commentaries and Translations.* Tehran: Society for the Appreciation and Cultural Works and Dignitaries.

Heynick, F. (2002). *Jews and Medicine: An Epic Saga.* Jersey City, NJ: KTAV Publishing House.

Husain, M. (2004). *Islam's Contribution to Science.* New Delhi: Anmol Publications.

Ibn Sina, A.A. (1951). *Rag Shenasi ya Resaleh dar Nabz [Pulsology, or, Treatise on Pulse]* (Persian) (ed. S.M. Meshkat). Tehran: Selsele Intisharat-e Anjomane Asare Melli.

Legge, H., Norton, M., and Newton, J.L. (2008). Fatigue is significant in vasovagal syncope and is associated with autonomic symptoms. *Europace* 10: 1095–1101.

Loukas, M., Lam, R., Tubbs, R.S. et al. (2008). Ibn al-Nafis (1210–1288): the first description of the pulmonary circulation. *Am. Surg.* 74: 440–442.

Madelung, W. (1975). The minor dynasties of northern Iran. In: *The Cambridge History of Iran, Vol. 4: From the Arab Invasion to Saljugs* (ed. R.N. Frye), 198–249. Cambridge: Cambridge University Press.

Mahmoodzadeh-Sagheb HR, Heidary Z, Nouri MH. Ibn Sina, Maktab va Nazaryyehay-e Jahanshenasy-eo [Avicenna and his Existence Doctrine and Ideology] (Persian). The Collection of Articles from the International Symposium of Avicenna. Bu-Ali Sina Scientific and Cultural Foundation: Hamadan; 2004.

Meri, J.W. (ed.) (2006). *Medieval Islamic Civilization: An Encyclopedia*, vol. 2. New York: Taylor & Francis.

Morley, H. (1853). *Anatomy in Long Clothes. Fraser's Magazine for Town and Country, November Issue.* London: J.W. Parker and Son.

Nabipour, I., Burger, A., Moharreri, M.R., and Azizi, F. (2009). Avicenna, the first to describe thyroid-related orbitopathy. *Thyroid* 19: 7–8.

Nafisi, S. (2005). *Zendegi, Kar, Andisheh va Rozegar-e Pur Sina [The Life, Work, Ideas and Days of Avicenna]* (Persian). Tehran: Asatir Press.

Nasr, S.H. (1993). *An Introduction to Islamic Cosmological Doctrines: Conceptions of Nature and Methods Used for Its Study by the Ikhwan Al-Safa, Al-Biruni, and Ibn Sina.* New York: State University of New York Press.

O'Dowd, M.J. (2001). *The History of Medications for Women: Materia Medica Woman.* New York: Parthenon Publishing Group.

Osler W. The evolution of modern medicine: a series of lectures delivered at Yale University on the Silliman Foundation in April 1913. http://www.gutenberg.org/files/1566/1566-h/1566-h.htm. Last accessed July 13, 2018.

Pachter, H.M. (1951). *Magic into Science: The Story of Paracelsus.* Scranton, PA: Henry Shuman.

Parvin, M. (2006). *Avicenna and I: The Journey of Spirits.* Bethesda, MD: Ibex Publishers.

Prioreschi, P. (2003). *A History of Medicine Volume 5: Medieval Medicine.* Omaha, NE: Horatius Press.

Ramsden, E.H. (1963). *Letters of Michelangelo, Translated from the Original Tuscan*, vol. 1. Stanford, CA: Stanford University Press.

Reisman, D.C. and Al-Rahim, A.H. (eds.) (2003). *Before and After Avicenna: Proceedings of the First Conference of the Avicenna Study Group.* Boston: Brill.

Sarbazi, M. (1994). *Zendegi-e Abu Ali Sina [The Life of Abu Ali Sina]* (Persian). Tehran: Toseey-e Ketabkhanehay-e Iran.

Savage-Smith, E. (1997). Europe and Islam. In: *Oxford Western Medicine: An Illustrated History* (ed. I. Loudon), 40–53. New York: Oxford University Press.

Shoja, M.M. and Tubbs, R.S. (2007a). The disorder of love in the Canon of Avicenna (AD 980–1037). *Am. J. Psychiatry* 164: 228–229.

Shoja, M.M. and Tubbs, R.S. (2007b). The history of anatomy in Persia. *J. Anat.* 210: 359–378.

Spence, L. (2003). *An Encyclopaedia of Occultism*. New York: Dover.
Willis, R. (1847). *The Works of William Harvey, MD. Translated from the Latin with a Life of the Author*. London: C. and J. Adlard.

From the Mongol Invasion of Persia to the Foundations of Modern Anatomy

Abd Al Kader M. Ali and Abdel-reheem Ead. 2007. http://www.levity.com/alchemy/islam06.html. Last accessed July 13, 2018.
Abdel-Halim, R.E. and Abdel-Maguid, T.E. (2003). The functional anatomy of the uretero-vesical junction. A historical review. *Saudi Med. J.* 24: 815–819.
Al-Qattan, M.M. (2006). History of anatomy of the hand and upper limb. *J. Hand. Surg. [Am.]* 31: 502.
Arjah, A., Hadiyan, F., Soltanifar, S., and Chehrekhand, Z. (1992). *Bibliography of the Medical Manuscripts in Iran* (Persian). Tehran: National Library of the Islamic Republic of Iran.
Azizi, M.H. (2005). Dr. Jacob Eduard Polak (1818–1891): The pioneer of modern medicine in Iran. *Arch. Iranian. Med.* 8: 151–152.
Batirel, H.F. (1999). Early Islamic physicians and the thorax. *Ann. Thorac. Surg.* 67: 578–580.
Belen, D. and Aciduman, A. (2006). A pioneer from the Islamic Golden Age: Haly Abbas and spinal traumas in his principal work, The Royal Book. *J. Neurosurg Spine.* 5: 381–383.
Blair B. The medical manuscripts of Azerbaijan, unlocking their secrets. 1997. http://www.azer.com/index.html. Last accessed July 13, 2018.
Bodjnordi, K.M. (1996a). *The Great Islamic Encyclopedia*, vol. 2 (Persian). Tehran: Foundation of Islamic Encyclopedia.
Bodjnordi, K.M. (1996b). *The Great Islamic Encyclopedia*, vol. 3 (Persian). Tehran: Foundation of Islamic Encyclopedia.
Bodjnordi, K.M. (1996c). *The Great Islamic Encyclopedia*, vol. 7 (Persian). Tehran: Foundation of Islamic Encyclopedia.
Choulant, L. (1852). *Geschichte Und Bibliographie Der Anatomischen Abbildung Ncach Ihrer Beziehung Auf Anatomische Wissenschaft Und Bildende Kunst*. Leipzig: Rudolph Weigel.
Durant, W. (1954). *Our Oriental Heritage*. New York: Simon and Schuster.
Elgood, C. (1952). *A Medical History of Persia and the Eastern Caliphate*. London: Cambridge University Press.
Elgood, C. (1970). *Safavid Medical Practice/The Practice of Medicine, Surgery and Gynaecology in Persia Between 1500 AD and 1750 AD*. London: Luzac.
Farhadi, M., Behzadian Nejad, G., and Bagbanzadeh, A. (1994). *Papers of the International Congress of the History of Medicine in Islam and Iran*, vol. 1 (Persian). Tehran: Iranian Institute for Science and Research Expansion.
Farhadi, M., Behzadian Nejad, G., and Bagbanzadeh, A. (1996). *Papers of the International Congress of the History of Medicine in Islam and Iran*, vol. 2 (Persian). Tehran: Iranian Institute for Science and Research Expansion.
Goodrich, J.T. (1997). Neurosurgery in the ancient and medieval worlds. In: *A History of Neurosurgery in its Scientific and Professional Contexts* (ed. S.H. Greenblatt). Park Ridge, IL: American Association of Neurological Surgeons.
Goodrich, J.T. (2004). History of spine surgery in the ancient and medieval worlds. *Neurosurg. Focus.* 16: E2.
Hamarneh, S. (1970). Contributions of 'Ali al-Tabari to ninth-century Arabic culture. *Folia Orient.* 12: 91–101.
Hekmat H. The story of contemporary anatomy in Iran (Persian). In: Abstract Book of 7th Iranian Congress of Anatomical Sciences. Kashan University of Medical Sciences: Kashan; 2006.

Lewis, B. (2004). *The Middle East, 2000 Years of History from the Rise of Christianity to the Present Day*. London: Phoenix Press.

Mahdavi, S. (2005). Shahs, doctors, diplomats and missionaries in 19th century Iran. *Bull. Br. Soc. Middle East. Stud.* 32: 169–191.

Majidzadeh, Y. (2001). *Ancient Mesopotamia: History and Civilization (Art and Architecture)*, vol. 3. Tehran: Iran University Press.

Matini, J. (1965). *Hidayat al-Mutaallimin fi al-Ttib by Abu Bakr Rabi ibn Ahmad al-Akhawaini al-Bukhari* (Persian). Mashhad: Meshed University Press.

Meyerhof, M. (1935). Ibn an-nafis (13th century) and his theory of the lesser circulation. *Isis* 23: 100–120.

Miller, A.C. (2006). Jundi-Shapur, bimaristans, and the rise of academic medical centers. *J. Royal Soc. Med.* 99: 615–617.

Moharreri, M.R. (2005). *Zakhireye Kharazmshahi* (Persian). Tehran: The Iranian Academy of Medical Science.

Moore, K.L. (1986). Scientist's interpretation of references to embryology in the Quran. *J. IMA* 18: 15–16.

Nabipour, I. (2003). Clinical endocrinology in the Islamic civilization in Iran. *Int. J. Endocrinol. Metab.* 1: 43–45.

Naderi, S., Acar, F., Mertol, T., and Arda, M.N. (2003). Functional anatomy of the spine by Avicenna in his eleventh century treatise Al-Qanun fi al-Tibb (The Canons of Medicine). *Neurosurgery* 52: 1449–1453.

Nadjmabadi (1987). *History of Medicine in Iran*, vol. 2 (Persian). Tehran: Tehran University Press.

Naficy, S. (2005). *Pur Sina* (Persian). Tehran: Asatir.

Nagamia, H.F. (2003). Islamic medicine history and current practice. *JISHM* 2: 19–30.

Naji MR. The Islamic History and Civilization in the Samanid Realm (Persian). Symposium on the Samanid Civilization, History and Culture Press: Tehran; 1999.

Newman, A.J. (1998). Tashrih-e Mansuri: human anatomy between the Galenic and prophetic medical traditions. In: *La Science dans le Monde Iranien* (ed. Z. Vesel, H. Beikbaghban and B. Thierry), 253–271. Tehran: Institut Francais de Recherche en Iran.

Newman, A.J. (2003). Baqir al-Majlisi and Islamicate medicine: Safavid medical theory and practice re-examined. In: *Society and Culture in the Early Modern Middle East, Studies on Iran in the Safavid Period* (ed. A.J. Newman), 371–396. Leiden: Brill.

Noury MS. First world-famous Iranian physician: Pur Sina. 2005. https://iranian.com/2012/09/12/first-world-famous-iranian-physician-pur-sina/. Last accessed July 13, 2018.

Persaud, T.V.N. (1984). *Early History of Human Anatomy. From Antiquity to the Beginning of the Modern Era*. Springfield, IL: C. C. Thomas.

Price M. History of ancient medicine in Mesopotamia & Iran. 2001. http://www.iranchamber.com/history/articles/ancient_medicine_mesopotamia_iran.php. Last accessed July 13, 2018.

Rust, J.H. (1982). Animal models for human diseases. *Perspect. Biol. Med.* 25: 662–672.

Savage-Smith, E. (1995). Attitudes toward dissection in medieval Islam. *J. Hist. Med. Allied Sci.* 50: 67–110.

Savage-Smith E. Mansur ibn Ilyas.Tashrih-i badan-i insan [Anatomy of the Human Body] (Iran, ca. 1390). 2005. http://www.nlm.nih.gov/exhibition/historicalanatomies/mansur_bio.html. Last accessed July 13, 2018.

Shoja, M.M. and Tubbs, R.S. (2007). The disorder of love in the Canon of Avicenna. *Am. J. Psychiatry* 164: 228–229.

Singer I, Jelinek E. Polak, Jakob Eduard. http://www.jewishencyclopedia.com/articles/12235-polak-jakob-eduard. Last accessed July 13, 2018.

Soroushian M. Persien, das Land und Seine Bewohner; Ethnograpische Schilderungen. 2004. http://www.zoroastrian.org.uk/vohuman/Article/Persien%20das%20Land%20und%20Seine%20Bewohner%20Ethnograpische%20Schilderungen.htm. Last accessed July 13, 2018.

Souayah, N. and Greenstein, J.I. (2005). Insights into neurologic localization by Rhazes, a medieval Islamic physician. *Neurology* 65: 125–128.

Tan, S.Y. (2002). Medicine in stamps Rhazes (835–925 AD). Medical scholar of Islam. *Singap. Med. J.* 43: 331–332.

Troupeau, G. (1995). The first treatise on diet: the Kitab Hawass al-agdiyah de Yuhanna ibn Masawayh. *Med. Secoli.* 7: 121–139.

Wakim, K.G. (1944). Arabic medicine in literature. *Bull. Med. Libr. Assoc.* 32: 96–104.

Weisser, U. (1980). The embryology of Yuhanna ibn Masawaith. *J. Hist. Arabic. Sci.* 4: 9–22.

Wright, D. (1985). *The Persians Amongst the English.* London: I.B. Tauris.

Zaehner RC. Zurvanism. 2006. http://www.iranchamber.com/religions/articles/zurvanism3.php. Last accessed July 13, 2018.

Baghdad

El Ghrari H *Architects of the Scientific Thought in Islamic Civilization: Hallmarks from the Biographies of Muslim Scholars in Various Ages.* Translated by Haouaria A. Imprimerie Beni Snassen: Berkane; 2003.

Garbutt, N. (1996). Ibn Jazlah: the forgotten Abbasid gastronome. *J. Econ. Soc. Hist. Orient* 39: 42–44.

Graziani, J.S. (1980). *Arabic Medicine in the Eleventh Century as Represented in the Works of Ibn Jazlah.* Karachi: Hamdard Academy.

Shoja, M.M. and Tubbs, R.S. (2007a). The history of anatomy in Persia. *J. Anat.* 210: 359–378.

Shoja, M.M., Tubbs, R.S., Ardalan, M.R. et al. (2007b). Anatomy of the cranial nerves in medieval Persian literature: Esmail Jorjani (1042–1137) and "the treasure of the khwarazm shah.". *Neurosurgery* 61: 1325–1330.

Tubbs, R.S., Shoja, M.M., Loukas, M., and Oakes, W.J. (2007). Abubakr Muhammad Ibn Zakaria Razi, Rhazes (865–925 AD). *Childs Nerv. Syst.* 23: 1225–1226.

5

North America

Compared to the rich history of anatomy from other parts of the world, the history of anatomy in the Americas is brief. However, in a relatively short period of time, much has come from the anatomical sciences in this region. Although there are many who could be included in a discussion of influential anatomists from North America, we have chosen two who give a flavor of nineteenth-century anatomy as taught and practiced in the United States.

5.1 United States

5.1.1 Henry Jacob Bigelow

Henry Jacob Bigelow (1818–1880) (Figure 5.1) was born in Boston, Massachusetts, the eldest of five children of to Mary Scollay and Jacob Bigelow. His father was a well-respected man who received his Doctorate of Medicine from the University of Pennsylvania in 1810 and served as Professor of Materia Medica at Harvard Medical School from 1815 to 1865; between 1846 and 1863, he was President of the American Academy of Arts and Sciences. Jacob is said to have had very similar qualities to his father. He began his formal education at a school for children and spent a year at Thayer's, a reputable private school, before attending high school at the Boston Public Latin School. He was admitted to Harvard College at the age of 15. Bigelow graduated in 1837, deciding to pursue a career in medicine. He studied under his father in a preceptorship for three years, and attended lectures given by his friend Oliver Wendell Holmes at Dartmouth Medical School. He was appointed house student of surgery at the Massachusetts General Hospital from 1838 to 1839, and the degree of Doctor of Medicine was conferred on him by Harvard Medical School in 1841. During 1842, he traveled to Paris, at that time the surgical hub of the world. During the two years he spent there, Bigelow regularly attended lectures given by Sir James Paget, the famous British surgeon. On returning to Boston at the age of 28, he was appointed visiting surgeon at Massachusetts General Hospital. He immediately took an interest in the study of surgical pathology and became one of the leading microscopists in the United States. During these early years of his career, he opened a "Charitable Surgical Institution" with Dr. Henry Bryant, which was located in the basement of the First Church on Chauncey Place. Bigelow married Susan Sturgis in May 1847 and had a son, William Sturgis. Susan died in 1853, and William, who also became a physician, died in 1926.

In 1849, Bigelow became Professor of Surgery at Harvard Medical School, where for 33 years he was admired by students and colleagues for his superb skills in surgical techniques and teaching. In 1846, Bigelow published "Insensibility during Surgical Operations Produced by Inhalation," a groundbreaking paper that changed the practice of surgery forever.

History of Anatomy: An International Perspective, First Edition. R. Shane Tubbs, Mohammadali M. Shoja, Marios Loukas and Paul Agutter.

Figure 5.1 Henry Jacob Bigelow (1818–1890). Quarter-plate daguerreotype or half-plate daguerreotype cut down. (*Source:* Courtesy of Harvard Art Museum/Fogg Museum, loan from the Massachusetts General Hospital Archives and Special Collections.)

On October 16, 1846, he witnessed the first public demonstration of the use of ether as a surgical anesthetic at Massachusetts General Hospital. The ether was administered by Dr. William Morton, a local dentist, and the surgery was performed by Dr. John Warren, a colleague. The procedure involved making an incision a few inches long near the lower aspect of the mandible of a patient named Gilbert Abbott. Mr. Abbott stated after the surgery that "the pain was considerable, though mitigated" and that it was "as though the skin had been scratched with a hoe." Bigelow postulated that there must have been a defect in the method of inhalation, as he found subsequent procedures to be completely successful. He studied ether further and engaged in experiments with several other subjects, which formed the content of his famous paper. Bigelow recounted the mental states and vital signs of various patients before, during, and after surgery involving the administration of ether vapor. His early experiments concerned short procedures, including tooth extractions, but quickly expanded to more invasive practices, such as tumor removal. One featured case, which Bigelow stated in his paper would "confirm ... the great utility of this process," was the use of ether on a young girl called Alice Mohan, who was undergoing amputation of her right leg above the knee. The procedure was performed by Dr. Hayward on November 7, 1846. Alice was rendered insensible after five minutes of inhaling ether vapor, and regained consciousness shortly after the procedure was finished. On being asked whether her leg had been removed or not, she was unsure of the answer. Through this procedure and several others, Bigelow concluded that these pivotal discoveries would lay the foundation for a much more humane and comfortable way to perform surgery. He developed a follow-up paper two years later entitled "Ether and Chloroform: A Compendium of their History, Surgical Use, Dangers, and Discovery," which described his own cases further and proved that ether and chloroform would render patients insensible during surgery. The paper also outlined the dangers associated with overusing these substances and noted that the patient's pulse was the best indicator of these. In an additional publication, "Surgical Anesthesia: Addresses and other Papers," Bigelow again stressed the importance of monitoring the frequency and force of the pulse while the patient was anesthetized, and cautioned that administration of anesthesia should only be carried out by experienced physicians. Although he was not the pioneer of ether, Bigelow undoubtedly contributed to its widespread use in operating rooms.

Bigelow is renowned for innovation in the field of orthopedic and urologic surgery. His notable publications include the *Manual of Orthopedic Surgery*, for which he won the 1844 Boylston Prize. This was described by Rutkow as the first comprehensive summary of the practice of orthopedic surgery in the United States, and a detailed account of the common practices prevalent in France at the time. In 1849, Bigelow was appointed Professor of Surgery at Harvard Medical School, where he remained for 33 years. He was admired by students and colleagues for his superb surgical skills and teaching technique.

In 1869, Bigelow published "The Mechanism of Dislocation and Fracture of the Hip, with the Reduction of the Dislocation by the Flexion Method," which summarized his extensive research concerning hip dislocation. In this paper, he described the structure and function of the iliofemoral ligament thoroughly for the first time (Bigelow and Bigelow 1894). This ligament extends from the anterior inferior iliac spine to the anterior inter-trochanteric line of the femur. It is commonly referred to as the "Y ligament," because Bigelow had noticed that its shape was similar to the inverted letter "Y" (due to the two fasciculi that extend from it). Bigelow described how through dissection of the ligament, he had observed the unique anatomical arrangement of its fibers and noted its strength. He characterized it as having a critical function in resisting hyperextension of the lower limb while walking, as the ligament forcibly holds the femur in its socket. Bigelow then related this ligament to the pathology of hip dislocations. He was the first to describe these dislocations in terms of their being regular and irregular. According to Bigelow, regular dislocations occur when one or both parts of the iliofemoral ligament remain intact, allowing the head of the femur to remain close to the acetabulum and stay fixed in a relatively constant position. In irregular dislocations, both parts of the ligament are ruptured so that the head of the femur can be displaced in multiple directions, eliciting a variety of inconstant signs. Such signs have implications for the reduction of dislocation of the hip joint, which is a medical emergency that must be diagnosed and treated promptly to avoid the high risk of avascular necrosis. Bigelow contested previous treatment suggestions taught by Sir Astley Cooper, which had the limb elevated in an extended position. He noted that this caused the ligament to become taut due to gravity. Bigelow explained that having the hip in flexion under the anesthetic effects of ether allowed the ligament to become relaxed, facilitating reduction of the dislocation. Furthermore, he described the vital importance of the iliofemoral ligament in the treatment, stating that "the whole manipulation must be conducted in reference to it."

In 1875, Bigelow published a paper that had lasting effects in the field of orthopedics. This paper, entitled "The True Neck of the Femur: Its Structure and Pathology," describes the calcar femorale, which is now known as Bigelow's septum. This structure is a strong tissue at the anterior portion of the lesser trochanter that has a crucial role in strengthening the neck of the femur. Bigelow discussed femoral neck fractures and suggested differences between an impacted and an unimpacted (loose) fracture. Furthermore, he explained the various methods of diagnosing and treating each, and stressed the importance of recognizing the differences to achieving the best possible outcome for the patient.

Bigelow's impact on urologic surgery is also noteworthy, particularly his work concerning the development of a procedure for bladder stone removal. This operation involves crushing and removing a urinary bladder stone in a single procedure. Previously, surgeons had crushed the stone and removed the majority of debris at a later date. Bigelow invented his method out of concern for possible trauma of the bladder due to persistence of dissolved fragments of stones. This caused much discomfort to patients, as the fragments would usually exit on their own. Bigelow found a way to incorporate ether into the procedure to allow the insertion of a catheter through the urethra to flush out pieces of stone. In his paper, "Lithotrity by a Single Operation," published in 1878, Bigelow described and included pictures of his creation of a more effective lithotrite and a large irrigation tube that could be used to remove debris from the bladder immediately after dissolving the stone. This innovative technique significantly lowered the mortality rate caused by surgery and offered a more efficient alternative for patients, who had a better postoperative outcome.

Being an avid scholar, and taking a keen interest in new and unique cases, Bigelow was curious about the famous case of Phineas Gage, a railroad worker who survived a traumatic skull injury. While Gage was working, an explosion occurred and an iron bar shot through his head, entering at the angle of the lower jaw, piercing through the left frontal lobe, and exiting through the top of his skull (Harlow 1848; Bigelow and Bigelow 1894). Bigelow brought Gage to Boston, with the help of his attending physician Dr. John Harlow (who was the first to publish the events concerning the remarkable story), where Gage remained under his care for several weeks. While there was much skepticism concerning the legitimacy of Gage's story, since it appeared that he was relatively uninjured by the accident, Bigelow demonstrated that the bar had traversed through Gage's head in a way that did not cause fatal injury. He traced and drilled a track through a skull to prove this. In his paper, "Dr. Harlow's Case of Recovery from the Passage of an Iron Bar through the Head," Bigelow described his findings and his accounts of meeting several individuals who had witnessed the accident. Bigelow testified that he became wholly convinced by the case and of the fact that it had been possible for Gage to survive the event, given the circumstances of how the bar had passed through his skull and brain. After Mr. Gage passed away, several years later, his body was examined and Bigelow's explanations of the accident were verified. This case caused immense public interest and changed the field of neuropsychiatry, as it provided evidence for behavioral syndromes caused by frontal lobe impairment.

One aspect of Bigelow's career that could serve to discredit him is his relatively slow acceptance of the concept of Listerism, the use of antiseptic techniques during surgical procedures. In his publication, "Orthopedic Surgery and other Medical Papers," Bigelow defended his position on the matter, stating, "It flatters neither the vanity nor the scientific sense to exorcise an invisible enemy with something very like a censer." However, he did eventually accept Listerism, as he continued, "But after the two years' experience I have accepted the new doctrine with most of its details."

With his major contributions to the medical field, Dr. Henry J. Bigelow has embedded himself in history as an innovative and highly skilled physician. After resigning from his position of surgeon at the Massachusetts General Hospital in 1886, Bigelow retired to his country home in Newton, Massachusetts. He died on October 30, 1880, at the age of 72, from complications pertaining to pyloric stenosis (Figure 5.2). More than a century after his death, his work continues to be recognized and admired, and is the foundation of routine procedures performed around the world. With his thorough study and promotion of ether, and his invention of groundbreaking surgical techniques, he has left a mark on modern medicine and will be remembered in the scientific community as a brilliant mind.

5.1.2 Oliver Wendell Holmes

5.1.2.1 Early Life

Oliver Wendell Holmes (1809–1894) (Figure 5.3) was born in Cambridge, Massachusetts, one of four children of Calvinist minister and Yale graduate Abiel Holmes and Sarah Wendell, who were married in 1801, the same month Thomas Jefferson was inaugurated. Oliver was the grandson of Captain David Holmes, a soldier and physician of the French and Indian War and the Revolutionary War.

Figure 5.2 Henry Jacob Bigelow (1818–1890). (*Source:* From Bigelow and Bigelow 1894.)

Figure 5.3 Oliver Wendell Holmes, Sr. Note the microscope, which Holmes introduced into the study of medicine in the United States. Note also Holmes' signature at the bottom of the photograph.

From Phillips Academy in Andover, he entered Harvard and received BA and MD degrees. He studied law briefly (Dane Law School, Cambridge, MA) before beginning his medical training, although the reason behind this change in direction was not clear to him. Following his graduation, he began writing poetry. Regarding his anatomical training in medical school, Holmes is quoted as saying:

> I have been going to Massachusetts General Hospital and slicing and slivering the carcasses of better men and women than I ever was myself or am like to be. It is a sin for a puny little fellow like me [he was 5′3″] to mutilate one of your six-foot men as if he were sheep, but vive la science.

From 1833 to 1835, Holmes continued his medical studies in Paris at La Charité, Hotel Dieu, and La Pitié Salpêtrière. During this time, he studied with Pierre Charles Alexandre Louis (1787–1872), a proponent of the philosophy that a physician's role is simply to facilitate nature's healing of the patient, and Holmes was also to opine this notion in his practice. Under Louis, Holmes became an expert in auscultation. While attending clinics in La Pitie and Hotel Dieu, he attended lectures by greats such as Dupuytren, Velpeau, Roux, Larrey, and Lisfranc, and performed multiple dissections, paying approximately 50 cents per cadaver.

In 1837, Holmes taught physiology and pathology at the Tremont Street Medical College with his colleague, famed anatomist and surgeon Henry Bigelow. In 1838, Holmes was appointed Professor of Anatomy at Dartmouth College in Hanover, a post he held for two years. He spent four months per year at Dartmouth, and was paid a salary of $400. Soon after starting there, he met Amelia Lee Jackson, the daughter of Judge Charles Jackson of the Massachusetts Supreme Court. They were married on June 15, 1840 at King's Chapel on Tremont Street. The next year, Amelia gave birth to a son, Oliver Wendell Jr., the future chief justice of the Massachusetts Supreme Court and associate justice of the US Supreme Court. In 1843, a daughter, Amelia, was born, and in 1846, a son, Edward Jackson, who would become a lawyer.

5.1.2.2 Teaching at Harvard

In 1847, the newly elected president of Harvard College, Edward Everett (1794–1865), a former congressman, governor of Massachusetts, ambassador to Great Britain, United States senator, and vice-presidential candidate in the campaign of 1860, appointed Holmes professor and chair of anatomy and physiology. Holmes followed John Collins Warren, who had served in this post from 1815 to 1847. Interestingly, Holmes would be the one to articulate the skeleton of Dr. Warren, who donated his body to the school on his death.

Holmes was well liked by students, and was referred to as "Uncle Oliver." One colleague recalled:

> He enters [the classroom] and is greeted by a mighty shout and stamp of applause. Then silence, and there begins a charming hour of description, analysis, anecdote, harmless pun, which clothes the dry bones with poetic imagery, enlivens a hard and fatiguing day with humor, and brightens to the tired listener the details for difficult though interesting study.

Holmes' training in Paris led him to teach students the importance of the anatomico-pathological basis of disease (Figure 5.4). He was considered one of the finest lecturers of his day, and his lectures were the best attended at Harvard, including presentations given during the noon hour and the last lecture of the day (Lassek 1958). He was said to have an animated

Figure 5.4 One of the many anatomical preparations made by Holmes, illustrating ankylosis of the sacrum and last three lumbar vertebrae.

style of delivery, sparkling wit, and an "obvious fund of knowledge [that] kept the weary and possibly hypoglycemic students reasonably alert." As an example of his style, Holmes described the female pelvis by stating, "Gentlemen! This is the triumphal arch under which every candidate for immortality has to pass." Holmes was also known to sing to his students in the dissection laboratory.

As a teacher, Holmes was vitally aware of the concerns of students. Hints of this are present in his speech given to the medical class entering Harvard in 1867:

> Some years ago I ventured to show in an introductory lecture how very small a proportion of the anatomical facts taught in a regular course, as delivered by myself and others, had a practical bearing whatever on the treatment of disease. How can I, how can any medical teacher justify himself in teaching anything that is not likely to be of practical use to a class of young men who are to hold in their hands the balance in which life and death, ease and anguish, happiness and wretchedness are to be daily weighed?

Holmes continued:

> But we cannot successfully eliminate and teach by itself that which is purely practical. The easiest and surest way of acquiring facts is to learn them in groups, in systems, and systemized knowledge is science ... Scores of proverbs show you can remember two lines that rhyme better than one without the jingle ... If the memory gains so much by mere rhythmical association, how much more will it gain when isolated facts are brought together under the laws and principles, when organs are examined in their natural connections, when structure is coupled with function, and healthy and diseased actions are studied as they pass one into the other! There is a great and accelerating accumulation of medical facts today. Yet it may be somewhat presumptuous to assume that the intellectual task now confronting entering students is much more formidable than

> the challenges of a century ago. Remember that the facts of today have merely superseded the "facts" of the past. To the student I would say, that however plain and simple may be our teaching, he must expect to forget much which he follows intelligently in the lecture room. But it is not the same as if he had never learned it. A man must get a thing before he can forget it. There is a great world of ideas we cannot voluntarily recall, they are outside the limits of the will. But they sway our conscious thought as the unseen planets influence the movements of those within the sphere of vision ... Some of you must feel your scientific deficiencies painfully after your best efforts. But everyone can acquire what is most essential.

Although Holmes preferred systemic anatomy, he commented on the popularity of regional (surgical) anatomy by stating:

> The second new method of studying the human structure, beginning with the labors of Scarpa, Burns, and Colles, grew up principally during the first third of this century. It does not deal with organs, as did the earlier anatomist, nor with tissues, after the manner of Bichat. It maps the whole surface of the body into an arbitrary number of regions, and studies each region successively from the surface to the bone, or beneath it. This is what regional, or, as it is sometimes called, surgical anatomy, does for the surgeon with reference to the part on which his skill is to be exercised. It enables him to see with the mind's eye through the opaque tissues down to the bone on which they lie, as if the skin were transparent as the cornea, and the organs it covers translucent as the gelatinous pulp of a medusa.

5.1.2.3 Inventor, Innovator, and Discoverer

Holmes contributed to medicine and anatomy. For example, when ether was first used at Massachusetts General Hospital on October 16, 1846, by dentist William G. T. Morton, it was Holmes who wrote to Dr. Morton suggesting that he coin the term "anesthesia." He helped introduce the use of stethoscopes in the United States following his training in Paris. Furthermore, he developed a handheld apparatus for viewing photographs, which he called the stereoscope.

Holmes began to offer instruction in microscopic anatomy at the Tremont Street Medical School in the late 1840s, and offered instruction in the use of the microscope to medical students at Harvard from 1855. He was one of the first Americans to introduce microscopy into a medical curriculum, a discipline that had begun after his medical studies had been concluded. In an address to the Boston Microscopical Society in 1877, Holmes said:

> My dealing with the instrument has been principally as a teacher, and not of microscopy as a specialty, but as a fractional portion of long-extended courses on anatomy, delivered to large classes. The most I could hope for was to teach them the rudiments of histology, and more especially to give them knowledge enough to make them wish for more. I have therefore aimed at having perfectly and easily manageable instruments, at selecting the more important and interesting objects, and at making everything as plain as practicable, knowing well that if a mistake in looking through a microscope is within the bounds of possibility, the young student will be certain to make it.

With his histological interests, Holmes documented nucleated cells in cancellous bone. Grossly, Holmes described the masseteric fossa on the human mandibular ramus, which is better developed in the Carnivora. He also described and named several small muscles that radiate from the axis, to which he gave the poetic name "Stella musculosa nuchae" (Figure 5.5). Regarding these anatomical findings, Holmes modestly said:

> But this scanty catalogue is only an evidence that one may teach long and see little that has not been noted by those who have gone before him.

He prepared approximately 100 anatomical specimens for the medical school, of which 37 have survived. These preparations range from comparative to human anatomical specimens (Figure 5.6). An under-recognized fact is that Holmes published papers concerning puerperal fever and methods for its prevention six years prior to Semmelweis' publication on the topic. As Dean of Harvard Medical School, he tried unsuccessfully to admit white women and free black men. However, Holmes was only willing to teach anatomy to women in separate classes and dissecting rooms from men.

Figure 5.5 Preparation made by Holmes of the muscles attaching to a cervical vertebra.

Figure 5.6 Preparation made by Holmes illustrating the internal aspect of the nasal cavity and nasopharynx.

Defending anatomy as a discipline, which was considered antiquated by some, Holmes declared that human anatomy was much the same study as it had been in the times of Vesalius and Fallopius. Therefore, he considered that it should bear equal importance in a medical curriculum of the nineteenth century. Interestingly, Holmes commented that he believed that there could be a relationship between the rise of great medical teachers and leaders and periods of social/political upheaval. His examples included Galen and the Roman emperor Marcus Aurelius, and Vesalius and Martin Luther.

He chided physicians who tried to pursue academics while maintaining a clinical practice:

> "I suppose I must go and earn this — guinea," said a medical man who was sent for while he was dissecting an animal. I should not have cared to be his patient. His dissection would do me no good, and his thoughts would be too much upon it. I want a whole man for my doctor, not a half one. I would have sent for a humbler practitioner, who would have given himself entirely to me, and told the other – who was no less a man than John Hunter – to go on and finish the dissection of his tiger.

5.1.2.4 Later Life

In Holmes' later life, medical interests and teaching gave way to literary pursuits. He won multiple prizes for writing medical essays and poetry. Literary friends included Emerson, Longfellow, Hawthorne, Whittier, Motley, and Lowell. A poem written by Holmes that illustrates not only his poetic ability but also his interest in anatomy runs:

> So the stout fetus, kicking and alive,
> Leaps from the fundus for his final dive
> Tired of the prison where his legs were curried
> He pants, like Rasselas, for a wider world.
> No more to him their wonted joys afford
> The fringed placenta and the knotted cord.

At the celebration of the Medical School's centennial, Holmes said:

> During the long period in which I have been a Professor of Anatomy in this Medical School, I have had abundant opportunities of knowing the zeal, the industry, the intelligence, the good order and propriety, with which this practical department has been carried on. The labors superintended by the Demonstrator and his assistants are in their nature repulsive, and not free from risk of disease ... The student is breathing an air which unused senses would find insufferable. He has tasks to perform which the chambermaid and the stable-boy would shrink from undertaking ... But we know that the great painters, Michael Angelo, Leonardo, Raphael, must have witnessed many careful dissections, and what they endured for art our students can endure for science and humanity.

Holmes retired from the Parkman Professorship of Anatomy in 1882, after 35 years of teaching (Figure 5.7). He was awarded an LLD degree from Harvard and made professor emeritus. In 1886, he was given honorary degrees from Oxford, Cambridge, and Edinburgh, and made acquaintance with Louis Pasteur. He died on October 7, 1894 at the age of 85. King's Chapel in Boston, where Holmes worshiped, erected a plaque in his memory. This memorial notes Holmes' achievements and refers to him as a "Teacher of Anatomy, Essayist and Poet." Furthermore,

Figure 5.7 Photograph taken on the occasion of Dr. Oliver Wendell Holmes' retirement from teaching anatomy at Harvard, and just after the dedication of the new facility on Boylston Street. Holmes is seen to the far left of the photo.

inscribed on his memorial is a quote from Horace's *Ars Poetica: Miscuit Utile Dulci*: "He mingled the useful with the pleasant." Most remembered for his literary works, it should not be forgotten that he taught for over three decades to hundreds of future physicians and made contributions to medicine with original descriptions and the implementation of new teaching methods (e.g., microscopy).

Bibliography

Henry Jacob Bigelow

Anonymous (1965). Henry Jacob Bigelow (1818–1890) of the Y ligament. *JAMA Editorials* 191: 141–142.

Bigelow, H.J. (1846). Insensibility during surgical operations produced by inhalation. *Boston Med. Surg. J.* 35: 309–317.

Bigelow, H.J. (1848). *Ether and Chloroform: A Compendium of their History, Surgical Use, Dangers, and Discovery*. Boston: David Clapp, Printer.

Bigelow, H.J. (1850). Dr. Harlow's case of recovery from the passage of an iron bar through the head. *Am. J. Med. Sci.* 16: 13–22.

Bigelow, H.J. (1900a). *Orthopedic Surgery and other Medical Papers*. Boston: Little, Brown, and Co.

Bigelow, H.J. (1900b). *Surgical Anesthesia: Addresses and other Papers*. Boston: Little, Brown, and Co.

Bigelow, H.J. (1900c). *I. The Mechanism of Dislocation and Fractures of the Hip. II. Litholapaxy;or Lithotrity with Evacuation*. Boston: Little, Brown, and Co.

Bigelow, H.J. and Bigelow, W.S. (1894). *A Memoir of Henry Jacob Bigelow*. Boston: Little, Brown, and Co.

Ellik, M. (1965). Henry Jacob Bigelow (1818–1890). *Investig. Urol.* 3: 217–219.
Hall, D.P. (1961). Our surgical heritage. *Am. J. Surg.* 101: 824–825.
Harlow, J.M. (1848). Passage of an iron rod through the head. *Boston Med. Surg. J.* 39: 383–393.
Harrington, T.F. and Mumford, J.G. (1905). *The Harvard Medical School: A History, Narrative and Documentary*, vol. II. New York: Lewis Publishing Co.
Holmes, O.W. (1890). Henry Jacob Bigelow. *Am. Acad. Arts Sci.* 26: 339–351.
Malenfant, J., Robitaille, M., Schaefer, J. et al. (2011). Henry Jacob Bigelow (1818–1890): his contributions to anatomy and surgery. *Clin. Anat.* 24: 539–543.
Neylan, T.C. (1999). Frontal lobe function: Mr. Phineas Gage's famous injury. *J. Neuropsychiatr. Clin. Neurosci.* 11: 280–283.
Peltier, L.F. (1997). The Classic: the true neck of the femur: its structure and pathology – Henry Jacob Bigelow (1818–1890). *Clin. Orthop. Relat. Res.* 344: 4–7.
Rutkow, I.M. (1987). American Surgical Biographies. *Surg. Clin. North Am.* 67: 1153–1180.
Rutkow, I.M. (1988). *The History of Surgery in the United States: 1775–1900.* San Francisco: Norman Publishing.
Smith, M.J.V. (1979). Henry Jacob Bigelow (1818–1890). *Urology* 14: 317–322.
Yang, E.C. and Cornwall, R. (2000). Initial treatment of traumatic hip dislocations in the adult. *Clin. Orthop. Relat. Res.* 377: 24–31.

Oliver Wendell Holmes

Anonymous (1894). Medical tributes to Dr. Holmes. *Boston Med. Sci J.* 131: 423–425.
Bergey, G.K. (1977). Oliver Wendell Holmes: the professor and autocrat addresses medical matters. *Yale J. Biol. Med.* 50: 681–688.
Dowling, W.C. (2006). *Oliver Wendell Holmes in Paris: Medicine, Theology, and the Autocrat of the Breakfast Table.* Hanover: University Press of New England.
Dunn, P.M. (2007). Oliver Wendell Holmes (1809–1894) and his essay on puerperal fever. *Arch. Dis. Child. Fetal Neonatal Ed.* 92: F325–F327.
Felts, J.H. (2002). Henry Ingersoll Bowditch and Oliver Wendell Holmes: stethoscopists and reformers. *Perspect. Biol. Med.* 45: 539–548.
Hartwell, E.M. (1881). *The Study of Anatomy, Historically and Legally Considered.* Boston: Tolman & White.
Holmes, O.W. (1883). *Medical Essays (1842–1882).* Boston: Houghton, Mifflin and Company.
Knox, J.H.M. (1907). The medical life of Oliver Wendell Holmes. *Johns Hopkins Hosp. Bull.* 18: 45–51.
Lassek, A.M. (1958). *Human Dissection. Its Drama and Struggle.* Springfield, IL: Charles C. Thomas.
Lindskog, G.E. (1974). Oliver Wendell Holmes "miscuit utile dulci.". *Yale J. Biol. Med.* 47: 277–290.
Morse, J.T. (1896). *Life and Letters of Oliver Wendell Holmes*, vol. 1. London: Sampson Low, Marston & Co.
Otis, E.O. (1909). The medical achievements of Dr. Holmes. *Boston Med. Surg. J.* 161: 951–957.
Shoemaker, H.N. (1953). The contemporaneous medical reputation of Oliver Wendell Holmes. *N. Engl. Q.* 26: 477–493.
Small, M.R. (1962). *Oliver Wendell Homes*, Twayne's United States Authors Series. New York: Twayne Publishers.
Sullivan, W. (1972). *New England Men of Letters.* New York: The Macmillan Company.
Tubbs, R.S., Shoja, M.M., Loukas, M., and Carmichael, S.W. (2012). Oliver Wendell Holmes, Sr. (1809–1894): physician, jurist, poet, inventor, pioneer, and anatomist. *Clin. Anat.* 25: 992–997.

Index

History of Anatomy: An International Perspective, First Edition. R. Shane Tubbs, Mohammadali M. Shoja, Marios Loukas and Paul Agutter.

b

d

e

n

o

Printed and bound by CPI Group (UK) Ltd, Croydon, CR0 4YY
16/04/2025
14658536-0001